Some Modern Mathematics
For Physicists and Other Outsiders

(Volume 1)

The Author

Paul Roman (Ph.D., Eötvös University, Budapest; Cand. Phys. Sci., Hungarian Academy of Sciences) has been Professor of Physics at Boston University since 1960. Prior to that, he held positions in Manchester, England and Budapest, Hungary. He also traveled and worked in other countries, including Italy, Germany, the Soviet Union, India, and Mexico. His primary professional interests are in theoretical physics, especially the theory of elementary particles, symmetries, quantum theory, field theory, and mathematical physics. He is an elected member of several learned societies and is listed in various "Who's Who." Dr. Roman's long list of scientific publications includes three well-known textbooks: *Theory of Elementary Particles, Advanced Quantum Theory,* and *Quantum Field Theory.*

Some Modern Mathematics For Physicists and Other Outsiders

An Introduction to Algebra, Topology, and Functional Analysis
(Volume 1)

Paul Roman
Professor of Physics, Boston University

Pergamon Press Inc.

New York · Toronto · Oxford · Sydney · Braunschweig

PERGAMON PRESS INC.
Maxwell House, Fairview Park, Elmsford, N.Y. 10523

PERGAMON OF CANADA LTD.
207 Queen's Quay West, Toronto 117, Ontario

PERGAMON PRESS LTD.
Headington Hill Hall, Oxford

PERGAMON PRESS (AUST.) PTY. LTD.
Rushcutters Bay, Sydney, N.S.W.

PERGAMON GmbH
D - 3300 Braunschweig, Burgplatz 1

Library of Congress Cataloging in Publication Data

Roman, Paul.
    Some modern mathematics for physicists and other
outsiders.

        Bibliography: v.2, p.
        1. Algebra. 2.   Topology. 3.   Functional analysis.
I.   Title.
QA155.5.R66   1975   510   74–1385

Vol. 1:
ISBN 0-08-018096-5   (pbk.)
ISBN 0-08-018097-3

Vol. 2:
ISBN 0-08-018134-1
ISBN 0-08-018133-3   (pbk.)

Printed in the United States of America

*Dedicated to
the 70th birthday of
Professor Eugene P. Wigner*

Contents of Volume 1

Contents of Volume 2

x

Preface

In these days of rapid proliferation of textbooks, any preface to a new book must begin with an apology, if not with a well-documented justification, concerning the raison d'être of this new volume. This is particularly true if the author ventures into a field which is not one of his primary competence. In the present case, my apology is quite simple. After many years of research in elementary particle theory and related topics, I realized that my "standard" mathematical background is inadequate to keep up with modern developments in theoretical physics. This realization was followed by a long period of hard study when I tried to dig out from the mountains of existing literature those concepts and tools without which I could no longer continue to be productive. I also realized that my graduate students should be spared this horrendous task and I introduced a new, one-year course on what, somewhat euphemistically, may be called "modern" mathematics. The outcome of these efforts is the present treatise.

It is obvious that a pragmatic survey of some relevant chapters of algebra, topology, measure theory, and functional analysis would not serve any useful purpose were it not unified by some central theme and presented in a manner that makes the hard work of absorbing the material not only useful but also truly enjoyable. In modern mathematics such a unifying viewpoint presents itself quite naturally: it is the pursuit of *structure*. Of course, a substantial segment of prospective learners will be impatient to gather, as fast as possible, readily available tools rather than to desire enchantment with beauty. Consequently, I tried hard to strike a healthy balance between structural investigations on one hand and practical theorems and methods on the other. I believe that this is one of the features which distinguishes the present volume from the host of other books, written both by professional specialists and by theoretical physicists. Another, and in my opinion, equally important feature is that I really started out from "scratch" and attempted to pave the way smoothly from elementary concepts to highly sophisticated and involved material. *The only prerequisite for the successful use of this book is a standard familiarity with basic calculus.* A superficial acquaintance with the elements of linear algebra (and perhaps with a few not quite elementary topics of classical analysis) will help, but is not essential.

Naturally, to some extent this book has the character of "selected topics." However, almost all topics covered in the earlier chapters *will be used* later on, and a strong thread of continuity ties the topics together. The attentive critique will observe that occasionally statements and examples are repeated, instead of referring the reader to earlier sections. This is done on purpose, so as to ease the student's work.

It was my firm policy always to proceed from the general to the specific. I took this decision not only so as to conform with the spirit of "modern mathematics," but also because I am convinced that, eventually, this approach is easier to digest and provides a much more stable and time-enduring knowledge than does the laborious method which starts with examples, proceeds to special cases and only then builds up the general theory.

On the other hand, I did not attempt to follow systematically the "Definition–Lemma–Theorem–Proof–Corollary–Remarks" sequence of most professional mathematics texts. Especially in the earlier chapters, many theorems are presented informally, "bringing out" the theorem by a series of observations rather than by first stating it and then supplying the detailed proof. Furthermore, many theorems (even some important ones) are stated without proof, in particular if the proof is atypical, lengthy and/or highly technical. From about Chapter 4 onward, both the rigor and the formal manner of presentation are increased, so that there is a certain amount of unevenness in style. This is the result of a conscious pedagogical decision, since I felt that the yet unexperienced reader should be at first spared the somewhat bleak succession of formal developments and should make fast progress in grasping basic structural features.

I believe that this book may fill the needs of most theoretical physicists (especially of those interested in quantum theory, high energy physics, relativity, modern statistical physics), many research engineers, and even other scientists who are concerned with structural problems, such as systems analysis. For students in these fields, this work is essentially a graduate level text. It is also possible that professional students of mathematics, in their earlier stage of studies (sophomore or in some cases junior undergraduate level), may profit from this volume inasmuch as it gives a survey of standard topics that they will have to study, eventually, in considerable depth.

This book may be used for self-study, since it is self-contained and, frankly speaking, originated from the author's self-study. For those who are already familiar with certain topics, this treatise may be used as a reference or quick refresher of once-learned but forgotten subjects. Pleasant experience has shown that the book will well serve as a textbook for a two- or three-semester course. Chapters 3 and 4 (with the essentials of the preceding two chapters) may be used for a short course or seminar in modern algebra. Similarly, Chapters 5 and 6 serve as a (somewhat superficial) guide to topology, and Chapters 7 and 8 to measure and integration. Those who wish to teach a leisurely one-semester course on the basics of functional analysis (especially Hilbert space theory) may start (for an already primed audience) with Chapter 9 and conclude with Chapter 12, with some material from Chapter 13 added on. Chapter 13, on the other hand, may be a good supplement for those who wish to penetrate deeper in the already absorbed field of Hilbert space operator theory. In connection with Chapters 12 and 13 I would like to mention that, having in mind the needs of the quantum theorists, I gave considerable attention to questions of domains and to unbounded operators, which is a topic unduly neglected in most introductory and even intermediate level texts.

In consequence of many factors, it was impossible to consistently show "applications" or even to indicate the areas where the discussed theorems and methods are particularly useful or necessary. I am aware that an occasional muttering about quantum theory does not indicate the physical importance of the discussed concept. However, this is a book on mathematics and I fervently hope that the student who patiently made his way through it will be able to understand any contemporary paper or book in the frontier areas of theoretical physics and to use with confidence any original and professional mathematical source that is needed to enlarge the knowledge he gained from this volume.

It was unavoidable that many topics, as important as those discussed in the text, had to be completely omitted from consideration. I particularly regret the omission of an introduction to topological (especially Lie) groups and their representation theory. Many topics of functional analysis that I discussed would have been well illustrated and put to direct use by the inclusion of a chapter on nonsingular Fredholm integral equations, to say the least. Commuting sets of operators and their spectral representation, complete sets of commuting operators, von Neumann algebras and related topics would have been also most desirable to be included, had space permitted this. I also regret the total omission of topics on differentiable structures. But I believe the interested reader is well equipped to study now these topics on his own. The annotated reading list in Appendix III may serve as a guide for him where to turn next.

PAUL ROMAN

Boston, Massachusetts

Organization of the Book

The logical connection of topics covered in this book is symbolized by the chart below. The numbers in the boxes refer to chapter numbers.

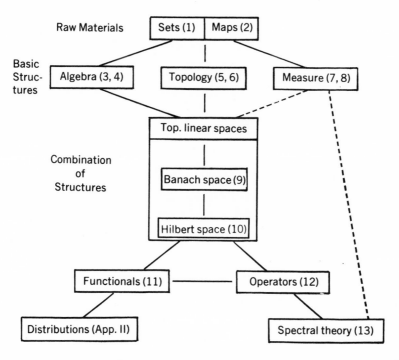

Each chapter consists of several sections and some sections have one or more subsections. Especially in the earlier parts of the book, the beginning paragraphs of a chapter or of a section may be quite long, without carrying any special number. (The Table of Contents will be helpful to locate subdivisions.) Subsection b of Section 4 in Chapter 13 is referred to by the symbol 13.4b, etc. Definitions and theorems are numbered according to their standing in a subsection or section. Thus, Theorem 13.4b(3) indicates the third theorem in Subsection 13.4b and Definition 13.4b(3) stands for the third definition in the same subsection. In the beginning part of a section (before a subsection, if any, is reached) the notation would be Theorem (or Definition) 11.2(3). Occasionally, one will find related theorems such as, say, Theorems 12.3b(3a) and 12.3b(3b) in succession. Figures and tables are numbered successively in each chapter. Thus, Fig. 13.3 refers to the third figure in the entire Chapter 13. Equations are numbered only if frequent reference is made to them or (especially in the beginning of the book) when special interest is drawn to them, for pedagogical reasons. Equations are numbered consecutively through each chapter. Thus, Eq. (13.4) refers to the fourth displayed and numbered equation in Chapter 13. Occasionally, in proofs, examples, or short discussions, for ease of reference during the discussion, equations are labeled by Greek letters. Thus, Eq. (γ) or simply (γ) refers to a so labeled, displayed equation in the immediate vicinity. There are many illustrative examples in the text. They are labeled in each subsection (or section), in alphabetic order, by Greek letters. Thus, we may refer to Example γ in Section 13.4 or Example α in Subsection 13.4a.

A special comment on problems is in order. Each subsection, most beginning parts of sections, and sometimes even the introductory parts of chapters are followed by a selection of problems. For example, Problem 13.4b–8 would refer to the eighth problem at the end of Subsection 13.4b, whereas Problem 12.2–3 is the third problem at the end of the initial part of Section 12.2 (just before Subsection 12.2a starts). Needless to say, the problems form an integral and even crucial part of the book. There are over 600 problems in toto and the reader is strongly urged to attempt all of them—but at the very least, half of them. (Experience shows, it can be done, within one academic year!) Some of the problems are in the same vein as the examples in the text, i.e. they are simple illustrations of theorems or definitions. Many other problems are designed to challenge the reader's understanding on the deeper level and to test his recall of earlier results. Finally, a certain proportion of problems is intended to extend the material that was covered in the text or to introduce some new concepts which, for one reason or another, could not be fitted smoothly into the main text. Occasionally, these problems introduce minor theorems which are then used in the main text, sometimes repeatedly. In many problems, and not only in the more difficult ones, hints are given to facilitate the solution. This is meant as an encouragement and the student is advised to first ignore the hints and find his own argument. For the successful tackling of the problems, no external reading whatsoever is necessary.

A final comment on references. In accord with standard custom in the mathematical textbook literature (and because of impractibility), no references to original publications are made. However, on occasions when we either omitted the proof of an important theorem or when details were neglected, we call the reader's attention to some standard text where he can look up the details. Such a reference will be indicated by a sentence like "See Helmberg[16], p. 215," which refers to Helmberg's book numbered as [16] in Appendix III of this treatise. Occasionally reference is made, quoting complete bibliographical data, to books that are not listed in Appendix III.

P. R.

Publisher's Note: For convenience to the student, this work has been divided into two volumes. Appendix II, which appears in Volume 2, does not appear in Volume 1 because it has no application there.

Acknowledgments

Departing from standard academic practice, I wish to use this occasion for expressing my gratitude, esteem, and indebtedness to several persons who, even though they were in no way directly connected with the project of this book, had a profound and formative influence on my mathematical education and on my attitude to mathematics.

First in chronological order, I wish to pay tribute to the memory of my Junior High School teacher, DR. J. MENDE. He perished tragically during the last days of the siege of Budapest, in January 1945. Next I recall the image of a friend of my early youth, the polyhistor P. T. MESSER. He was brutally murdered in the Spring of 1945, in Mauthausen, at the tender age of 21 years, before his talents became known to the world. It was only a few months later that my inspiring teacher, PROF. B. VON KERÉKJÁRTO, the well-known topologist of the 1930's, departed from this world, in consequence of illness and malnutrition suffered in the last phase of World War II. It was he who first revealed to me the fascinating interplay of analysis and algebra and, more generally, who made me aware of the unity of mathematics. Finally, even though I never had the opportunity to be his pupil or to work under his guidance, I wish to acknowledge the strong influence of Nobel Laureate PROF. E. P. WIGNER of Princeton University on my attitude to the role of mathematics in physics. I shall never forget the summer of 1946 when, amongst the ruins of my hometown, I first read a battered copy of his classic volume *Gruppentheorie und ihre Anwendung auf die Quantenmechanik der Atomspektren*, which was lent to me by my esteemed physics professor, the late K. F. NOVOBÁTZKY.

Turning to the conventional part of these acknowledgments, I would like to express my sincere thanks to a number of colleagues without whose interest, enthusiasm, and keen criticism this work would never have grown from a set of amateurish lecture notes into a (hopefully, more professional) book. To start with, I pay my tribute to MR. A. SIEGEL who, as a mathematics–physics junior at Boston University, attended my lectures in 1971–1972 and read, with unparalleled sharpness, the entire manuscript, discovering and correcting an alarming number of errors. Most helpful was also my graduate student and teaching assistant, MR. P. L. HUDDLESTON, who in the period 1970–1972 contributed a great deal to the eradication of mistakes and inconsistencies from the earlier version of the manuscript. Several other students who took the course, in particular MR. S. M. WOLFSON and MR. S. I. FINE, deserve most sincere thanks. Last but not least, I record my warmest gratitude to my friend, PROF. B. GRUBER of Southern Illinois University who carefully read the final manuscript and suggested numerous corrections and improvements.

P.R.

Introduction

Four opinions on mathematics:

1. The negativist: "Mathematics is the science where one does not know what one is speaking about, nor if what one says is true." (B. RUSSELL)
2. The humanist: "Mathematics is one of those subjects which obliges one to make up his mind." (L. G. DES LAURIERS)
3. The artist: "Mathematics is not a question of calculation perforce but rather the presence of royalty: a law of infinite resonance, consonance and order." (LE CORBUSIER)
4. The modern descriptionist: "Mathematics is the logical study of *relationships* among certain entities, and not the study of their *nature*." (J. DIEUDONNÉ)

Since the days of Galileo it is commonly accepted that mathematics is the language of science and that the laws of nature are expressible in the language of mathematics. Equally important but apparently less appreciated, however, is the notion that mathematics plays also a more fundamental, more sovereign role in the formulation of scientific ideas. The point can be made that mathematics *per se* is the necessary vehicle for formulation, comparison, description, verbalization, and structuring of all scientific and logical thought. As E. P. Wigner repeatedly emphasized, "... the mathematical language has more to commend it than being the only language which we can speak; ... it is, in a very real sense, the correct language." Indeed, it appears that the human mind cannot deal directly with reality. Therefore, as H. Eyraud puts it, "in its continuous effort to understand nature, to grasp and to express its laws, the human mind has been led to construct an edifice of symbols and operations which constitutes mathematics."

This fundamental role of mathematics in human scientific thought is rather surprising and even mystifying. A partial clarification is perhaps provided if one considers the purpose and aim of modern mathematicians. It appears that, in their view, *the primary goal of mathematics is the study of structures.* This attitude, which developed only during the last 50 or 70 years, is radically different from the approach of classical mathematics. Classical mathematics was essentially "constructionist" in its nature. In order to prove the existence of an object (a function, a number, etc. satisfying certain conditions), it was thought necessary to give a procedure for constructing this particular object. In contrast, the objects of modern mathematics are abstract symbols, identified by "description" only. These objects are grouped together and their *relationships* are studied. The "truth" of a relationship is ascertained by *rules* which demand merely the examination of the *form* of the relevant connections between the objects.

The preceding paragraph was not intended to give a professional evaluation of the edifice of modern mathematics.† It, however, may help one to understand the close tie that exists between the exploration of laws of nature and the working of modern mathematics. Both the natural sciences and mathematics are ultimately concerned with the study of structures.

In the following we give a brief review of mathematical structures and their interdependence.

The basic raw material of mathematical (and more generally, scientific) thought is the concept of a *set*. A set is a completely amorphous, structureless object. Sets can be compared and related by the fundamental notion of a *map*. Speaking figuratively in terms of a linguistic analogy, sets are the "objects" and "subjects," whereas maps play the role of "verbs."

Sets can be given various *structures*. A set endowed with some structure is usually called a *system* or sometimes a *space*.

At the present time, the human mind can conceive of only two basic types of structure:‡

    (a) Algebraic,
    (b) Analytic.

---

†Indeed, I am surely not qualified for such a task. The interested reader may find a penetrating analysis of these questions in the essay by J. Dieudonné, "Modern axiomatic methods and the foundations of mathematics," p. 251 in Volume II of *Great Currents of Mathematical Thought*, (F. Le Lionnais (Ed.), Dover, 1971.

‡These and all other concepts mentioned in this Introduction will of course be defined and explored more rigorously in the subsequent chapters. In the present overview, we give only a brief, intuitive description.

Algebraic structures are essentially those which permit the *composition* of two (or more) elements, leading to another element of the set. Analytic structures, on the other hand, have two major subclasses. In the case of *topological structures* we are concerned with relations between elements that can somehow be characterized by the concept of *neighborhoods*. The second subclass of analytic structures, which can be called *measure spaces*, are used to formulate the notion of *extent* (area, volume, mass, charge, etc.). Briefly speaking, analytic structures subsume all familiar geometrical systems.

Naturally, a given set may possess more than one kind of structure. As a matter of fact, the more structure a set carries, the more interesting it is, and its study leads to a richer theory. Much of the day-to-day work of contemporary mathematics consists precisely of interrelating various structures in an intimate and mutually interdependent manner. *Functional analysis*, for example, superimposes a topological and/or measure structure on algebraic systems and may also be called "topological algebra." On the other hand, one may consider the algebraic composition of certain given topological structures and is thus led to what became known as *algebraic topology*.

The concept of maps (functions) attains a special importance in connection with structured sets. This is so because then *the mapping will relate different structures*. For example, maps on algebraic structures lead to the detailed theory of algebraic systems, such as group theory. Maps on topological spaces reveal many theorems of classical analysis and their generalizations. Maps on measure spaces give rise to the theory of integration. The study of maps on spaces which have combined analytic and algebraic structures leads to the riches of modern functional analysis and includes topics like functionals, operators, Hilbert space theory, generalized functions, integral equations, and much more.

We now set out on our journey to systematically explore the structures of mathematics.

PART ONE

The Raw Materials of Mathematics

1

Sets

The concept of a set is the most general and most fundamental notion of mathematics. Precisely because of its utmost generality, it is extremely difficult to give a logically unassailable definition of a set. G. Cantor, the creator of set theory said that *a set is "a collection of definite and distinct objects of our intuition or of our thought."* Thus, instead of the term "set," we often use the equivalent expressions "ensemble," "collection," or even "class" of objects. Correspondingly, we may talk about the set of students in a specified room, about the set of all integers, about the set of all circles in the plane, and so on.

The apparent simplicity of this notion is deceptive. Indeed, a careless use of the above definition quickly leads to logical deficiencies and even to so-called "paradoxes." For many decades, a bitter war was fought between mathematicians of different schools with the aim of eliminating the enormous difficulties. It is not within our power to even touch upon the now accepted framework of "axiomatic set theory," which itself is a fundamental chapter in the foundations of mathematics.

On the other hand, as far as we are concerned, we can circumvent the pitfalls by a simple consideration. We shall never be interested in "abstract" sets. All the sets we are going to consider will always form "parts" of some "bigger" set, whose existence we merely *postulate.* Thus, we decide to work in the comforting environment of a given "universe of discourse." More technically, in every specific study we shall assume, either explicitly or implicitly, the existence of some *universal set S*, and all sets $A, B, \ldots, X, Y, \ldots$ we consider, will be collections of objects which are taken from the universal set S. For example, if we are interested in the geometry of the plane, our stipulated universal set "consists of all points of the plane" and the various sets that we study are different geometrical configurations such as circles, squares, etc.

As a rule, we denote sets by capital letters and *members* or *elements* of a set by lowercase letters. If x is an element of the set X (i.e. if it belongs[†] to X), we write $x \in X$. The notation $x \notin X$, on the contrary, means that x is not a member of X.

It is now fairly straightforward to unambiguously characterize a set. Let S be a given universal set and let P denote some "recognizable" property which some elements of S might have. Then we may specify a set X by collecting all elements of S which share the property[‡] P. We then write

$$X = \{x \,|\, x \in S \quad \text{and} \quad x \quad \text{has the property} \quad P\}. \tag{1.1}$$

and we read this formula as the sentence "The set X consists of elements x such that x belongs to S and has property P." Here the symbol x represents a "generic element" of X and the vertical line $|$ is short for the term "such that." Since the existence and nature of the universal set is usually obvious, we often abbreviate the notation and simply write

$$X = \{x \,|\, x \quad \text{has property} \quad P\}. \tag{1.1a}$$

For example, let the universal set be the set $\mathbf{N}$ of all natural numbers (0, 1, 2, etc.). Then

$$\mathbf{E}^+ = \{x \,|\, x \in \mathbf{N} \quad \text{and} \quad x = 2y \quad \text{for some} \quad y \in \mathbf{N}\}$$

defines the set of all even nonnegative numbers (0, 2, 4, etc.). For short, we just write

$$\mathbf{E}^+ = \{x \,|\, x = 2y \quad \text{for some} \quad y \in \mathbf{N}\}.$$

In many cases, an alternative notation for sets may be used which consists simply by enumerating the elements. For example,

$$\mathbf{n} = \{0, 1, 2, \ldots, n\}$$

denotes the set of the first n natural numbers.[§] Even if it is impossible to

---

†Since we are not attempting a rigorous, axiomatic study of set theory, we shall use loose terms (like "belongs") without any further a-do.

‡We may also say that we collect those elements of S "for which the proposition P is true."

§The universal set is not indicated. We tacitly assume this to be, say, the set of all natural numbers; but many other choices are possible, e.g., the set of reals.

"write down" all elements, convention and habit often permit the use of this notation. For example,

$$\mathbf{N} = \{0, 1, 2, \ldots\}$$

will clearly indicate to everyone the set of all natural numbers.

We remark here that, obviously, it does not matter in what order we list the elements of a set. Thus, for example,

$$\{a, b, c\} = \{a, c, b\} = \{b, a, c\} = \{b, c, a\} = \{c, a, b\} = \{c, b, a\}.$$

This is also a good opportunity to remind the reader that, by definition, a set contains each of its elements exactly once, and no repetitions occur.

Occasionally we meet sets which contain a single element, say $x \in S$. Such a set is called a *singleton* and is denoted by $\{x\}$. The singleton *set* $\{x\}$ must not be confused with the *element* x of S.

It is logically possible (and, in fact, desirable) to define a set which contains no elements whatsoever. This set is called the *empty set* and is denoted by the symbol $\emptyset$. One way to precisely define $\emptyset$ is to select a property P which does not pertain to any element of the (given) universal set† S. Then Eq. (1.1) will describe the empty set.‡ For example, let P be the proposition $x \neq x$. Then

$$\emptyset = \{x \mid x \neq x\}$$

describes the empty set.

We have one final, important general remark. Given a universal set S, it is perfectly legitimate to define sets whose elements are themselves sets. In order to avoid the cumbersome term "a set of sets," we often use the phrase "collection of sets" or "class§ of sets." For example, let S be the Euclidean plane. We define X to be the class of all equilateral triangles T. Note that each equilateral triangle T is itself a set, namely a well-defined set of points of S. However, when we talk about X, we are not interested in the fact that each element T of X is made up of individual points of S. Each T must be considered in its own right. The attentive reader will notice that tacitly we abandoned our universal set S and assumed a new universal set $\hat{S}$, which can be thought of, say, as consisting of all plane triangles.

---

†In other words, we make a proposition P which is false for every $x \in S$.

‡It can be shown that this definition is unique, i.e. there exists only one empty set.

§Strictly speaking, the terminology "class" should be avoided because in the axiomatic theory of sets it is used in another technical sense.

PROBLEMS

1-1. Which of these sets is the empty set? (Assume $S = \{\text{real numbers}\}$.)
(a) $A = \{x \,|\, x^2 - 9 \quad \text{and} \quad 2x = 4\}$,
(b) $B = \{x \,|\, x^2 \neq x\}$,
(c) $C = \{x \,|\, x + 8 = 8\}$,
(d) $D = \{x \,|\, x^2 < 0\}$.

1-2. Let $\emptyset$ denote, as always, the empty set. Explain the meaning of the following symbols:

$$\{\emptyset\}, \quad \{\{\emptyset\}\}, \quad \{\emptyset, \{\emptyset\}\}.$$

1.1 OPERATIONS WITH SETS

In this section we shall familiarize ourselves with the basic manipulations of sets. Without always explicitly saying so, we assume that all sets $A, B, \ldots$ are defined in reference to some specified universal set S.

Two sets are said to be *equal* if they have exactly the same elements. We write $A = B$. A set A is said to be a *subset* of a set B iff† $x \in A$ implies that $x \in B$. We also say that "A is included in B." We denote set inclusion by writing $A \subset B$. Sometimes it is advantageous to write this the other way around, i.e. in the form $B \supset A$ (B includes A).

It is important to note that *our definition of set inclusion permits the special case that the two sets are actually equal.* If $A \subset B$ but $A \neq B$, then we often emphasize this by saying that A is a proper subset‡ of B.

It is clear that $A = B$ iff $A \subset B$ *and* $B \subset A$. As a matter of fact, the only way to prove that A equals B is to show that A is a subset of B and B is a subset of A.

It is trivial but important to observe that, given any set X, we have $X \subset S$ and furthermore $\emptyset \subset X$. Also, if $A \subset B$ and $B \subset C$, then $A \subset C$.

Often one considers the class of all possible subsets of a given (universal) set S. This set of sets is called the *power set* of S and will be denoted by $\mathscr{P}(S)$. Thus,

$$\mathscr{P}(S) = \{A \,|\, A \subset S\}. \tag{1.2}$$

For example, if $X = \{1, 2\}$, then

---

†We remind the reader that the ugly shorthand "iff" stands for "if and only if."

‡Many texts reserve the symbol $\subset$ for the case of *proper* inclusion and use, for the general case, the symbol $\subseteq$. Both conventions have their advantages and disadvantages. The reader should be aware that our notation $A \subset B$ will always mean that either A is a proper subset of B or A is identical with B.

$$\mathscr{P}(X) = \{\{1, 2\}, \{1\}, \{2\}, \emptyset\}.$$

Note that $\emptyset$ and S itself always belong to $\mathscr{P}(S)$.

In the following we shall discuss methods which are used frequently to define new sets from one or more given sets.

Let A be an arbitrary set. The set of all elements x which belong to the universal set S but do not belong to A, is called the *complement* of A and will be denoted by A^c. Thus (cf. Fig. 1.1a),

$$A^c = \{x \,|\, x \in S \quad \text{and} \quad x \notin A\}. \tag{1.3}$$

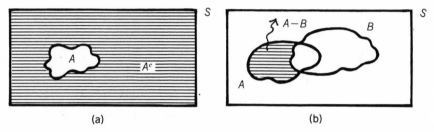

(a)   (b)

Fig. 1.1.

It should be obvious that

$$(A^c)^c = A, \tag{1.4a}$$

$$\emptyset^c = S, \tag{1.4b}$$

$$S^c = \emptyset. \tag{1.4c}$$

Furthermore,

$$A \subset B \quad \text{implies} \quad B^c \subset A^c. \tag{1.4d}$$

To give an example, let S be the set of complex numbers and let A be the set of reals. Then A^c is the set of all nonreal numbers, i.e. the set of numbers $z = x + iy$ with $y \neq 0$.

Let now A and B be two arbitrary sets. We define their *difference* $A - B$ to be the set whose elements are those elements of A that are not contained in B. Thus,

$$A - B = \{x \,|\, x \in A \quad \text{and} \quad x \notin B\}, \tag{1.5}$$

as is illustrated in Fig. 1.1b. For example, let S be the set of all students at Boston University. Let A be the set of all students attending course

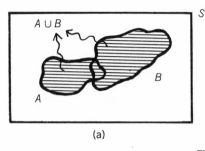 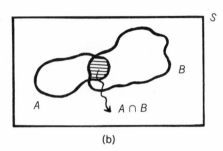

(a)   (b)

Fig. 1.2.

PY703. Let B be the set of all girls attending Boston University. Then $A - B$ is the set of all boys attending course PY703.

For obvious reasons, $A - B$ is sometimes called the complement of B relative to A. Indeed, we also have

$$A^c = S - A.$$

But it must be emphasized that $A - B$ makes sense even if B is not a subset of A.

Now we turn out interest to the basic composition rules of sets, unions, and intersections. Let A and B be arbitrary sets. We define their *union*, to be denoted by $A \cup B$, as the set which contains exactly all elements of A and B together. Thus,

$$A \cup B = \{x \mid x \in A \quad \text{and/or} \quad x \in B\}. \tag{1.6}$$

The union is illustrated† in Fig. 1.2a. In a similar way, the *intersection* $A \cap B$ of A and B is defined as the collection of elements that are common to A and B. Thus,

$$A \cap B = \{x \mid x \in A \quad \text{and} \quad x \in B\}, \tag{1.7}$$

as illustrated in Fig. 1.2b. If A and B have no common elements, i.e. if $A \cap B = \emptyset$, then we say that A and B are *disjoint*.

We list below the fundamental properties which govern unions and intersections. The reader should have no difficulty to verify these rules directly from the definitions. It will be also useful if he illustrates these rules by drawing diagrams. We find:

$$A \cup B = B \cup A, \tag{1.8a}$$

---

†This illustration reveals why in Eq. (1.6) the word "and/or" is used rather than just "or." The overlapping part of A and B gives elements of $A \cup B$ which are both in A *and* B.

$$A \cap B = B \cap A, \tag{1.8b}$$

$$A \cup (B \cup C) = (A \cup B) \cup C, \tag{1.9a}$$

$$A \cap (B \cap C) = (A \cap B) \cap C, \tag{1.9b}$$

$$A \cup (B \cap C) = (A \cup B) \cap (A \cup C), \tag{1.10a}$$

$$A \cap (B \cup C) = (A \cap B) \cup (A \cap C). \tag{1.10b}$$

For reasons to be discussed in Chapter 3, the relations (1.8a, b) are referred to as the laws of commutativity, Eqs. (1.9a, b) are the associative laws, and Eqs. (1.10a, b) are two laws of distributivity. In addition to these laws, the reader should memorize the rather trivial relations

$$A \cup A = A, \tag{1.11a}$$

$$A \cap A = A, \tag{1.11b}$$

$$A \cup A^c = S, \tag{1.12a}$$

$$A \cap A^c = \emptyset. \tag{1.12b}$$

In many applications, the so-called De Morgan laws come in very handy. They connect unions and intersections with complementation:

$$(A \cup B)^c = A^c \cap B^c, \tag{1.13a}$$

$$(A \cap B)^c = A^c \cup B^c. \tag{1.13b}$$

The formation of unions and intersections can be generalized for the case of more than two sets. Let us have a collection of sets. Suppose that the members of this collection are labeled† by some index α. Thus, a typical element of the collection will be denoted by A_α and we may symbolize, somewhat unconventionally, the whole collection by $\{\ldots A_\alpha \ldots\}$. We now define the union of all sets A_α by

$$\bigcup_\alpha A_\alpha = \{x \,|\, x \in A_\alpha \quad \text{for at least one } \alpha\} \tag{1.14}$$

and, similarly, we define the intersection of all sets A_α by

$$\bigcap_\alpha A_\alpha = \{x \,|\, x \in A_\alpha \quad \text{for all} \quad \alpha\}. \tag{1.15}$$

It is clear that if $\alpha = 1, 2$ (i.e., if $\{\ldots A_\alpha \ldots\} = \{A_1, A_2\}$), then these definitions give back, as a special case, Eqs. (1.6) and (1.7), respectively.

†We are somewhat loose here in our presentation and intuitively anticipate the notion of an "indexed family of sets" which will be discussed in Section 2.5.

Incidentally, we note that if $\{\ldots A_\alpha \ldots\}$ consists of finitely many or countably many[†] members, then it will be our custom to use, instead of α, a Latin label, such as k. In that case, we also use often the more detailed notations

$$\bigcup_{k=1}^{n} A_k \quad \text{or} \quad \bigcup_{k=1}^{\infty} A_k \quad \text{and} \quad \bigcap_{k=1}^{n} A_k \quad \text{or} \quad \bigcap_{k=1}^{\infty} A_k.$$

But it should be borne in mind that the above generalized definitions for unions and intersections have unrestricted validity, even if "the index α is not discrete."

We give an example of the above notions. Let A_k be the set of real numbers x such that $-1/k \leq x \leq +1/k$. Consider the collection $\{\ldots A_k \ldots\}$, $k = 1, 2, \ldots$. Then

$$\bigcup_{k=1}^{\infty} A_k = A_1, \qquad \bigcap_{k=1}^{\infty} A_k = \{0\}.$$

The composition rules (1.8) through (1.10) are easily generalized to unions and intersections of arbitrary collections of sets. For example,

$$B \cap (\bigcup_{\alpha} A_\alpha) = \bigcup_{\alpha} (B \cap A_\alpha).$$

The generalized De Morgan laws read

$$(\bigcup_{\alpha} A_\alpha)^c = \bigcap_{\alpha} A_\alpha^c, \tag{1.16a}$$

$$(\bigcap_{\alpha} A_\alpha)^c = \bigcup_{\alpha} A_\alpha^c. \tag{1.16b}$$

Apart from the formation of unions and intersections, there exists another, rather different and very important method of constructing new sets from given sets. To discuss this method, we first need the concept of an *ordered pair*. An ordered pair (a, b) consists of two objects a and b written in this order.[‡] The ordered pairs (a, b) and (a', b') are said to be equal iff $a = a'$ and $b = b'$. (From this follows that $(a, b) \neq (b, a)$ unless the objects a and b are identical.)

Let now A and B be two arbitrary (nonempty) sets. We define the *Cartesian product* $A \times B$ of these two sets as the collection of all ordered pairs (a, b) which can be formed by taking a first term from A and a second term from B. Thus,

$$A \times B = \{(a, b) | a \in A \quad \text{and} \quad b \in B\}. \tag{1.17}$$

[†]The precise meaning of these terms will be clarified in Section 2.4.

[‡]This somewhat intuitive notion of an ordered pair can be given a purely set-theoretic definition, but at this stage we do not wish to burden the reader with questions of rigor. An alternative definition of an ordered pair (or n-tuple) will be given in Section 2.5.

In general, $A \times B$ is "bigger" than A or B, because if A has n elements and B has m elements, then $A \times B$ has nm elements. It should be noted that $A \times B \neq B \times A$.

We give now simple examples:

Example α.   Let $A = \{0, 1\}$ and $B = \{0, 2\}$. Then

$$A \times B = \{(0, 0), (0, 2), (1, 0), (1, 2)\},$$

$$B \times A = \{(0, 0), (0, 1), (2, 0), (2, 1)\}.$$

Example β.   Let A be the set of real numbers x such that $a \leq x \leq b$ and let B be the set of real numbers y such that $c \leq y \leq d$. In the familiar manner of Cartesian "analytic geometry" we may visualize these sets as portions of the "X-axis" and "Y-axis," respectively. The product set $A \times B$ can then be thought of as a portion of the "XY-plane," and is visualized† in Fig. 1.3.

It is of course possible to take the Cartesian product $A \times A$ of a set A with itself. For example, the Cartesian product of the set **R** of all real numbers with itself is the set of all ordered pairs (x, y) of real numbers.

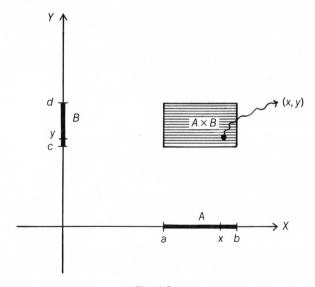

Fig. 1.3.

†Note that $B \times A$ would be visualized as a portion of the "YX-plane."

We denote $\mathbf{R} \times \mathbf{R}$ by $\mathbf{R}^2$ and usually we identify it with the "coordinate plane."

The *diagonal* of $A \times A$ is the subset of points $(a, b) \in A \times A$ such that $a = b$. Thus,

$$\text{Diag } A \times A = \{(a, a)|a \in A\} \subset A \times A. \tag{1.18}$$

The definition of the Cartesian product of two sets is easily extended to the case of several factors. This is achieved by the almost self-evident notion of *ordered n-tuples*† $(a, b, c, \ldots, n)$. If $A_1, A_2, \ldots, A_n$ are arbitrary (nonempty) sets, then one defines

$$A_1 \times A_2 \times \cdots \times A_n = \{(a_1, a_2, \ldots, a_n)|a_k \in A_k; \quad k = 1, 2, \ldots, n\}. \tag{1.19}$$

Instead of $A_1 \times \cdots \times A_n$ we sometimes write $\Pi_{k=1}^n A_k$. If $A_1 = A_2 = \cdots = A_n \equiv A$, then we often write for their product A^n.

It is possible to generalize the formation of Cartesian products for arbitrary (not even necessarily countable) collections of sets. However, we must postpone the discussion of this procedure until the end of Section 2.5.

PROBLEMS

1.1-1. Let $X = \{a, b, c\}$. Which of the following statements are true and which are false?
(a) $a \in X$,   (b) $a \subset X$,   (c) $\{a\} \in X$,   (d) $\{a\} \subset X$.

1.1-2. Find the power set of $X = \{3, \{1, 4\}\}$.

1.1-3. Show that any one of the conditions $X \subset Y$, $X \cap Y = X$, and $X \cup Y = Y$ implies the other two conditions.

1.1-4. Prove the important relation $A - B = A \cap B^c$.

1.1-5. Let $A = \{1, 2, 3, 4\}$, $B = \{3, 4, 5, 6\}$. Find $A - B$ and $B - A$.

1.1-6. Prove that $B - A$ is a subset of A^c, and that $B - A^c = B \cap A$.

1.1-7. Prove the following theorems:
(a) $A \subset B$ iff $A - B = \emptyset$,
(b) $A - B = A$ iff $A \cap B = \emptyset$,
(c) $A - (A - B) = A \cap B$,
(d) $A \cap (B - C) = (A \cap B) - (A \cap C)$.

1.1-8. Let $\mathcal{P}(X)$ denote the power set of X. Prove the following:
(a) $\mathcal{P}(X) \cap \mathcal{P}(Y) = \mathcal{P}(X \cap Y)$,
(b) $\mathcal{P}(X) \cup \mathcal{P}(Y) \subset \mathcal{P}(X \cup Y)$.

†The method of induction is used here implicitly. The ordered triple of a, b, c is looked upon as an ordered pair whose first component is itself an ordered pair. Thus, $(a, b, c) = ((a, b), c)$. Then $(a, b, \ldots, n) = ((a, b, \ldots, n - 1), n)$.

1.1-9. Let $X = \{a, b\}$, $Y = \{2, 3\}$, and $Z = \{3, 4\}$. Find $(X \times Y) \cap (X \times Z)$.

1.1-10. Show that $A \times B = \emptyset$ iff either A or B is the empty set.

1.1-11. Let $X = A \times B$ and $Y = C \times D$ and assume that neither X nor Y is empty. Show that

$$X \subset Y \quad \text{iff} \quad A \subset C \quad \text{and} \quad B \subset D.$$

1.1-12. Prove the following theorem: If $A \times B = C \times D \neq \emptyset$, then $A = C$ and $B = D$. (*Hint*: Use the result of the preceding problem.)

1.1-13. Show that

$$A \times B = \bigcup_{\beta} C_\beta,$$

where, for each $\beta \in B$, we defined the set C_β by $C_\beta = A \times \{\beta\}$.

1.1-14. Prove the following distributive laws:
(a) $A \times (B \cup C) = (A \times B) \cup (A \times C)$,
(b) $A \times (B \cap C) = (A \times B) \cap (A \times C)$,
(c) $A \times (B - C) = (A \times B) - (A \times C)$.

1.2   RELATIONS IN SETS

The usual statement that a set has no *a priori* structure should not be construed to mean that there are no "relations" between its elements. In this subsection we shall study precisely such relations.

From the purely logical point of view, a relation between elements of a given set X may be introduced as follows. Let R be a proposition involving ordered pairs of elements of X which may be either true or false. If, for a particular pair (a, b) the proposition R is true, then we say that "a stands in relation R to b." We indicate this by writing $a\,R\,b$. For example, if X is the set of reals and R is the proposition "$a - b$ is a negative number," then any a and b for which this is true are related to each other by the proposition R. It is customary and convenient to use a special symbol for this relation, and we write $a < b$ which we read as "a is smaller than b."†

It is a very important and useful fact that the concept of relation may be defined in a purely set-theoretic language. This is made possible by the circumstance that, as we emphasized at the beginning of this chapter, sets are specified in terms of selecting elements by propositions. Let X be a set and let R be a relation in it, as defined above. Then, clearly, the ordered

---

†In passing we note that we are interested only in what one sometimes calls "binary" relations. It is possible to define ternary, etc. relations. An example would be the relation of parenthood between people: a, b, c are related by parenthood if c is the offspring of a and b.

pairs (a, b) of related elements form a certain subset of $X \times X$. Conversely, if we choose an arbitrary subset of $X \times X$, i.e., select a collection of ordered pairs (a, b) that satisfy some proposition, we thereby specify a relation in X. Thus, we can adopt the following *definition*: A relation in the set X is a specified subset R of $X \times X$. We say that "a stands in relation R to b" if $(a, b) \in R \subset X \times X$. We denote this circumstance by writing $a R b$.

This definition identifies the concept of a relation R with a certain set R of ordered pairs and this identification is emphasized by the notation.

Example α.   Let X be the set of reals, let R be the subset of $X \times X$ given by

$$R = \{(a, b)|a - b \quad \text{is a negative number}\}.$$

The corresponding relation is that of "a is smaller than b."

Example β.   Let $\mathbf{Z}$ be the set of integers, let $R \subset \mathbf{Z} \times \mathbf{Z}$ be given by

$$R = \{(a, b)|a \neq 0, \quad b = na \quad \text{for some integer} \quad n\}.$$

Then the relation $a R b$ means: "a divides b."

The notion of a relation may be generalized so as to connect elements of different sets. If X and Y are two sets, then "a relation from X to Y" is a subset R of $X \times Y$ and we write $x R y$ iff $(x, y) \in R$.

Example γ.   Let M be the set of all men, W be the set of all women. Let R be the set of ordered pairs (m, w) where m and w are a man and a woman wedded to each other. Clearly, $R \subset M \times W$. The relation $m R w$ is that of "being married."

Example δ.   Let T be the set of all triangles, Q be the set of all squares. Let (t, q) be a pair of a triangle and square which have the same area. The collection R of all such pairs is a subset of $T \times Q$ and defines the relation of "having equal area."

Many relations have interesting general properties. We shall use the following *terminology*:

(a) A relation R in X is *reflexive* if $a R a$ for every $a \in X$, i.e. if every element stands in relation R to itself. This implies that $(a, a) \in R$, i.e. R contains the diagonal of $X \times X$.

(b) A relation R is *symmetric* if $a R b$ implies $b R a$.

(c) A relation R is *transitive* if $a R b$ and $b R c$ imply that $a R c$.

(d) A relation R is *antisymmetric* if $a R b$ and $b R a$ imply that $a = b$.

Example ε.   Let X be the set **Z** of all integers. The relation "m divides n" is reflexive and transitive but neither symmetric nor antisymmetric. The relation "$m^2 + m = n^2 + n$" is reflexive, symmetric, and transitive, but not antisymmetric. The relation "$m < n$" is transitive and antisymmetric,† but not reflexive and not symmetric. The relation "m and n have opposite signs" is symmetric, but not reflexive, not transitive, and not antisymmetric.

In considering properties of relations, one should remember the trivial fact that, in general, for a given couple a and b, neither $a\,R\,b$ nor $b\,R\,a$ need hold and in fact, for a given a, there may not exist any b such that $a\,R\,b$ holds. Conversely, it may happen that *every* given element a is related to some, maybe even infinitely many elements b.

PROBLEMS

1.2-1. Let R be the relation from $A = \{2, 3, 4, 5\}$ to $B = \{3, 6, 7, 10\}$ defined by the proposition "x divides y." Express R in terms of the "solution set" of R, i.e. find the relevant subset R of $A \times B$.

1.2-2. Which of the properties "reflexive," "symmetric," "transitive," "antisymmetric" applies to the following relations in the set of reals:
(a) $|x| \leq |y|$,
(b) "$x - y$ is a multiple of 2π,"
(c) $x^2 + y^2 = 1$.

1.2-3. Let R and R' be two relations in a set X. Show that:
(a) If R and R' are both symmetric, then $R \cup R'$ is symmetric.
(b) If R is reflexive and R' is arbitrary, $R \cup R'$ is reflexive.

1.2a. Equivalence Relations

The most important type of relation that one encounters frequently in all branches of pure and applied mathematics is called equivalence relation.

DEFINITION 1.2a(1).   *A relation in a set X is said to be an equivalence relation iff it is reflexive, symmetric, and transitive.*

It is customary to use the special symbol "$\sim$" for an equivalence relation (in place of the general symbol R). Thus, if $a, b, c, \ldots \in X$ and $\sim$ is an

---

†The latter property is "vacuously satisfied": there is no pair (a, b) such that $a < b$ and $b < a$ hold simultaneously.

equivalence relation in X, then

  (i) $a \sim a$ for all $a \in X$,
  (ii) $a \sim b$ implies $b \sim a$,
  (iii) $a \sim b$ and $b \sim c$ imply $a \sim c$.

In some cases it is more convenient to use the letter E for denoting an equivalence relation. Thus, we shall sometimes write $x \, E \, y$ instead of $x \sim y$.

Example α.   Let X be any set, let "$a = b$" mean the relation "a is identical to b," i.e. consider the relation of equality in the usual sense.† It is trivial to verify that "$=$" is an equivalence relation.

Example β.   Let X be the set of all triangles in the plane. The relation "triangle a is congruent to triangle b" is an equivalence relation in X.

Example γ.   Let A and B be two arbitrary sets. Define in $X = A \times B$ the relation $(a, b) \, E \, (a', b')$ iff $a = a'$. This equivalence relation is illustrated in Fig. 1.4: the point (a, b) is equivalent to all points (a, b') which lie on the same "vertical line."

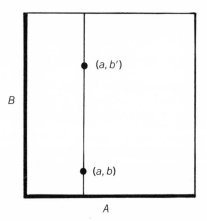

Fig. 1.4.

Example δ.   Define in the set $\mathbf{Z}$ of integers the relation "$a \sim b$ iff $a - b$ is an even integer." A little calculation shows that this is an equivalence relation.

---

†In set-theoretic language, this relation is given by $R = \operatorname{Diag} X \times X$.

One reason for the importance of equivalence relations is that they permit the separation of the set into convenient subsets. In order to discuss this topic we need a preliminary concept. First, we shall say that a collection $\{\ldots A_\alpha \ldots\}$ of sets is a *cover* of the set X if $\cup_\alpha A_\alpha = X$. A cover whose members A_α are pairwise disjoint (i.e. $A_\alpha \cap A_\beta = \emptyset$ if $\alpha \neq \beta$) is called a *partition* of X. Thus, denoting a partition of X by π, we can characterize it conveniently as follows:

$$\pi = \{A_\alpha | A_\alpha \subset X, \quad \text{and each } x \in X \text{ belongs exactly to one } A_\alpha\}. \quad (1.20)$$

For example, a partition for the integers $\mathbf{Z}$ is given by $\pi = \{\mathbf{N}^+, \mathbf{N}^-, \{0\}\}$, where $\mathbf{N}^+(\mathbf{N}^-)$ is the set of positive (negative) integers. Another partition of $\mathbf{Z}$ would be $\pi = \{\mathbf{E}, \mathbf{O}\}$, i.e. into even and odd numbers. Yet another partition of $\mathbf{Z}$ is the one where each number is a separate set of π, i.e. $\pi = \{\ldots \{-2\}, \{-1\}, \{0\}, \{1\}, \{2\}, \ldots\}$.

We are now prepared to formulate the crucial fact about equivalence relations:

THEOREM 1.2a(1).   *Any equivalence relation in a set X leads to a unique partition of X. Conversely, any given partition of X defines an equivalence relation on X.*

In order to see the validity of these assertions, we first introduce the notion of an *equivalence class*. Suppose $\sim$ is an equivalence relation in X. Let x be a fixed (but arbitrary) element of X. Consider the collection of all elements $y \in X$ which are related to the given x. This collection (a subset of X) is called the equivalence class of x (or the equivalence class containing x, or sometimes the equivalence class generated by x). A standard notation for this class is $[x]$, so that

$$[x] = \{z | z \in x \quad \text{and} \quad z \sim x\}. \quad (1.21)$$

Conversely, any subset of X which, for some x, has the form of the r.h.s. of Eq. (1.21), is called an equivalence class for the given equivalence relation. The particular x which generates this equivalence class is often referred to as a representative element of the class.

Now suppose $[x]$ and $[y]$ have one common element, i.e. $z \in [x]$ and also $z \in [y]$. Then $z \sim x$ and $z \sim y$. Using both the symmetry and transitivity of the equivalence relation, we conclude that $x \sim y$. If now z' is another element of $[x]$, i.e. $z' \sim x$, then, by transitivity $z' \sim y$. Hence, $z' \in [y]$. In a similar manner it follows that $z'' \in [y]$ implies $z'' \in [x]$. In other words, if $[x]$ and $[y]$ have one common element, then all elements are common, i.e. $[x] = [y]$, i.e. any element of the equivalence class $[x]$ fully determines this class.

On the other hand, suppose $[x] \neq [y]$. Then they cannot have *any* common element. For, suppose the opposite, i.e. assume that there exists an element u such that $u \in [x]$ and $u \in [y]$. Then, by the previous result, $[x] = [y]$, contrary to assumption. Thus, $[x] \neq [y]$ implies $[x] \cap [y] = \emptyset$.

Finally, because of reflexivity, every element $z \in X$ belongs to *some* equivalence class, namely the one which it generates.†

Summarizing the above observations we see that the collection $\{\dots [x] \dots\}$ of all distinct‡ equivalence classes forms precisely a partition for X: each $x \in X$ belongs exactly to one $[x]$ and the union of all $[x]$ gives the set X. This proves the first part of Theorem 1.2a(1).

The proof of the second part is even simpler. Let π be a partition $\{\dots A_\alpha \dots\}$. Define the relation "$x \sim y$ iff both x and y belong to the same A_α." The reader will have no difficulty to verify that this is indeed an equivalence relation in X. He will also see that the corresponding equivalence classes are precisely the sets A_α of π.

The merit of our theorem is that it tells us that there is no essential distinction between a partition and an equivalence relation: either one determines the other.

The equivalence classes determined by some given equivalence relation E on X often deserve attention *as a collective*. In this way we are led to the concept of a *quotient set*:

DEFINITION 1.2a(2).   *Let E be an equivalence relation in a set X. The collection of all (distinct) equivalence classes induced on X by E is called the quotient set of X modulo E and is denoted§ by X/E. Thus,*

$$X/E = \{C | C \subset X \quad \text{and} \quad C = [x] \quad \text{for some} \quad x \in X\}. \qquad (1.22)$$

In a less detailed notation we may also write

$$X/E = \{\dots [x] \dots\}.$$

Comparing this definition with the first part of Theorem 1.2a(1), we see that the partition π induced by a given equivalence relation E is precisely

---

†It is perfectly possible that, for a specific element z, the class $[z]$ consists of a single element, namely z itself.

‡The term "distinct" is necessary since, as we saw above, we may take any element of an equivalence class as a generating element and so in our notation $[x]$ and $[y]$ may refer to the same class.

§Occasionally one also sees the awkward notation $X/\sim$.

the quotient set X/E, i.e.

$$\pi = \frac{X}{E}. \tag{1.23}$$

Conversely, the study of the second part of the theorem tells us that every given partition π can be looked upon as a quotient set X/E for some unique equivalence relation† E.

Quotient sets are one of the most powerful tools of modern mathematics. They have been called the "secret weapon" of the mathematician. We give a few illustrative examples.

Example ε. Let X be the set of ordered pairs (p, q) of integers with $q \neq 0$. We define "$(p, q)\ E\ (p', q')$ iff $pq' = p'q$." The reader will easily verify that this is an equivalence relation. The equivalence class $[r]$ generated by the element $r = (p, q)$ is the set of all pairs (np, nq), with $n = \pm 1, \pm 2, \ldots$. At this point we note that our equivalence relation, in common language, defines equality of fractions: the fraction $r = p : q$ is equal to the fraction $r' = p' : q'$ iff $pq' = p'q$. We also recall that the fractions $p : q$ and $np : nq$, even though not identical, are considered to be "equal." In the present context we see that this simply means that they belong to the same equivalence class. It should now be obvious that the quotient set X/E is nothing but the set of all rational numbers. We may look upon this set as a partition of all fractions into distinct rational numbers. Each rational number is, strictly speaking, an equivalence class $[r]$.

Example ζ. Let π be the partition of the integers $\mathbf{Z}$ into even and odd numbers. The relation "$x\ E\ y$ iff both x and y are even or both are odd" is easily checked out to be an equivalence relation. There are two equivalence classes: all even numbers belong to one of these, all odd numbers belong to the other. We have $\mathbf{Z}/E = \{\mathbf{E}, \mathbf{O}\}$.

Example η. Consider the equivalence relation defined in Example γ above. The equivalence class $[(a, b)]$ generated by the element (a, b) consists of the points of the "vertical line" passing through the point (a, b), cf. Fig. 1.4. (Formally, it can be described by setting $[(a, b)] = \{a\} \times B = \{(a, h) | h \in B\}$.) The quotient set X/E is the collection of all vertical lines. The set $X = A \times B$ is partitioned precisely into these vertical lines; each point (a, b) belongs to exactly one such line and vertical lines have no common points (i.e. X is the union of all disjoint vertical lines).

---

†Namely the relation $x\ E\ y$ meaning "x and y belong to the same member A_α of π."

The reader is urged to analyze in a similar manner the equivalence classes and quotient sets defined by the equivalence relations of Examples **α**, **β**, and **δ** above.

PROBLEMS

1.2a-1. Let X be an arbitrary set. Define in the power set $\mathcal{P}(X) = \{A \mid A \subset X\}$ the relation "$A \, R \, B$ iff $A \subset B$." Is this an equivalence relation?

1.2a-2. Let **N** be the set of natural numbers. Let E be the relation in $\mathbf{N} \times \mathbf{N}$ defined by "$(a, b) \, E \, (c, d)$ iff $a + d = b + c$." Prove that this is an equivalence relation. Determine the partition which it induces and determine the quotient set $(\mathbf{N} \times \mathbf{N})/E$.

1.2a-3. Let (x, y) denote the points of the plane and define $(x, y) \, E \, (x', y')$ if $x - x'$ and $y - y'$ are integers. Prove that E is an equivalence relation and show that the quotient set by E may be described as, i.e. is equivalent to the set of points on a torus. (*Remark*: the surface of a torus consists of pairs (φ, θ) with $0 \leqslant \varphi < 2\pi$, $0 \leqslant \theta < 2\pi$.)

1.2a-4. Define the relation E in the unit circle (*not* the disc!) by stipulating that for two points p, q of the circle $p \, E \, q$ if $p = q$ or if they are diametrically opposite. Show that E is an equivalence relation and that the corresponding quotient set can be described as the projective line. (*Remark*: The projective line is the union of the real line and *one* "ideal point," called the "point at infinity.")

1.2b. Order Relations

Next in importance to equivalence relations are the so-called order relations.

DEFINITION 1.2b(1).   *A relation in a set X is said to be an order relation if it is reflexive, antisymmetric, and transitive.*

It is customary to use the special symbol "$\leqslant$" for order relations. Thus, if $a, b, c, \ldots \in X$ and $\leqslant$ is an order relation in X, then

  (i) $a \leqslant a$ for all $a \in X$,
  (ii) $a \leqslant b$ and $b \leqslant a$ imply $a = b$,
  (iii) $a \leqslant b$ and $b \leqslant c$ imply $a \leqslant c$.

The notation "$a \leqslant b$" is usually read as "a is less than or equal to b," because the most familiar order relation is precisely the standard ordering of the real numbers. (This is defined by "$a \leqslant b$ iff $b - a$ is a nonnegative number.") Sometimes it is convenient to write, instead of $a \leqslant b$, the phrase $b \geqslant a$ ("b is greater than or equal to a"), but it means precisely the same as $a \leqslant b$.

If $a \leqslant b$ and $a \neq b$, we usually write $a < b$ ("a is strictly less than b").

A set X with an order relation $\leqslant$ defined in it is called a *partially ordered set* or simply a *poset*.† If, in addition, for every pair, $a, b \in X$ either $a \leqslant b$ or $b \leqslant a$, we speak of a *totally ordered set* or *chain*.‡ The term "poset" will be used to include the latter as a special case.

Example α.   The set of real numbers with the usual ordering is a chain.

Example β.   If $\mathcal{P}(X)$ is the power set of an arbitrary set X, then set inclusion $A \subset B$ is an order relation on $\mathcal{P}(X)$. But the order is not total, since for two given sets A, B we need not have either $A \subset B$ nor $B \subset A$. Thus, $\mathcal{P}(X)$ is a poset but not a chain.

Example γ.   Let $\mathbf{N}^+$ be the set of positive integers. Define "$a \leqslant b$ iff a divides b." This order relation makes $\mathbf{N}^+$ a poset, but not a chain.

Example δ.   Order relations can be defined by diagrams. We indicate the members of a given set by appropriate symbols and agree to connect two elements x and y by an arrow from x to y iff $x < y$ and if there is no z such that $x < z < y$. For example, let $x = \{a, b, c, d, e\}$. Then the diagram

defines the following ordering: $b < a$, $c < a$, $d < b$, $d < c$, $e < c$ and, because of transitivity, also $d < a$, $e < a$. (There is no arrow between d and a or between e and a, because there exists, in each case, at least one intervening element.) Note that the order defined on X by this diagram is not a total order since e and d are not comparable: neither $e \leqslant d$ nor $d \leqslant e$. Likewise, b and c are not comparable. In the diagram of a *chain*, every pair of elements is connected somehow by an unbroken line of arrows (hence the terminology). It is clear that every diagram defines an order relation and conversely, every finite poset (and sometimes even a countably infinite poset) may be assigned a diagram. For example, the order relation "x divides y" in the set $X = \{1, 2, 3, 4, 5, 6\}$ is fully described by the diagram

$4 \nwarrow \quad 6$

$\quad \nwarrow 2 \nearrow \quad \nwarrow 3 \quad 5$

$\quad \nwarrow 1 \nearrow$

---

†Many people consider posets as the simplest case of algebraic *structures*. We, however, reserve this notion for sets with composition laws, cf. Chapter 3, and consider order relations *a par* with any other relations.

‡Thus, in a chain every couple of elements is *comparable*.

In the following we must acquaint ourselves with a simple but somewhat confusing terminology used in connection with posets.

Suppose that the poset X has an element f such that $f \leq x$ for every $x \in X$. For obvious reasons, we call f the *first element* † of X. Similarly, if there exists an element l such that $x \leq l$ for all $x \in X$, then l is called the *last element* ‡ of X. From antisymmetry it easily follows that both the first and last element, if it exists, is unique. First and/or last elements need not exist. For example, the set $\mathbf{N}$ of natural numbers (with the standard ordering) has a first element (namely 0) but no last element. The set $\mathbf{Z}$ of all integers (with the standard ordering) has neither a first nor a last element. The set $\mathscr{P}(X)$ with $A \subset B$ as order relation (cf. Example $\boldsymbol{\beta}$) has both a first element, $\emptyset$, and a last element, X.

The notion of first (last) element must not be confused with that of a minimal (maximal) element. An element m of a poset is called a *minimal element* if there is no element in X which is strictly smaller than m. (This means that m is minimal iff $x \leq m$ implies $x = m$.) Similarly, μ is called a maximal element if there is no element in X which is strictly greater than μ. (This means that μ is maximal iff $\mu \leq x$ implies $x = \mu$.) The situation may be well visualized by the diagram describing the order relation of Example $\boldsymbol{\delta}$. There, we have no first element (because d and e cannot be compared) but we have two minimal elements, d and e. On the other hand, a is both a last element and a maximal element.

Next we discuss the very important notion of *bounds*.

Let X be a poset and A some subset of X. (We permit the special case that $A = X$.) An element $b \in X$ is called a *lower bound* of the subset A iff for every $a \in A$, we have $b \leq a$. Note that a subset A need not have any lower bound, and even if it has one, this b need not belong to A (it can be some element of X that is not contained in A). Furthermore, a given subset A may have several lower bounds. Denote, for a moment, the set of all lower bounds of A by the symbol A_*. Suppose A_* has a last (greatest) element.§ We call this element the *infimum* of A (often the confusing term "greatest lower bound" is used). The standard notation for this element is $\inf(A)$. Thus,

$$\inf(A) \leq a \quad \text{for all} \quad a \in A \quad \text{and} \quad \inf(A) \geq b,$$

where b is any lower bound of A.

In complete analogy, an element β is called an *upper bound* of the

†Also called smallest or least element.
‡Also called largest or greatest element.
§We recall that if this exists, it is unique.

subset A iff for every $a \in A$, we have $a \leq \beta$. Similar remarks about the upper bounds apply as for lower bounds. Denoting the set of upper bounds by A^*, we agree to call its (unique) first (smallest) element (if it exists) the *supremum* of A (often the term "least upper bound" is used). With the standard notation $\sup(A)$ we then have:

$$\sup(A) \geq a \quad \text{for all} \quad a \in A \quad \text{and} \quad \sup(A) \leq \beta,$$

where β is any upper bound of A.

If $A \subset X$ has an inf (sup), it is said to be bounded from below (above). If both exist, we simply say that "A is bounded."

We give now a few examples; the reader is advised to make up a few more, so as to assimilate these confusing ideas.

Example ε.   Let $A \subset X$ consist of two comparable elements, $A = \{a, b\}$ and suppose $a \leq b$. Then $\inf(A) = a$, $\sup(A) = b$.

Example ζ.   Let A be a subset of the reals x such that $\rho \leq x \leq \sigma$. Then $\inf(A) = \rho$, $\sup(A) = \sigma$, and both belong to A. If, however, B is the subset of reals x such that $\rho < x < \sigma$, then again $\inf(A) = \rho$, $\sup(A) = \sigma$, but neither of them belongs to A.

Example η.   Let $\mathbf{Q}$ be the set of rationals and let $A \subset \mathbf{Q}$ with elements q defined by $2 < q^2 < 3$. There are infinitely many lower and upper bounds (namely rationals r such that $r < \sqrt{2}$ or $r > \sqrt{3}$) but these sets have no last or first element, respectively, hence there is no inf or sup.

Example ϑ.   Let $\mathbf{R}$ be the set of *real* numbers and let $A \subset \mathbf{R}$ consist of all *rational* numbers x subject to the condition $2 < x^2 < 3$. Then there *is* an $\inf(A)$ and a $\sup(A)$, namely the numbers $\sqrt{2}$ and $\sqrt{3}$. Note that $\inf(A)$ and $\sup(A)$ do *not* belong to A. (Compare this example with the previous one!)

Example κ.   Let $X = \{a, b, c, d, e, f, g\}$ and take $A = \{d, e, c\}$. Let the order in X be defined by the diagram

The only lower bound of A is f (which does *not* belong to A). Clearly, $\inf(A) = f$. Upper bounds of A are the elements a, b, and c. (The last

belongs to A, the first two do not.) Since the first element of the set $\{a, b, c\}$ is c, we have $\sup(A) = c$.

We note here that, *for subsets of the real numbers*, with the standard ordering, often a special terminology is used. If $A \subset \mathbf{R}$ is a *finite* set, the terms $\min(A)$ and $\max(A)$ are used instead of inf and sup. Even if $A \subset \mathbf{R}$ is an *infinite* set but $\inf(A)$ or $\sup(A)$ exists *and belongs to A*, the words minimum and maximum are frequently employed.

We list here, without proof, a few useful simple theorems:

THEOREM 1.2b(1).   *If both* $\inf(A)$ *and* $\sup(A)$ *exist, then* $\inf(A) \leqslant \sup(A)$.

THEOREM 1.2b(2).   *If* $A \subset B \subset X$ *and* $\inf(A)$, $\inf(B)$ *exist, then* $\inf(A) \geqslant \inf(B)$. *Similarly,* $\sup(A) \leqslant \sup(B)$ *(provided both exist).*

THEOREM 1.2b(3).   *If* $A \subset X$ *has a first element f, then* $\inf(A) = f$. *Similarly, if A has a last element l, then* $\sup(A) = l$.

There is one more concept to discuss. Suppose X is a poset with the property that for every pair $a, b \in X$ of elements the set $A = \{a, b\}$ has both an inf and a sup.† Then we call X a *lattice*. Obviously, every chain is a lattice, but there are many lattices that are not totally ordered.

Example λ.   The set of reals with the standard ordering is a lattice, with $a \wedge b = \min(\{a, b\})$ and $a \vee b = \max(\{a, b\})$.

Example μ.   The set $X = \{a, b, c, d\}$ ordered by the diagram

is a lattice. Note that a and c are not comparable but nevertheless $\{a, c\}$ has both an inf and a sup; indeed, $\inf(\{a, c\}) = d$ and $\sup(\{a, c\}) = b$. A counterexample is given when the same set is ordered by the diagram

Here, for example, $\{d, c\}$ has no sup, in fact it has no upper bound at all.

---

†It is customary to call $\inf(\{a, b\})$ the *meet* of a and b and denote it by $a \wedge b$. Similarly, $\sup(\{a, b\})$ is called the *join* of a and b and is denoted by $a \vee b$.

With this we leave for a while the topic of elementary set theory. Several further basic notions can be explored only after we familiarize ourselves with maps.

PROBLEMS

1.2b-1. Let $A = \{a, b, c\}$ be ordered by the following rule:

Determine the family F of all (nonempty) totally ordered subsets of A. Make F a poset by set inclusion and construct the order diagram of F.

1.2b-2. Let $\mathbf{Q}$ be the set of rationals, with the natural order. Consider the subset $A = \{x \mid x \in \mathbf{Q}, 8 < x^3 < 15\}$. Is A bounded from above? Is A bounded from below? Does $\sup(A)$ exist? Does $\inf(A)$ exist?

1.2b-3. Let $B = \{2, 3, 4, 5, 6, 8, 9, 10\}$ be ordered by "x is a multiple of y." Construct the ordering diagram. Find all maximal and all minimal elements. Is there a first element? Is there a last element? Has the subset $\{2, 4, 6, 8\}$ a sup or an inf?

1.2b-4. Let $X = \mathbf{N} \times \mathbf{N}$, where $\mathbf{N}$ is the set of natural numbers, be ordered by the following rule: $(a, b) \le (a', b')$ iff $a \le a'$ and $b \le b'$. Has the poset a first, last, minimal, maximal element? Has it an inf and a sup? Is it a lattice?

1.2b-5. Given the set $A = \{1, 2, 3, 4\}$. Construct all partitions π_i of this set. Define in the set of all partitions the order relation: $\pi_i \le \pi_k$ if every set in π_i is a subset of one set in π_k. Draw the corresponding ordering diagram and show that the ordered set of the π_i is a lattice.

2

Maps

Let X and Y be two arbitrary sets. *A map f is a rule which assigns to each element $x \in X$ one and only one element $y \in Y$.* We say that f is a map from X to Y and write

$$f: X \to Y, \qquad \text{or sometimes} \quad X \xrightarrow{f} Y.$$

The term "function" will be used as a synonym of map. In special circumstances the technical terms operator, transformation, functional are used to designate a particular map.

The set X is called the *domain* of the function f and, for emphasis, is often denoted by D_f. We also say that f is *defined* on D_f. The set Y is called the *codomain* of f. It is possible that the domain and codomain coincide. Then we have a map $f: X \to X$, from X to X.

The specific element $y \in Y$ which, by f, is assigned to the specific element $x \in X$, is referred to as the *image* of x under f, and we denote this by writing

$$x \mapsto y \quad \text{or} \quad f(x) = y.$$

We sometimes read this as "x is mapped on y." Care should be taken to distinguish between the symbols f and $f(x)$. The former merely denotes the "rule" of assignment; the latter is nothing but a specific element of the codomain set. Actually, $f(x)$ is often called "the value of f assumed at x."

We now give a selection of simple examples.

Example α. Familiar to everyone are the functions whose domain and codomain is the set of real numbers (or subsets thereof). Such "real valued functions on the set of reals" may be given by algebraic rules, like

$x \mapsto x^2$ (also written as $f(x) = x^2$); or by "transcendental" means (such as power series), e.g., $x \mapsto \sin x$; or even by prescriptions which cannot be written down in the form of a "formula." For example, the map $f: \mathbf{R} \to \mathbf{R}$ defined by

$$f(x) = \begin{cases} 1 & \text{if } x \text{ is a rational number} \\ 0 & \text{if } x \text{ is irrational,} \end{cases}$$

is a perfectly good definition of a function. Equally familiar are the "complex valued functions on the set of complex numbers," where both the domain and codomain is the set of complex numbers.

Example β. Let $X = T$ be the set of all triangles in the plane. Let $Y = \mathbf{R}$ be the set of all real numbers. We may define the function $f: T \to \mathbf{R}$ by the rule "assign to each triangle its area."

Example γ. Let again $X = T$ be the set of all triangles in the plane and let $Y = K$ be the set of all circles in the plane. The rule "assign to each triangle the corresponding inscribed circle" defines a function $f: T \to K$.

Example δ. Let X be the set of inhabitants of Naples who possess a telephone. The telephone directory of Naples is a rule f that assigns to every phone-owner a real number. Thus, $f: X \to \mathbf{R}$ is a perfectly well-defined function.

Example ε. The last three examples show the extreme generality of the concept of a map. However, not every rule of assignment is a map. Let $X = Y = \mathbf{R}$, and consider the rule "$x \mapsto y$ iff $y^2 = x$." This is not a map since, for example, it assigns to $x = 4$ *two* elements of Y, namely $+2$ and -2. The definition of a map *insists* on the uniqueness of the assignment. Thus, we see that the so-called "multiple valued functions" of elementary calculus are *not* functions.†

The definition of a map, as given above, is somewhat intuitive and one may object that it involves undefined concepts such as "rule." It is therefore of great importance that one can give a rigorous, purely set theoretic definition of a function. To start with, suppose we are given a map $f: X \to Y$ (as defined above). The set

$$\Gamma(f) = \{(x, y) | x \in X, \quad y \in Y, \quad \text{and } y = f(x)\} \tag{2.1}$$

is called the *graph of the function f*. Clearly, $\Gamma(f) \subset X \times Y$. Recalling then our terminology from Section 1.2, we see that the graph of a function is a

---

†There are special methods (Riemann surfaces) which permit the reduction of such "bad" assignments to acceptable functions.

relation from X to Y. Of course, not every relation from X to Y is the graph of some function. The condition for this to hold is that in the ordered pair (x, y) the term x is arbitrary and the term y must be unique if x is given, i.e. to every $x \in X$ there must be related one and only one $y \in Y$.

These considerations suggest that we *identify a function and its graph.* Thus, we are led to the following:

DEFINITION 2(1).   *A function $f\colon X \to Y$ is a set of ordered pairs (x, y)* (i.e. *a subset of $X \times Y$) such that* EACH $x \in X$ *appears as the first term in* ONE AND ONLY ONE *ordered pair.*

Figure 2.1a shows the graph of a function $f\colon A \to Y$. Figure 2.1b represents a relation from A to B which is not the graph of any function. As a further example, let $X = \{1, 2, 3\}$, $Y = \{\alpha, H, \square\}$. We may define a

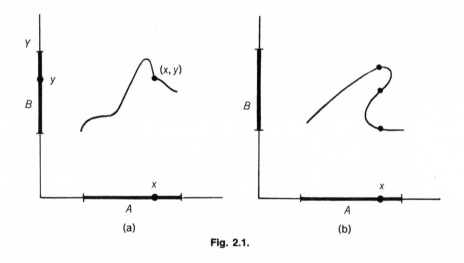

Fig. 2.1.

function $f\colon X \to Y$ as the set $f = \{(1, H), (2, \alpha), (3, \square)\}$. Its graph is given in the diagram of Fig. 2.2. Another function $g\colon X \to Y$ would be $g = \{(1, H), (2, \alpha), (3, \alpha)\}$. But the set $\{(1, H), (2, \alpha), (2, \square)\}$ is not a function.

We now discuss some standard terminology. The *range* of a function $f\colon X \to Y$ is the set $\{\ldots f(x) \ldots\}$ of images under the mapping, and is denoted by R_f. Sometimes the notation $f(X)$ is used and referred to as the

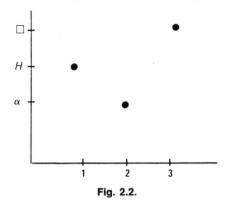

Fig. 2.2.

"image of the set X." It should be clear that R_f is a subset of the codomain Y with which it should not be confused.† For example, in Fig. 2.1a, Y is the codomain and B is the range.

In general, the range is a *proper* subset of the codomain (cf. Fig. 2.3). Then we often say that f maps X "into" Y. However, in many important

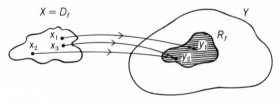

Fig. 2.3.

cases $R_f = Y$. We then say that the mapping is *onto* Y, or that f is a *surjection*. If f is a surjection, then every $y \in Y$ is the image of *at least* one $x \in X$. Therefore, the equation $f(x) = y$ has at least one solution x for every given $y \in Y$.

Looking the other way, we observe that, in general, the definition of a map permits that several x are mapped on the same y (cf. Fig. 2.3). If, however, every y in R_f is the image of exactly one $x \in X$, then the map is said to be *one-to-one*, or f is called an *injection*. This means that

---

†The primary concept in the definition of the function is the codomain. The range can be ascertained only after f is specified.

$f(x) \neq f(x')$ if $x \neq x'$. Therefore, if f is an injection, the equation $f(x) = y$ has exactly one solution x for every $y \in R_f$ with R_f a subset of Y.

Finally, the "best" functions are those which are both *one-to-one and onto*. In this case f is called a *bijection*.† If $f: X \to Y$ is a bijection, then the equation $f(x) = y$ has exactly one solution x for every $y \in Y$.

We wish to make the reader conscious of the triviality that the pairs of expression

onto ——— surjective

one-to-one ——— injective

one-to-one and onto ——— bijective

are synonymous adjectives.‡

To give an example, the map $f: \mathbf{R} \to \mathbf{C}$ from the reals to the complex numbers, defined by $x \mapsto i|x|$ is neither injective nor surjective. The map $x \mapsto ix$ is injective but not surjective. The map $f: \mathbf{R} \to \mathbf{R}$ defined by $f(x) = x^3$ is bijective.

We can now unambiguously define *equality of functions*. We say that $f = g$ iff f and g have the same domain, the same codomain, and $f(x) = g(x)$ for all $x \in X$. (The criterion on the codomains is necessary in order to avoid ambiguities. For example, f defined as a function on the reals to the reals by $f(x) = x^2$ is not equal to the function g defined as a function on the reals to the positive reals by $g(x) = x^2$. The function g is surjective, but f is not.)

Next we review terminology pertaining to some frequently met types of function. The map

$$1_X : X \to X \quad \text{given by} \quad x \mapsto x \quad \text{for all} \quad x \in X$$

is called the *identity map* on X. It is bijective. The map

$$C : X \to Y \quad \text{given by} \quad x \mapsto c \quad \text{for all} \quad x \in X,$$

with c a fixed element of Y, is called a *constant function*.

Often one speaks of "*a function of two variables*." This is merely a shorthand for the following. Let

$$f: X \times Y \to Z \quad \text{be given by some rule} \quad (x, y) \mapsto z.$$

†Some authors use the term "bi-uniform map."

‡We shall use the older and newer terminology interchangeably. Following present custom, we are inclined to use the modern terms (right column) in algebraic topics and the older terms (left column) in topologic topics.

It is then customary to write $f(x, y) = z$ rather than $f((x, y)) = z$. Functions of several variables are defined in a similar way.

An important class of such functions is the following. Let

$$p: X \times Y \to X \quad \text{be given by} \quad (x, y) \mapsto x.$$

We call p a *projection* onto the X-axis. Similarly,

$$q: X \times Y \to Y \quad \text{given by} \quad (x, y) \mapsto y$$

is called the projection onto the Y-axis.

Suppose we have a map $f: X \to Y$ and let $A \subset X$. We say that f *induces* the map

$$\varphi: A \to Y \quad \text{given by} \quad \varphi(a) = f(a) \quad \text{for all} \quad a \in A.$$

Usually φ is called the *restriction* of f to A. Note that $D_\varphi = A \subset X$, codomain of φ = codomain of f, but in general $R_\varphi \neq R_f$. It is important to note that the graph of the restriction φ of f is a subset of the graph of f. Indeed, one often denotes the restriction of a function by writing $\varphi \subset f$.

The converse concept is that of an *extension*. If $\varphi: A \to Y$ and $X \supset A$, $Z \supset Y$, then *any* function $f: X \to Z$ with the property that $f(x) = \varphi(x)$ for $x \in A$, is called an extension of φ from A to the domain X.

PROBLEMS

2-1. Let $A = \{a, b, c\}$ and $B = \{1, 0\}$. Construct all possible functions from A to B. (Use diagrams.) State which of these are surjective, injective, bijective.

2-2. Let $X = \{a, b, c, d\}$ and define $f: X \to X$ by $f(a) = a$, $f(b) = c$, $f(c) = a$, $f(d) = a$. What is the range (or image) of f?

2-3. When is a constant function an injection and when is it a surjection? When is it a bijection?

2-4. Let $X = \{1, 2, 3, 4\}$ and $\mathbf{R}$ = the reals. Let $f: X \to \mathbf{R}$ be defined by $f(x) = x^2$. Find the graph $\Gamma(f)$ of f.

2-5. Let $f: A \to B$ be displayed on a "coordinate diagram" of $A \times B$. What geometrical property must f have if
(a) f is an injection,
(b) f is a surjection,
(c) f is a constant function?

2-6. Consider $f(x) = x$, where $x \geq 0$. Which of the following functions is an extension of f?
(a) $g(x) = |x|$ for all real x,
(b) $h(x) = 1_{\mathbf{R}}(x)$,
(c) $k(x) = x$, where $-1 \leq x \leq 1$.

2.1 COMPOSITE FUNCTIONS AND INVERSES

Let $g : A \rightarrow B$ and $f : B \rightarrow C$. We can define a new function, the *composite* $f \circ g$ of g with f, as follows:

$$f \circ g : A \rightarrow C \quad \text{given by} \quad (f \circ g)(a) = f(g(a)) \quad \text{for all} \quad a \in A. \quad (2.2)$$

The definition is visualized in Fig. 2.4. Note that the symbol $f \circ g$ makes

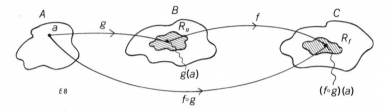

Fig. 2.4.

sense only because, by definition, the domain of f contains the range of g. Sometimes one writes fg in place of $f \circ g$, but this may lead to confusion. The reader should also be aware that, in general, $f \circ g \neq g \circ f$, and in fact one of the two objects may not be defined.

Composition of functions is easily visualized by *mapping diagrams*. The vertices of such diagrams are sets, which are connected by labeled arrows, indicating maps. A *path* is a sequence of maps. A diagram is said to be *commutative* if for any two sets X and Y in the diagram, any two paths from X to Y yield, by composition, the same function. For example, the diagram

$$A \xrightarrow{\;\;f\;\;} B$$
$$g \searrow \quad \nearrow h$$
$$C$$

is commutative if $f = h \circ g$. Of course, diagrams may be drawn even if they are not commutative. Then, in the above example, f simply denotes a map which is "independent" of g and h.

Composition of functions obeys the following two rules:

(a) Associativity:

$$(f \circ g) \circ h = f \circ (g \circ h), \quad (2.3a)$$

(b) Identity law:

$$\text{If} \quad f: A \to B, \quad \text{then} \quad 1_B \circ f = f \circ 1_A = f. \tag{2.3b}$$

The proof of these laws can be based on the inspection of the corresponding diagrams. Equation (2.3a) is represented by

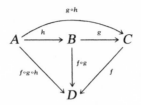

while Eq. (2.3b) corresponds to

$$1_A \subset A \xrightarrow{\;f\;} B \supset 1_B$$

Both diagrams are commutative.

A simple example of composite function is the following. Let $g: T \to \mathbf{R}$ be the map associating to every triangle its area (Example β on p. 27). Let $f: \mathbf{R} \to \mathbf{C}$ be the map which assigns to every real number x the complex number e^{ix}. Then $f \circ g: T \to \mathbf{C}$ associates to every triangle a point $e^{i\alpha}$ of the unit circle, where α is the area of the triangle. Note that the range of $f \circ g$ is a subset of $\mathbf{C}$.

Next we would like to define "inverse functions." The discussion of this topic is best prepared by illuminating the concept of a *set function*. This term simply refers to a function whose domain is a set of sets (i.e. a class, in our customary terminology). Thus, a set function maps *sets* onto elements of some arbitrary other set. A particularly interesting example of a set function arises in the following way. Let X and Y be arbitrary sets and let $f: X \to Y$. Define the set function†

$$f_*: \mathscr{P}(X) \to \mathscr{P}(Y)$$

by the following prescription: If A is an element of $\mathscr{P}(X)$ (i.e. $A \subset X$), then let

$$f_*(A) = B, \quad \text{where} \quad B = \{b \,|\, b \in Y \quad \text{and} \quad b = f(x) \quad \text{for some} \quad x \in A\}.$$

(Obviously, $B \subset Y$, hence $B \in \mathscr{P}(Y)$.) One says that f induces the set function f_*. Clearly, (cf. Fig. 2.5) B is the image set of A under f. For this reason, often the subscript $*$ is omitted, and one writes $f(A) = B$, but it should be kept in mind that f_* and f are *different* functions.

---

†$\mathscr{P}$, as always, stands for the power set.

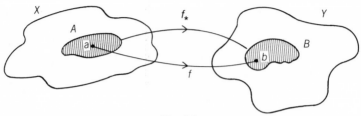

Fig. 2.5.

We now come to our main topic. Let $f: X \to Y$. Let $B \subset Y$. Find the subset $A \subset X$ consisting of all elements which are mapped by f into B, i.e. find A given as $A = \{x \mid x \in X \text{ and } f(x) \in B\}$. Clearly, A has the property that $f(A) \subset B$. This set A is called the *inverse image* of B, and is denoted by the symbol $f^{-1}(B)$. Doing this for every subset B of Y, we obviously define a set function

$$f^{-1}: \mathscr{P}(Y) \to \mathscr{P}(X). \tag{2.4}$$

It is customary to call f^{-1} the *inverse* of f. The situation is visualized in Fig. 2.6. One easily verifies the following relations:

$$f^{-1}(Y) = X, \tag{2.5a}$$

$$f^{-1}(\emptyset) = \emptyset, \tag{2.5b}$$

$$f^{-1}(\bigcup_{\alpha} B_{\alpha}) = \bigcup_{\alpha} f^{-1}(B_{\alpha}), \tag{2.5c}$$

$$f^{-1}(B_1 - B_2) = f^{-1}(B_1) - f^{-1}(B_2). \tag{2.5d}$$

It is important to observe that, whereas f^{-1} as defined above is a function from $\mathscr{P}(Y)$ to $\mathscr{P}(X)$, it is *not* a function from Y to X. Indeed, let $B = \{b\}$ consist of a single element $b \in Y$. In general, there will be *several* elements $a_1, a_2, \ldots \in X$ which are mapped on b. Furthermore, there may exist an element $b \in B$ such that *no element* $x \in X$ is mapped on b. (Then

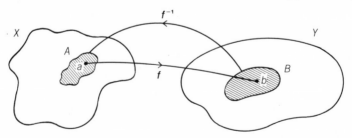

Fig. 2.6.

$f^{-1}(\{b\}) = \emptyset$.) However, suppose $f: X \to Y$ is bijective. Then our set function f^{-1} in effect yields a function from Y to X. In this case, somewhat confusingly for the beginner, we continue to use the same symbol f^{-1} to denote this new map. We set $f^{-1}(y)$ equal to that unique element $x \in X$ for which $f(x) = y$. The function

$$f^{-1}: Y \to X \tag{2.6}$$

so obtained we call the *inverse function* of f, as opposed to our previous set function (2.4) which we simply called the inverse of f.

In summary: whereas the inverse of f always exists, *the inverse function f^{-1} exists iff f is bijective*. In that case, f^{-1} is given by the rule $f^{-1}(f(x)) = x$ for all $x \in X$.

It is clear that if f is bijective, then the inverse function f^{-1} is also bijective. Hence, f^{-1} will have an inverse function and we see easily that $(f^{-1})^{-1} = f$. Thus, $f(f^{-1}(y)) = y$ for all $y \in Y$.

These observations can be formalized by

THEOREM 2.1(1). *If $f: X \to Y$ is a bijection (so that the inverse function $f^{-1}: Y \to X$ exists), then $(f^{-1} \circ f): X \to X$ is the identity function 1_X and $(f \circ f^{-1}): Y \to Y$ is the identity function 1_Y.*

We leave it as an exercise to prove the converse:

THEOREM 2.1(2). *Let $f: X \to Y$ and $g: Y \to X$. Then $g = f^{-1}$ if $(g \circ f): X \to X$ is 1_X and $(f \circ g): Y \to Y$ is 1_Y.*

These theorems may be expressed by saying that $f: X \to Y$ and $g: Y \to X$ are inverses of each other iff both diagrams

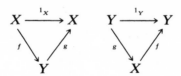

are commutative.

Let us illuminate the notion of inverses on two examples.

Example α. Let $f: \mathbf{R} \to \mathbf{R}$ be given by $x \mapsto x^2$. Let $B = \{x \mid 4 \leqslant x \leqslant 9\}$. Then $f^{-1}(B) = \{x \mid 2 \leqslant x \leqslant 3 \text{ or } -3 \leqslant x \leqslant -2\}$. Let $B = \{4\}$. Then $f^{-1}(B) = \{2, -2\}$. Finally, let $B = \{-3\}$. Then $f^{-1}(B) = \emptyset$. Obviously, f has no inverse function.

Example β. Let $f: \mathbf{R} \to \mathbf{R}$ be given by $x \to x^3$. Since this is a bijection, the inverse function exists and we usually write $f^{-1}(x) = \sqrt[3]{x}$.

As an application, we conclude this section with a brief discussion of the concept of *set isomorphism*. Suppose there are given two sets X and Y and there exists a bijective function $f: X \to Y$. We then often say that the two sets are isomorphic, or that f furnishes an isomorphism between X and Y. The importance of this notion hinges on the fact that *set isomorphism is an equivalence relation in the class of all sets.*† Indeed, 1_X is a bijection; $f \circ g$ is a bijection if f and g are bijections; and, as we pointed out above, f^{-1} is a bijection together with f. Therefore, every set is isomorphic to itself; if X is isomorphic to Y and Y isomorphic to Z then X is isomorphic to Z; and finally, if X is isomorphic to Y, then Y is isomorphic to X. Thus, denoting set isomorphism by the symbol $\simeq$, we have

$$X \simeq X,$$
$$X \simeq Y \quad \text{and} \quad Y \simeq Z \quad \text{imply} \quad X \simeq Z,$$
$$X \simeq Y \quad \text{implies} \quad Y \simeq X.$$

Isomorphic sets may be identified in all their set-theoretic properties. The only difference between isomorphic sets is in the concrete nature of their elements. These statements follow from the obvious observation that *isomorphic sets are in a one-to-one correspondence.*‡ For example, the sets $X = \{1, 2, 3\}$ and $Y = \{\triangle, \square, \alpha\}$ are isomorphic, because the map

$$f(1) = \triangle, \qquad f(2) = \square, \qquad f(3) = \alpha$$

is a bijection.§ A more instructive example is the following. Let $f: X \times Y \to Y \times X$ be given by $(x, y) \mapsto (y, x)$. This is clearly a bijection, hence $X \times Y \simeq Y \times X$. This "identification" of $X \times Y$ and $Y \times X$ does not mean, of course, that they are the same sets: they consist of different elements, yet, as sets, they "behave" identically. A trivial example of two nonisomorphic sets is the pair $X = \{1, 2\}$ and $Y = \{\triangle, \square, \alpha\}$. It is not possible to find a bijection between X and Y.

---

†When we talk of the "class of all sets," we tacitly assume that there exists some universal set to which all the sets we talk about belong.

‡"One-to-one correspondence" is an expression often used to denote the existence of a bijective map.

§There are, of course, five more bijections between X and Y.

PROBLEMS

2.1-1. Prove that the composite function $g \circ f : A \to C$ is an injection if both $f : A \to B$ and $g : B \to C$ are injections. Prove that the same holds when the word "injection" is replaced by "surjection."

2.1-2. Let $f : \mathbf{R} \to \mathbf{R}$ and $g : \mathbf{R} \to \mathbf{R}$ (where $\mathbf{R}$ means the reals) be defined by

$$f(x) = x^2 + 3x + 1, \qquad g(x) = 2x - 3.$$

Find formulae which define the composite functions $g \circ f$, $f \circ g$, $g \circ g$, $f \circ f$.

2.1-3. Let $f : X \to Y$. Show that f is an injection if there exists a map $g : Y \to X$ such that $g \circ f = 1_X$. Show that f is a surjection iff there exists a map $h : Y \to X$ such that $f \circ h = 1_Y$.

2.1-4. Let A, B, C, D be arbitrary sets and $\beta : A \to C$ and $\gamma : B \to D$ arbitrary maps. Let $\alpha : A \to B$ be a surjection and $\delta : C \to D$ an injection. Suppose the diagram

is commutative. Show that there exists a unique map $\varphi : B \to C$ such that the diagram

is completely commutative.

2.1-5. Let $f : A \to A'$ and $g : B \to B'$. The "Cartesian product function" $f \times g$ is defined as the map $f \times g : A \times B \to A' \times B'$ specified by the assignment $(a, b) \mapsto (f(a), g(b))$. Let $p : A \times B \to A$ and $q : A \times B \to B$ be the standard projections to the "axes" (i.e. $p(a, b) = a$, $q(a, b) = b$) and let p' and q' denote the projections $p' : A' \times B' \to A'$, $q' : A' \times B' \to B'$. Consider the diagram

$$
\begin{array}{ccccc}
A & \xleftarrow{\ p\ } & A \times B & \xrightarrow{\ q\ } & B \\
{\scriptstyle f}\downarrow & & {\scriptstyle f \times g}\downarrow & & \downarrow{\scriptstyle g} \\
A' & \xleftarrow{\ p'\ } & A' \times B' & \xrightarrow{\ q'\ } & B'
\end{array}
$$

and show that each "square" commutes.

2.1-6. Let $f: A \to B$ and $g: B \to C$ possess inverse functions $f^{-1}: B \to A$ and $g^{-1}: C \to B$. Prove that the composite $g \circ f: A \to C$ has an inverse function which is given by $f^{-1} \circ g^{-1}$.

2.1-7. Let $f: A \to B$. What is $f^{-1}(B)$?

2.1-8. Let $f: \mathbf{R} \to \mathbf{R}$ be defined by $f(x) = \sin x$. Find $f^{-1}(\{0\})$, $f^{-1}(\{2\})$, $f^{-1}([-1, 1])$. Has f an inverse function?

2.2 EQUIVALENCE RELATIONS AND MAPS

In this section we shall considerably deepen our insight into the nature and importance of equivalence relations.

Let $f: X \to Y$. Define in X a relation R by the following rule: $a R b$ iff $f(a) = f(b)$. It is a trivial matter to check that R is an equivalence relation. Thus we have

THEOREM 2.2(1). *Any given function f defines an equivalence relation in its domain. Equivalent elements are those whose images under f coincide.*

The equivalence relation generated by a given function f is called the *equivalence kernel* of f and is denoted by ker (f). Figure 2.7 illustrates the

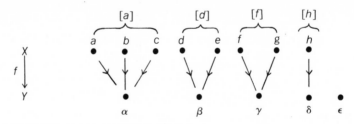

Fig. 2.7.

situation on a simple example. In passing we note that if f is injective, then each equivalence class generated by ker (f) consists of a single element of X.

Conversely, suppose now that we are given an equivalence relation E in some set X. We recall that this gives rise to the partition $\pi = X/E$, whose elements are the equivalence classes $[x]$. Let us assign to each $x \in X$ the equivalence class $[x]$ to which it belongs. In this manner we define a function

$$p_E: X \to X/E \quad \text{given by} \quad x \mapsto [x]. \tag{2.7}$$

This function is so important that it has a special name: it is called the *canonical mapping* or *canonical projection* of X onto X/E. (It is obvious that p_E is a surjection, since every $[x]$ is the image of at least one element x.)

A further study reveals that $\ker(p_E)$ is precisely the originally given equivalence relation E. Indeed, suppose $p_E(x) = p_E(y)$. Then x and y must belong to the same equivalence class, hence $x \, E \, y$. Conversely, if $x \, E \, y$, then $p_E(x) = [x]$ and $p_E(y) = [y] = [x]$, so $p_E(x) = p_E(y)$.

We summarize these results:

THEOREM 2.2(2). *Any given equivalence relation E in X defines a surjective function $p_E \colon X \to X/E$, specified by* Eq. (2.7), *with* $\ker(p_E) = E$.

We may paraphrase the essential content of this theorem by saying that any equivalence relation E in a set X is the kernel of some function p_E with $D_{p_E} = X$.

The ideas described above are simple but, as we shall see in much of our later work, have profound consequences. We give now only two, elementary illustrative examples.

Example α.   Let $f \colon T \to \mathbf{R}$ be the map described in Example β on p. 27, associating to every triangle its area. According to Theorem 2.2(1), this map gives rise to an equivalence relation among triangles: all triangles with the same area are equivalent.

Example β.   Consider the equivalence relation E described in Example γ of Subsection 1.2a which relates to each point (a, b) of the set $A \times B$ all points on the "vertical line" passing through (a, b), cf. Fig. 1.4. We recall that the corresponding equivalence classes are precisely the "vertical lines." Hence, the canonical map p_E is the function from $X = A \times B$ to the set of vertical lines, assigning to each point (a, b) the vertical line that passes through it.

The most important consequence of the preceding considerations is that they permit the *representation of any given function as the composite of functions* with simpler properties. Let $f \colon X \to Y$ be an arbitrary function between arbitrary sets. Consider the equivalence relation $\ker(f)$ generated by f. The quotient set $X/\ker(f)$ consists of equivalence classes $[x]$, each $[x]$ being a subset of X such that they have the same image under f. The canonical map $p \colon X \to X/\ker(f)$ is given by $x \mapsto [x]$, and we recall that p is a *surjection*. Now define the function

$$g \colon X/\ker(f) \to Y \quad \text{given by} \quad [x] \mapsto f(x). \tag{2.8}$$

This map is often called the *induced map* (induced by f). Note that in Eq. (2.8) $f(x)$ stands for the image under $f: X \to Y$ of *any* element z of $[x]$. The function g is certainly an *injection*, because if $[x_1] \neq [x_2]$, then $f(x_1) \neq f(x_2)$, as follows from the definition of $[x]$. Now, consider the composite map $g \circ p: X \to Y$. Using Eqs. (2.7) and (2.8), we find $(g \circ p)(x) = g(p(x)) = g([x]) = f(x)$. Thus, we have the identity $f = g \circ p$. In other words, *we succeeded in representing f as the composite of a surjection and an injection.* The situation is fully described by saying that the diagram

$$X / \mathrm{ker} \ (f)$$

is commutative, with p surjective and g injective. For later reference, we formalize this result:

THEOREM 2.2(3).   *Any function $f: X \to Y$ can be decomposed as $f = g \circ p$, where p is the surjective canonical map of X onto $X / \mathrm{ker} \ (f)$ and g is the injective induced map defined by* Eq. (2.8).

It is clear that if $f: X \to Y$ happens to be surjective (i.e. if $f(X) \equiv R_f = Y$), then g will be not only injective but also surjective, so that actually g is bijective. If this is not the case, we can use a simple trick. Let

$$i: f(X) \to Y \quad \text{be given by} \quad i(y) = y \quad \text{for all} \quad y \in f(X).$$

This injective function is clearly the restriction of 1_Y to the subset $f(X)$ of Y and is called very graphically an *insertion* (it inserts the range of f into the codomain of f). We can now set $g = i \circ b$, where†

$$b: X / \mathrm{ker} \ (f) \to f(X) \quad \text{given by} \quad [x] \mapsto f(x), \tag{2.8a}$$

and is obviously a bijection. Consequently, the given f is now decomposed as $f = i \circ b \circ p$, where p is surjective, b bijective, and i injective. This is visualized by the commutative diagram

$$X \xrightarrow{\ \ p \ \ } X / \mathrm{ker} \ (f) \xrightarrow{\ \ b \ \ } f(X) \xrightarrow{\ \ i \ \ } Y$$
$$f$$

---

† Clearly, b is essentially the same function as g but with a restricted codomain.

The crux of this consideration is that *any function f is essentially determined by a bijection b.*

The following example will illuminate the situation. Let $f\colon \mathbf{Z} \to \mathbf{R}$ be the map which assigns to each integer n the real number $x = n^2$. This map is neither surjective nor injective (note that $n^2 = (-n)^2$). We have $R_f \equiv f(\mathbf{Z}) = \{x \mid x = n^2 \text{ for some } n \in \mathbf{Z}\}$. The equivalence kernel of f is defined by the rule: "n is equivalent to n' iff $n^2 = n'^2$." Hence $[n] = \{n, -n\}$ and so

$$\mathbf{Z}/\ker(f) = \{\{n, -n\} \mid n \in \mathbf{Z}\}.$$

We have $p\colon \mathbf{Z} \to \mathbf{Z}/\ker(f)$ given by $n \mapsto \{n, -n\}$; then $b\colon \mathbf{Z}/\ker(f) \to (\mathbf{Z})$ is the map $\{n, -n\} \mapsto f(n) = n^2$; and the injection i is the trivial map $n^2 \in f(\mathbf{Z}) \mapsto (x = n^2) \in \mathbf{R}$. We clearly have $f = i \circ b \circ p$. The composition is illustrated in Fig. 2.8.

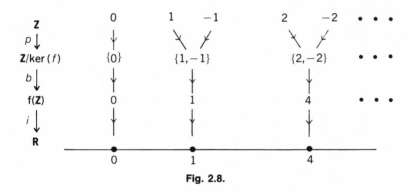

Fig. 2.8.

PROBLEMS

2.2-1. Let A and B be two sets, equipped with an equivalence relation E and F, respectively. Let $p\colon A \to A/E$ and $q\colon B \to B/F$ be the canonical projections onto the quotient sets. Prove that the function $p \times q\colon A \times B \to (A/E) \times (B/F)$ is surjective. Describe the equivalence kernel of $p \times q$. (*Note*: The product function $p \times q$ was defined in Problem 2.1-5.)

2.2-2. Let E be an equivalence relation in X. Let $f\colon X \to Y$ be any function such that $a \mathrel{E} b$ (with $a, b \in X$) implies $f(a) = f(b)$. Prove that there exists a unique function $g\colon X/E \to Y$ such that $f = g \circ p_E$. (*Note*: Theorem 2.2(3) may be considered as a corollary to the theorem of this problem.)

2.2-3. Let $p\colon A \times B \to A$ be the projection $(a, b) \mapsto a$. Describe the bijection b which, by Theorem 2.2(3) (and the subsequent discussion) determines the function p.

2.3 ORDERED SETS AND MAPS

In this section we discuss some simple concepts which combine the notions of order relations and of functions.

(*i*) *Bounded functions.* Let $f: X \to P$ be a function from an arbitrary set X into some poset P. The function f is said to have a lower bound if the subset $f(X) \subset P$ has a lower bound, i.e. if there exists a $p \in P$ such that $p \leqslant f(x)$ for all x. The upper bound of a function is defined similarly. In particular, if $f(X) \subset P$ has an infimum (supremum), then we say that *the function f has an infimum* (*supremum*) *in X*, and we denote this by writing

$$\inf_{x \in X} f \quad \text{and} \quad \sup_{x \in X} f.$$

Thus, if f is a function bounded from below, then

$$\inf_{x \in X} f \leqslant f(x) \quad \text{for all} \quad x \quad \text{and} \quad \inf_{x \in X} f \geqslant p,$$

where p is any lower bound of f (i.e. $p \leqslant f(x)$ for all x). Similarly, if f is a function bounded from above, then

$$\sup_{x \in X} f \geqslant f(x) \quad \text{for all} \quad x \quad \text{and} \quad \sup_{x \in X} f \leqslant g,$$

where g is any upper bound (i.e. $g \geqslant f(x)$ for all x).

If both inf f and sup f exist, the function f is said to be bounded. For example, if $f: X \to \mathbf{R}$ or $f: X \to \mathbf{C}$, where X is an arbitrary set, then f is a bounded function iff $|f(x)| \leqslant M$, where M is an arbitrary nonnegative number.

(*ii*) *Monotone functions.* Let A and B be both posets with order relations denoted by $\leqslant$ and $\leqslant'$, respectively. Let $f: A \to B$ be a map with the property that

$$x \leqslant y \quad \text{implies} \quad f(x) \leqslant' f(y).$$

We then say that the function f is *monotone*. The term "*nondecreasing*" is also used. If, in addition,

$$x < y \quad \text{implies} \quad f(x) <' f(y),$$

then f is called a *strictly monotone* or "increasing" function.

It is easily seen that if f is monotone and bijective, then it is automatically strictly monotone.

(*iii*) *Isomorphism of posets.* Suppose we have two posets A and B and suppose there exists a monotone function $f: A \to B$. Since f is an order preserving map, we say that it provides a *morphism of order* or a morphism of posets. Particularly important is the case when f is bijective. Then it preserves strict order and we say that we have an *isomorphism of order*. The posets A and B are called isomorphic posets or *similar posets*, and we write $A \cong B$.

It should be clear that isomorphism of order is a stronger statement than simple set isomorphism. Two posets are similar, $A \cong B$, if, first of all, they are equivalent (isomorphic) in the set-theoretic sense, $A \simeq B$, (i.e. there is a bijection f between them) and second, if the isomorphism f is strictly order preserving (i.e. if f has the property that $f(x) <' f(y)$ iff $x < y$).

For example, let **Z** be the set of integers and **E** be the set of even numbers, each ordered in the usual way. The bijective map $x \mapsto 2x$ is strictly order preserving, hence $\mathbf{Z} \cong \mathbf{E}$. On the other hand, the poset **Z** of all integers is not order isomorphic to the poset **N**. It is easy to find a simple *set* isomorphism, for example,

$$
\begin{array}{cccccccc}
0 & -1 & 1 & -2 & 2 & -3 & 3 & \ldots \\
\updownarrow & \updownarrow & \updownarrow & \updownarrow & \updownarrow & \updownarrow & \updownarrow & \\
0 & 1 & 2 & 3 & 4 & 5 & 6 & \ldots
\end{array}
$$

but there is no bijection that would also preserve the (usual) order.

Consider now an arbitrary class $\mathcal{K}$ of posets. It is an easy exercise to convince oneself that *isomorphism of order is an equivalence relation* in $\mathcal{K}$. Hence, similar posets may be identified not only in their set-theoretic, but also in their order properties. They differ only in the concrete nature of their elements. The class $\mathcal{K}$ of all posets is partitioned into (disjoint) classes of similar posets. Each member of such a class can be looked upon as "copy" of the same kind of poset, even though it is composed of different elements.

PROBLEMS

2.3-1. Prove that if A is a totally ordered set and $A \cong B$, then B is totally ordered.

2.3-2. Let A and B be isomorphic ordered sets. Prove that $a \in A$ is a first (last, minimal, maximal) element of A if its image $f(a)$ is a first (last, minimal, maximal) element of B.

2.4 CARDINAL NUMBERS

In this section we collect, without giving proofs and without even respecting requirements of strict logic, some basic facts concerning the "size" of sets.

Suppose we have two sets X and Y which are isomorphic in the pure set-theoretic sense (cf. p. 36). Since this means that there is a one-to-one correspondence between their elements, we shall say that two isomorphic sets have "equal size." In a more formal way, we often call isomorphic sets *numerically equivalent.*† Alternatively, we say that the two sets have the same *cardinal number.* If the set X is isomorphic to the set of symbols $\{1, 2, \ldots, n\}$, then we say that X is a *finite set* and has the cardinal number n. Thus, if X and Y are isomorphic finite sets, then they have the same cardinal number iff they have "the same number of elements."

Suppose a set X is numerically equivalent (i.e. isomorphic) to some subset of a set Y. Then we say that X is dominated by Y and write $X \preceq Y$. We now agree to the following definition: the cardinal number of X is less than or equal to‡ the cardinal number of Y iff $X \preceq Y$.

We may raise the question: can a *proper* subset of X be numerically equivalent to X itself? For finite sets, this is obviously impossible. However, it is perfectly possible for infinite sets. For example, the set of natural numbers N and its subset of even numbers E are numerically equivalent, since the map $f: N \to E$ given by $x \mapsto 2x$ is a bijection. The general situation is described by the celebrated *Schröder–Bernstein theorem* which tells us that if $X \preceq Y$ and $Y \preceq X$, then $X \simeq Y$. The converse statement is trivial.

Any set X which is dominated by N is obviously either a finite set or numerically equivalent to N. Conversely, if a set X is infinite, then $N \preceq X$. Thus, N is the infinite set which has the smallest cardinal number. We say that N is *countably infinite*§ and denote its cardinal number by $\aleph_0$ (aleph null). Any set which is isomorphic to N has, of course, cardinal number $\aleph_0$. Clearly, if X is an infinite set with the property that we can *list all its elements*, then X is isomorphic to N and so has cardinal number $\aleph_0$. Sets which are either finite or have a cardinal number $\aleph_0$ are called, for obvious reasons, *countable sets*¶ (or denumerable sets).

A nontrivial example of a countable infinite set is that of the positive rational numbers Q^+. Being accustomed to the standard ordering in Q^+, it is not easy to see how one possibly enumerates (lists) the elements of Q^+. Nevertheless, it can be done. Let us write down all positive fractions in this array:

---

†We recall that set isomorphism is an equivalence relation.
‡Note that Y *is* a subset of itself, even though not a proper one.
§The term "denumarably infinite" is also used.
¶We emphasize that we shall use this term always in this sense, i.e. it includes the possibility of finiteness. Some authors use the term "at most countable."

$$\frac{1}{1} \quad \frac{1}{2} \quad \frac{1}{3} \quad \frac{1}{4} \quad \cdots$$

$$\frac{2}{1} \quad \frac{2}{2} \quad \frac{2}{3} \quad \frac{2}{4} \quad \cdots$$

$$\frac{3}{1} \quad \frac{3}{2} \quad \frac{3}{3} \quad \frac{3}{4} \quad \cdots$$

$$\frac{4}{1} \quad \frac{4}{2} \quad \frac{4}{3} \quad \frac{4}{4} \quad \cdots$$

$$\vdots$$

Now, for example, let us list elements proceeding via "right to left diagonals":

$$\frac{1}{1}, \quad \frac{1}{2}, \quad \frac{2}{1}, \quad \frac{1}{3}, \quad \frac{2}{2}, \quad \frac{3}{1}, \quad \frac{1}{4}, \quad \frac{2}{3}, \quad \frac{3}{2}, \quad \frac{4}{1}, \quad \cdots$$

Every positive rational will occur in this list *at least* once. We throw out those that already appeared and thus get a correct listing

$$1, \quad \frac{1}{2}, \quad 2, \quad \frac{1}{3}, \quad 3, \quad \frac{1}{4}, \quad \frac{2}{3}, \quad \frac{3}{2}, \quad \cdots$$

This list indeed represents a bijection between $\mathbf{N}^+$ and $\mathbf{Q}^+$:

$$
\begin{array}{ccccc}
1 & 2 & 3 & 4 & 5 \quad \cdots \\
\updownarrow & \updownarrow & \updownarrow & \updownarrow & \updownarrow \\
1 & \frac{1}{2} & 2 & \frac{1}{3} & 3 \quad \cdots
\end{array}
$$

and so $\mathbf{Q}^+$ and $\mathbf{N}^+$ (or $\mathbf{N}$) are numerically equivalent.†

If X is dominated by Y but Y is not dominated by X (so that the Schröder–Bernstein theorem does not apply and X is not isomorphic to Y), we say that X is *strictly dominated* by Y and write $X \prec Y$. Obviously, $\mathbf{n} \prec \mathbf{N}$. One then may ask as to whether there are sets that strictly dominate $\mathbf{N}$, i.e. whether there exist infinite sets with higher cardinality than $\aleph_0$. The answer lies in the affirmative, and in fact it turns out that there is an endless hierarchy of cardinal numbers. The precise statement is expressed by *Cantor's theorem* which says that every set X is strictly dominated by its power set $\mathscr{P}(X)$, i.e. $X \prec \mathscr{P}(X)$. Thus, given a

†We may consider the listing as a new ordering of the rationals: $1 < \frac{1}{2} < 2 < \frac{1}{3} < 3 < \ldots$.

set X we easily can fabricate a new set which has higher cardinality. For example, if X is a finite set with n elements, its power set is easily seen to have 2^n elements.

The most important example of an infinite set which strictly dominates N is the set R of all real numbers. The familiar elementary proof of this statement† consists in assuming that the real numbers can be listed and then showing that whatever this listing was, one can construct a real number which differs from all in the list. Hence, $N < R$.

The real numbers form, therefore, an uncountably infinite set and we say that R has the cardinal number of the continuum. This is usually denoted by c (German c). It can be shown that c is actually the cardinal number of the power set of a set which has cardinal number $\aleph_0$. However, it is not known whether c is the "immediate successor" of $\aleph_0$, i.e. whether or not there is any set X such that $N < X < R$. Cantor's surmise was that there is no such set X; this statement is called the "continuum hypothesis." If this is true, then we could write $\aleph_1$ instead of c, indicating that it is the immediate successor of $\aleph_0$. The present status of this puzzle is that both the continuum hypothesis as well as its negation are compatible with the axioms of set theory.

It is not difficult to show by direct construction that the cardinal number of the plane is the same as that of the real line and in fact even R^n has also the cardinal number c of the continuum.

This last remark leads us to consider arithmetic rules for combining cardinalities. We mention only a few, which will be used later in our work.

If A and B are disjoint sets with finite cardinal numbers n and m, then, obviously, $A \cup B$ has cardinal number $n + m$. It is equally easy to see that if A is finite (cardinal number n) and B countably infinite, then $A \cup B$ has cardinal number $\aleph_0$. We indicate this by writing

$$n + \aleph_0 = \aleph_0.$$

Less obvious is the rule

$$\aleph_0 + \aleph_0 = \aleph_0,$$

and rather surprising is the following generalization:

If $\{\ldots X_k \ldots\}$ is a countable class of countable sets X_k, then $\cup_{k=1}^{\infty} X_k$ is itself a countable set.‡

On the other hand, one has the rules

---

†See any standard text, e.g., Simmons[36], p. 36.

‡We note here that this and several other theorems of this type are best proved by using the Schröder–Bernstein theorem.

$$\aleph_0 + c = c \quad \text{and} \quad c + c = c.$$

One may also inquire about the cardinality of the Cartesian product set $A \times B$ when the cardinality of A and B is known. For finite sets, the trivial answer is that $A \times B$ has cardinality *nm*. In obvious notation, we have the following generalizations to infinite sets:†

$$\aleph_0 \, \aleph_0 = \aleph_0,$$

$$\aleph_0 c = c,$$

$$c \, c = c.$$

We shall not discuss further laws of the cardinal number arithmetic. Nor do we consider sets with cardinality higher than c. Perhaps, we just give an example of a set that strictly dominates **R**: this is the set of all real valued functions on **R**.

PROBLEMS

2.4-1. Show that the set **Z** of all integers is countable. Then proceed to show that the set **Q** of *all* rational numbers is countable.

2.4-2. Let $X_1, X_2, \ldots, X_n$ be countable sets. Prove that $X_1 \times X_2 \times \ldots \times X_n$ is countable.

2.4-3. Prove that any countably infinite set X contains a proper subset which is numerically equivalent to X. In fact, try to prove the following generalization: every infinite set is numerically equivalent to a proper subset of itself. (*Hint*: The Schröder–Bernstein theorem will be helpful.)

2.4-4. Prove that any set X is strictly dominated by the set Y which consists of all functions from X to the set $\{0, 1\}$. (*Hint*: Show first that Y is isomorphic to $\mathscr{P}(X)$; then refer to Cantor's theorem.)

2.5 SEQUENCES AND FAMILIES

From his practical mathematical experience, the reader is no doubt familiar with the somewhat vague notions of sequences of elements and parametrized families of elements. In this section we show that these intuitively simple concepts are actually disguised forms of a certain type of function.

Let $\mathbf{N}^+ = \{1, 2, 3, \ldots\}$ be the set of positive integers and let X be an

---

†The first equality is not really new: it is equivalent to our previous statement that the countable union of countable sets is countable. An example of the third equality is our previous remark that $\mathbf{R}^2$ has the same cardinal number as **R**.

arbitrary set. Suppose we have some arbitrary map $f: \mathbf{N}^+ \to X$. This map *defines a sequence.* Usually we would write $f(k) = x$ for the image $x \in X$ that corresponds to the integer k under the mapping $k \mapsto x$. However, it is customary to write instead of $f(k)$ the symbol x_k and to call x_k the "kth" term of the sequence (x_k)." Thus, from the logical point of view, *a sequence is a special map. The notation x_k refers to the particular element $x \in X$ which is the image of the integer k under the map.* Very often the sequence is denoted by the more detailed symbol $(x_1, x_2, \ldots, x_k, \ldots)$. Sometimes, when no confusion can arise, we just talk about "the sequence x_k," even though strictly speaking x_k is the symbol for a particular term of the sequence.

It is very important to realize that a sequence in X (or a sequence of elements taken from X) is *not* a subset of X. The individual images x_k of the integers k that "make up" the sequence must be referred to as the *terms of the sequence*† and *not* the "members" or "elements" of the sequence. Clearly (and in contradistinction to sets), a sequence may have *repeated terms.* It may even happen that a sequence has only one *distinct* term. Indeed, if $f: \mathbf{N}^+ \to X$ is the constant map which assigns to every integer k the same particular element $\hat{x} \in X$, then the sequence is $(\hat{x}, \hat{x}, \hat{x}, \ldots)$ i.e. $x_1 = x_2 = \cdots = x_k = \cdots = \hat{x}$. It is also clear that when writing down a sequence, it *does* matter in which order we indicate the elements. For example, $(x_1, x_2, \ldots, x_k, \ldots)$ is different from $(x_2, x_1, \ldots, x_k, \ldots)$. We can talk meaningfully about the "first, second, $\ldots, k$th term" of a sequence.

On the other hand, the collection of all *distinct* terms of a sequence (x_k) is obviously a subset of X. To avoid confusion, we indicate this set by $\{\ldots x_k \ldots\}$, in conformity with our standard set-notation. It should be obvious that any sequence (x_k) contains countably infinite many terms, but the set $\{\ldots x_k \ldots\}$ whose elements are the distinct terms of the sequence, may have either countably infinite many or just finite many elements. It is also worthwhile to realize that if X is an infinite set of arbitrary cardinal number, the elements of any countable subset of X may serve as terms of some sequence.

In passing we note that we can replace in the definition of a sequence the set $\mathbf{N}^+$ by the set $\mathbf{N} = \{0, 1, 2, \ldots\}$ of all natural numbers. The only difference will be that the "first" element of the sequence will then be denoted by x_0 rather than by x_1.

---

†Sometimes the distinct terms x_k are called the *set of values* of the sequence (x_k), emphasizing the viewpoint that a sequence is really a map.

Very often we talk about a *subsequence* of a given sequence. Conceptually, by this we mean the restriction of $f: \mathbf{N}^+ \to X$ to some subset $\mathbf{K}$ of $\mathbf{N}^+$. Formally, we may say that a subsequence (x_{k_i}) of (x_k) is obtained from (x_k) if one removes some, maybe infinitely many terms, but still leaves behind infinitely many terms and does not alter their order.

To conclude the discussion of sequences, we remark that a sequence may be considered as a generalization of an ordered n-tuple. Indeed, let $f: \mathbf{n} \to X$ be a map from $\mathbf{n} = \{1, 2, \ldots, n\}$ to a set X. Denoting the image $f(k)$ by x_k, we have n images (some of them may coincide) and we can display them by the familiar symbol $(x_1, x_2, \ldots, x_n)$. Actually, n-tuples are sometimes called "finite sequences," but, whenever possible, we shall not use this terminology.†

The concept of a sequence can be considerably generalized. Let A be some set whose elements we shall denote by the generic symbol α. Let X be a given set. Let $f: A \to X$ be an arbitrary map. Instead of writing $f(\alpha) = x$ for the image $x \in X$ that corresponds to the element α under the mapping $\alpha \mapsto x$, we decide to use the notation x_α. In this way, for a given f and specified set A, we select from X an "*indexed family*." The *terms* of the family are labeled by members α of the set A. For this reason we call A an *indexing set*. The entire family in X (or the indexed family of elements taken from X) is denoted by the symbol (x_α). The individual terms of the family are the images x_α.

In summary, an indexed family is, from the conceptual point of view, a *map* from some indexing set A to the set X. Once again, a clear distinction must be made between a family (x_α) and the subset $\{\ldots x_\alpha \ldots\}$ of X. The latter is the collection of all distinct terms of the family. A family (x_α) may very well consist of one single term repeated many times, possibly even uncountably many times. Nevertheless, the corresponding set $\{\ldots x_\alpha \ldots\}$ will then have only one element.

The most familiar example of an indexed family (other than an n-tuple or a sequence) is what we usually call a "*continuous one-parameter family*" of objects. In this case, the indexing set is the set $\mathbf{R}$ of real numbers. The terms of the family are parametrized (labeled) by real numbers. From our present viewpoint, such a continuous one-parameter family (x_α) of terms taken from X is nothing else than a specified function $f: \mathbf{R} \to X$, where x_α stands for $f(\alpha) = x$. As a matter of fact, it is not uncommon to write $x(\alpha)$ instead of x_α. This usage may be confusing for the beginner.

---

†Thus, if we say "sequence," we shall mean an "infinite sequence," unless we specifically say "finite sequence."

Similarly, a continuous two-parameter family $(x_{\alpha,\beta})$ is a map from the indexing family $\mathbf{R} \times \mathbf{R}$ to X with $x_{\alpha,\beta}$ (or $x(\alpha, \beta)$) representing $f(\alpha, \beta) = x$, and so on.

One can define *subfamilies* in a way similar to our definition of subsequences. If M is a subset of the indexing set A, then the restriction of f to M is called a subfamily of (x_{α}) (with $\alpha \in A$) and may be denoted by writing (x_{μ}), $\mu \in M$. This restriction has M as the set of indices.

Finally, we comment on our notation. It will be our practice throughout this book to use a Greek subscript (like α) to label terms of an *arbitrary* family. In contrast, we reserve Latin indices (like k) for the labeling of terms of sequences or n-tuples. Of course, (x_{α}) *may* denote a sequence or an n-tuple.

Quite frequently one encounters indexed families (or sequences) whose terms are sets (taken from some universal set). Naturally, one then uses the notation (X_{α}) for the indexed family of sets and the symbol $\{\ldots X_{\alpha} \ldots\}$ for the class of sets whose elements are the terms of the family (X_{α}).

In order to illustrate the concepts that were introduced in this section, we shall now show how the procedure of forming Cartesian products of sets (cf. Section 1.1) can be generalized for classes of infinitely many sets.

Let $\{X_1, X_2, \ldots, X_n\}$ be a class of n sets. Let $X = \cup_{k=1}^{n} X_k$. Let f be some function from $\mathbf{n} = \{1, 2, \ldots, n\}$ to X which assigns to each integer $k = 1, 2, \ldots, n$ some specialized point x_k of the set X_k. Thus, $f(k) = x_k \in X_k$. From the viewpoint of the present section, we may identify f with an n-tuple $(x_1, \ldots, x_k, \ldots, x_n)$, where the term x_k is some element of the set X_k. By our definition (Eq. (1.19)), this n-tuple is an element of the product set $X_1 \times X_2 \times \cdots \times X_n$. Now, consider the collection of *all* functions $f: \mathbf{n} \to X$, defined by $f(k) = x_k \in X_k$. In other words, consider, from our present viewpoint, the collection of *all* ordered n-tuples $(x_1, \ldots, x_n)$. This collection is precisely the product set $X_1 \times \cdots \times X_n$. Thus, instead of Eq. (1.19), we have the following alternative definition: The product $X_1 \times \cdots \times X_n$ is the collection of all functions $f: \mathbf{n} \to \cup_{k=1}^{n} X_k$, where each f assigns to k some element x_k of the set X_k, $k = 1, 2, \ldots, n$.

This consideration leads to the following generalization. Let $\{\ldots X_{\alpha} \ldots\}$ be an arbitrary collection of sets, whose elements X_{α} are the terms of an indexed family (X_{α}) of sets. Denote the indexing set by A (i.e. $\alpha \in A$). Let $X = \cup_{\alpha} X_{\alpha}$. Let

$$f: A \to X = \bigcup_{\alpha} X_{\alpha}$$

be some function which assigns to each α some element x_{α} of the set X_{α}.

The collection of *all* such functions f will be called the Cartesian product $\Pi_\alpha X_\alpha$. We may condense this definition into the formula

$$\Pi_\alpha X_\alpha = \{f \,|\, f: A \to \cup_\alpha X_\alpha, \quad f(\alpha) = x_\alpha \in X_\alpha\}. \qquad (2.9)$$

Note that each f is really an indexed family (x_α), with terms taken from X in a specified way, namely one term exactly from each X_α. We also may say that every element of the product set $\Pi_\alpha X_\alpha$ is a function f which chooses exactly one member x_α from each indexed set X_α.

We illustrate the subject on an example. Let X_α be the "vertical line" erected at the point α of the real axis **R**, cf. Fig. 2.9. What is $\Pi_\alpha X_\alpha$? Clearly, we have $A = \{\ldots \alpha \ldots\} = \mathbf{R}$ and $\cup_\alpha X_\alpha = \mathbf{R}^2$. Let then $f: \mathbf{R} \to \mathbf{R}^2$ be a function which assigns to each real number a point on the vertical line passing through it. (Thus, $f(\alpha) = x_\alpha$, $f(\beta) = x_\beta$, as shown in the figure.) The family (x_α) defined by f is really the graph of a function $\varphi: \mathbf{R} \to \mathbf{R}$. Therefore, *the product $\Pi_\alpha X_\alpha$, being the collection of all f, can be identified with the class of all real valued functions on the real line.*

PROBLEMS

2.5-1. Show that the class of all (infinite) sequences of numbers with terms 0 and 1 only, is an uncountable set. (*Hint*: Use the result of Problem 2.4-4.)

2.5-2. Let (X_k) (with $k = 1, 2, \ldots, n$) be a finite family of sets. Let H be some

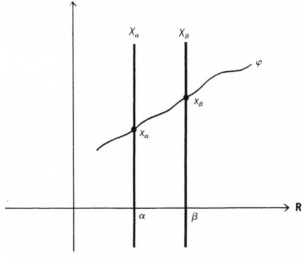

Fig. 2.9.

subset of the set $\mathbf{n} = \{1, 2, \ldots, n\}$. Let $P_H = \cup_{k \in H} X_k$ and $Q_H = \cap_{k \in H} X_k$. Let $\mathcal{A}_l$ be the class of sets whose elements are all subsets of $\mathbf{n}$ which have l elements. Show that

$$\bigcup_{H \in \mathcal{A}_l} Q_H \supset \bigcap_{H \in \mathcal{A}_l} P_H, \qquad \text{if } 2l \leq n + 1,$$

$$\bigcup_{H \in \mathcal{A}_l} Q_H \subset \bigcap_{H \in \mathcal{A}_l} P_H, \qquad \text{if } 2l \geq n + 1.$$

2.5-3. Prove that

$$(\bigcup_\alpha X_\alpha) \times (\bigcup_\beta Y_\beta) = \bigcup_{\alpha,\beta} (X_\alpha \times Y_\beta)$$

and that

$$(\bigcap_\alpha X_\alpha) \times (\bigcap_\beta Y_\beta) = \bigcap_{\alpha,\beta} (X_\alpha \times Y_\beta).$$

(*Note*: The notation $\cup_{\alpha,\beta}$ means $\cup_{(\alpha,\beta) \in A \times B}$ and similarly for $\cap_{\alpha,\beta}$.)

PART TWO

The Basic Structures of Mathematics

IIA: Algebraic Structures

3

Algebraic Composition Laws and Systems

The first mathematical knowledge about structures that a child acquires is the discovery that on the set of natural numbers there is defined a composition law (addition) which obeys certain rules. Our first search for structures will be a generalization of this experience.

Let A be a set with elements $a, b, c, \ldots$. An *algebraic composition law* on A is a rule which assigns to any ordered pair (a, b) an element c of A. Thus, a composition law on A is a map

$$k: A \times A \to A \quad \text{given by} \quad (a, b) \mapsto c.$$

Instead of writing $k(a, b) = c$, it is more practical to devise some special symbol, such as $\square$, or $\triangle$, or $*$, etc. for denoting the composition and we write $a \square b = c$, where $c \in A$ is the composite of $a \in A$ and $b \in A$. Strictly speaking, we should talk of a binary composition law, since it is clear that a map $N: A \times A \times \cdots \times A \to A$ could be considered as an "n-ary composition law." These, however, do not play an important role, so that the adjective "binary" is inessential in our future work.

Trivial examples of composition laws are, of course, addition or multiplication on the set of reals, and so on. A nontrivial example is the following. Let us denote by A^A the set of *all* functions from A to A, i.e. let $A^A = \{f | f: A \to A\}$. Let

$$\square: A^A \times A^A \to A^A$$

be defined by $g \square f = g \circ f$ for all $g, f \in A^A$, where $\circ$ denotes the composition of maps, as discussed in Section 2.1. Clearly, $\square$ is a

57

composition law on A^A and this explains why the term "composition of functions" was used in a more general setting.

What we discussed above is often called an *internal* composition law. A simple generalization is the concept of an *external* composition law. Let S (with elements $\alpha, \beta, \gamma \ldots$) and A (with elements $a, b, c \ldots$) be two *different* sets. Then a map

$$k : S \times A \to A \quad \text{given by} \quad (\alpha, a) \to b$$

is called an external composition law. Instead of $k(\alpha, a) = b$ we again write $\alpha \,\square\, a = b$. A familiar example is the scalar multiplication of vectors.† There is no real difference between internal and external composition laws, except for practical considerations. The former is a special case of the latter.

Binary composition laws (whether internal or external) are sometimes called binary *operations*. We may read $a \,\square\, b = c$ as "a operates on b yielding c." We shall avoid this terminology.

Suppose we have a set A for which a composition law is given. We then call the set together with the composition law an *algebraic structure* or *algebraic system*. An algebraic structure on A defined by an internal‡ composition law $\square$ is denoted by the symbol $(A, \square)$ and it is customary to say that "an algebraic structure is an ordered pair consisting of a set A and a composition law $\square$." Often, when no confusion can arise, we speak of "the structure A," even though A is really the symbol of the underlying set only. A given set A may carry, of course, different algebraic structures and we distinguish them by writing $(A, \square)$, $(A, \triangle)$, $(A, *)$, etc., in each case using a different symbol for the appropriate composition law. It often happens that a fixed set is endowed with two (or more) distinct composition laws. Then we write $(A, \square, *)$ and so on.

Let us introduce some standard terminology for denoting *special elements* of an algebraic system.

(α) A *unit element* (or identity element) for an internal composition law is an element $e \in A$ such that§

$$e \,\square\, a = a \quad \text{and} \quad a \,\square\, e = a \quad \text{for all} \quad a \in A.$$

If a unit element exists, then it is unique. Suppose that e and $\bar{e}$ are both

---

†We shall make this notion precise in Section 4.3.

‡If the composition law is external, then we may use the notation $(A, S, \square)$.

§If only one of the stated equations holds, we speak of a left (or right) unit element, respectively. These relaxed concepts are not very important, so that when we say "unit element," we shall always assume that it is a *two-sided* unit element.

units. Then we should have $e \,\square\, \bar{e} = \bar{e}$ and also $e \,\square\, \bar{e} = e$, so that $\bar{e} = e$, q.e.d. On the other hand, not every system has a unit. For example $(\mathbf{N}^+, +)$ (the set $\mathbf{N}^+$ of positive integers equipped with ordinary addition as composition law) has no identity element.

(β) A *regular element* for an internal composition law is any element $r \in A$ such that†

$$r \,\square\, a = r \,\square\, b \quad \text{implies} \quad a = b$$

and

$$a \,\square\, r = b \,\square\, r \quad \text{implies} \quad a = b$$

for any $a, b \in A$. For example, for the system $(\mathbf{R}, \cdot)$, i.e. ordinary multiplication on the set of reals, every element except the number zero is regular. If r is a regular element, then we often say that a "cancellation law" holds for r.

(γ) Suppose now that the system $(A, \square)$ has a unit element e. An element $a \in A$ is said to have an *inverse* if there exists an $a' \in A$ such that‡

$$a' \,\square\, a = e \quad \text{and} \quad a \,\square\, a' = e.$$

From the definition it is evident that the unit element e always has an inverse which coincides with e. Indeed, $e \,\square\, e = e$.

If an element a has exactly one inverse,§ it is standard practice to denote this inverse by the *symbol* a^{-1}.

To give trivial examples of inverses note that in the system $(\mathbf{R}, +)$ every element has an inverse, namely algebraic opposite (that is, $x^{-1} = (-x)$); whereas in $(\mathbf{R}, \cdot)$ every element except the number zero has an inverse (namely $x^{-1} = 1/x$). Attention should be called to the fact that regular elements must not be confused with elements that possess an inverse. For example, in $(\mathbf{N}^+, +)$ every element is regular, but no element has an inverse (since there is no unit element in the system).¶

To complete the above discussions, we remark that if we deal with an external composition law, we still may speak of a unit element ($\varepsilon \in S$ is a unit if $\varepsilon \,\square\, a = a$ for all $a \in A$) and of a regular element ($\rho \in S$ is regular if $\rho \,\square\, a = \rho \,\square\, b$ implies $a = b$).

The next topic to discuss is the terminology which refers to composition laws that have *special properties*.

†If only one of the stated conditions holds, we speak of a left (or right) regular element, respectively.

‡If only one of the two stated conditions holds, we call a' a left (or right) inverse of a, respectively.

§For a condition that guarantees uniqueness of an inverse, see Problem 3-4.

¶More complicated situations will be met later.

(a) An internal composition law on A is said to be *associative* if

$$a \,\square\, (b \,\square\, c) = (a \,\square\, b) \,\square\, c \quad \text{for all} \quad a, b, c \in A.$$

In that case, parentheses may be omitted from any chain of compositions, as is easily seen by induction.

(b) An internal composition law on A is *commutative* if

$$a \,\square\, b = b \,\square\, a \quad \text{for all} \quad a, b \in A.$$

(c) Suppose now that we have a system with two composition laws, $(A, \square, \triangle)$. We say that $\triangle$ is *left distributive* over $\square$ if

$$a \,\triangle\, (b \,\square\, c) = (a \,\triangle\, b) \,\square\, (a \,\triangle\, c)$$

and $\triangle$ is *right distributive* over $\square$ if

$$(a \,\square\, b) \,\triangle\, c = (a \,\triangle\, c) \,\square\, (b \,\triangle\, c)$$

for all $a, b, c \in A$. Of course, if $\triangle$ is distributive over $\square$ (either left or right or both), then it is not necessary that $\square$ be distributive over $\triangle$. For example, in $(\mathbf{R}, +, \cdot)$, multiplication is both left and right distributive over addition but addition is not distributive over multiplication $(a + bc \neq (a + b)\cdot(a + c))$.

By way of summary, we illustrate now *all* concepts which we introduced above on one single interesting example.† Let S be a universal set and consider the power set $\mathscr{P}(S)$, whose elements we denote by $A, B, C, \ldots$. Let us define two composition laws: union and intersection. $(\mathscr{P}(S), \cup, \cap)$ is obviously an algebraic system with two internal composition laws. Reviewing Section 1.1, we easily ascertain the following properties of our system. Both laws are associative and commutative. Each law is left and right distributive over the other. The unit element for $\cup$ is $\emptyset$ and the unit element for $\cap$ is S. For $\cup$, the only regular element is $\emptyset$ (because $A \cup B = A \cup C$ implies $B = C$ iff $A = \emptyset$). Similarly, for $\cup$, the only regular element is S. For $\cup$, the only element that has an inverse is $\emptyset$ and its inverse is itself. (This is so because we already know that $\emptyset$ is the unit, so we ask: for which elements A exist an A' such that $A' \cup A = A \cup A' = \emptyset$? The answer is: the required relations can be satisfied only if $A = \emptyset$ and then $A' = \emptyset$.) In a similar manner we ascertain that for the law $\cap$, the only element that has an inverse is S and its inverse is itself.

---

†More examples are given in the problems and will be met in the next chapter, when we study special systems.

We conclude this general discussion of algebraic systems with a consideration concerning the *restriction of a composition law* to a subset. Let $(A, \square)$ be an algebraic system and let $B \subset A$. We say that B *is closed under the law* $\square$, if for any pair $b, b' \in B$, we find that $b \square b' \in B$. If B is closed under $\square$, then $(B, \square)$ is an algebraic system in its own right, and may be called a *subsystem* of $(A, \square)$. For example, $(\mathbf{N}, \cdot)$ is a subsystem of $(\mathbf{R}, \cdot)$, because it is closed, since the product of two natural numbers is a natural number. But $(\mathbf{n}, \cdot)$ is not a subsystem of $(\mathbf{R}, \cdot)$, because it is not closed, since the product of two numbers k and l, both less than or equal n, may or may not be less than or equal n.

An important observation here is that even if $(A, \square)$ has a unit element, its subsystem $(B, \square)$ does not necessarily have one. For example, both $(\mathbf{E}, \cdot)$ and $(\mathbf{O}, \cdot)$ are subsystems of $(\mathbf{R}, \cdot)$, the first does not have a unit, the second has one (namely the number 1, which is the same as the unit of $(\mathbf{R}, \cdot)$). Similarly, other *properties of $(A, \square)$ may or may not carry over to a subsystem $(B, \square)$*.

We have here, finally, a comment on usage. When we say that $(A, \square)$ is an algebraic system, we of course imply that the composition law $\square$ can be performed for any pair $a, b \in A$, i.e. that A is closed under $\square$. In the older literature this was not the custom, and the characterization of a specific algebraic system started always with asserting the existence of a composition law "such that for every $a, b \in A$, the element $a \square b \in A$." This is completely unnecessary and will not be done in our future work.

PROBLEMS

3-1. Let $\square$ be the binary operation of least common multiple, defined on the set of positive numbers, i.e. $a \square b = $ l.c.m. of a and b. Is this combination law commutative? Is it associative? Has it an identity element? Which elements of the set have inverses, and what are they? Which elements are regular?

3-2. Let A^A be the set of all functions $f: A \to A$, and let $\circ$ denote the composition of functions. When is this composition law commutative? Is it associative? What is the unit element? Which elements have inverses and what are the inverses? Which elements are regular?

3-3. Consider the combination laws defined by $(a, b) \mapsto a$ and $(a, b) \mapsto b$ and show that they are associative.

3-4. Suppose the law $\square$ is associative and has a unit element e. Show that if an element x has an inverse x', this is unique and x is regular for the law $\square$.

3-5. Suppose the law $\square$ is associative and has a unit element e. Suppose that both x and y have inverses x' and y'. Show that $x \square y$ also has an inverse. (What is it?) Is $x \square y$ and/or $x' \square y'$ a regular element? (Use previous theorem.)

3-6. Let $(A, \square)$ be an algebraic system with a unit element e. Suppose the equation

$$(a \square b) \square (c \square d) = (a \square c) \square (b \square d)$$

holds for all choices of $a, b, c, d \in A$. Prove that $\square$ is both associative and commutative.

3-7. Let $\mathbf{Z}$ be the set of integers and let $(\mathbf{Z}, \square, \triangle)$ be the system where we define

$$a \square b = a + 2b \quad \text{and} \quad a \triangle b = 2ab.$$

($x + y$ and xy stand for ordinary sum and product.) Show that $\triangle$ is left distributive over $\square$. Do you need a direct calculation to show that $\triangle$ is also right distributive over $\square$? Is $\square$ distributive over $\triangle$?

3-8. Let $\square : A \times A \to A$ be an associative composition law on A, with a unit element e. Let B be the subset of A consisting of those elements which have an inverse. Prove that B is closed under $\square$.

3.1 MORPHISMS OF ALGEBRAIC SYSTEMS

Our purpose in this section is to establish relations between different algebraic systems, similarly as we found it profitable to look for relations between completely unstructured sets (set isomorphism, cf. the end of Section 2.1) or between different posets (morphisms of posets, cf. end of Section 2.3). This study will eventually lead to means of "identification" of seemingly different algebraic structures and thus it plays a crucial role in the exploration of systems.

Let $(A, \square)$ and $(A', \square')$ be algebraic systems. Then

$$f : (A, \square) \to (A', \square')$$

is said to be a *morphism between the two algebraic systems*, furnished by the map f, if this map f from A to A' has the property that

$$f(a \square b) = f(a) \square' f(b) \quad \text{for all} \quad a, b \in A. \tag{3.1}$$

Thus, a morphism "carries" the composition law $\square$ on A to the composition law $\square'$ on A', in addition to mapping elements $a, b, \ldots$ on elements $a', b', \ldots$. If $a \square b = c$ in the first system and if $a \mapsto a'$, $b \mapsto b'$, $c \mapsto c'$, then in the second system we have $a' \square' b' = c'$. In other words, the image of the composite of two elements is the composite of the corresponding images.

A simple example of a morphism is the following. Consider $(\mathbf{R}^+, \cdot)$, i.e. the set of all positive nonzero reals equipped with ordinary multiplication and $(\mathbf{R}, +)$, i.e. the set of *all* reals equipped with ordinary addition. The

map
$$\log: (\mathbf{R}^+, \cdot) \rightarrow (\mathbf{R}, +)$$

is a morphism, because $\log (a \cdot b) = \log a + \log b$. (Note that whereas $a, b > 0$ by definition, $a' = \log a$ and $b' = \log b$ can have arbitrary signs.) It may be interesting to observe in this particular example that in the first system the inverse of a is $1/a$ and the corresponding images are also inverses in the second system because $\log 1/a = -\log a$, which, for addition, is the inverse of $\log a$.

Suppose that we have two systems, each equipped with more than one (say, two) composition laws. Then a map

$$f: (A, \square, \triangle) \rightarrow (A', \square', \triangle')$$

may be a morphism for either one of the composition laws, or for both of them (or, of course, for neither one). For example, let $(\mathbf{N}, +, \cdot)$ be the set of natural numbers equipped with ordinary addition and multiplication, and let $(X, \oplus, *)$ be the set $X = \{0, 1\}$ on which the two composition laws $\oplus$ and $*$ are defined by the rules

$$0 \oplus 0 = 0, \qquad 0 * 0 = 0,$$
$$0 \oplus 1 = 1 \oplus 0 = 1, \qquad 0 * 1 = 1 * 0 = 0,$$
$$1 \oplus 1 = 0, \qquad 1 * 1 = 1.$$

Let f be the map from $\mathbf{N}$ onto X which assigns to each even natural number the element 0 and to each odd natural number the element 1 of X. This map is a morphism for both composition laws. This follows from the simple observation that if $e_1, e_2, \ldots$ are even numbers and $o_1, o_2, \ldots$ are odd numbers, then

$$\begin{array}{ll} e_1 + e_2 \text{ is even,} & e_1 \cdot e_2 \text{ is even,} \\ e_1 + o_1 \text{ is odd,} & e_1 \cdot o_1 \text{ is even,} \\ o_1 + o_2 \text{ is even,} & o_1 \cdot o_2 \text{ is odd.} \end{array}$$

Hence $f(e_1 + e_2) = 0$, $f(e_1 \cdot e_2) = 0$, so that $f(e_1 + e_2) = f(e_1) \oplus f(e_2)$, $f(e_1 \cdot e_2) = f(e_1) * f(e_2)$, and so on.

Morphisms may be *combined* with each other, i.e. two morphisms performed in succession give a new morphism. More precisely, let

$$f: (A, \square) \rightarrow (A', \square') \quad \text{and} \quad g: (A', \square') \rightarrow (A'', \square'')$$

be morphisms. Then

$$g \circ f: (A, \square) \rightarrow (A'', \square'')$$

is a morphism between the first and third system. Indeed,

$$(g \circ f)(a \,\square\, b) = g(f(a \,\square\, b)) = g(f(a) \,\square'\, f(b)) \equiv g(a' \,\square'\, b')$$
$$= g(a') \,\square''\, g(b') \equiv g(f(a)) \,\square''\, g(f(b))$$
$$= (g \circ f)(a) \,\square''\, (g \circ f)(b).$$

Morphisms are *classified* according to the mapping properties of the function f which provides the morphism. A morphism is called a

monomorphism if f is injective,

epimorphism or *homomorphism*† if f is surjective,

isomorphism if f is bijective.

Monomorphisms do not play an important role. Homomorphisms have already several interesting features and occur frequently in applications. Most important of all are, however, isomorphisms, as it shall transpire soon. To give elementary examples, we note that the map $f: (\mathbf{N}, +, \cdot) \to (X, \oplus, *)$ discussed above is a homomorphism (epi- but not mono-morphism), whereas the map log: $(\mathbf{R}^+, \cdot) \to (\mathbf{R}, +)$ is an isomorphism.

It is perfectly possible that we consider a map of an algebraic system *into itself* which preserves the composition law. We then use the term *endomorphism*. Thus,

$$f: (A, \square) \to (A, \square)$$

is an endomorphism if $f(a \,\square\, b) = f(a) \,\square\, f(b)$. If, in particular, f is bijective (i.e. if we have an isomorphic endomorphism), we use the term *automorphism*.‡

For example, in the system $(\mathbf{Z}, +)$ the map $x \mapsto 2x$ is an endomorphism (since $f(x_1 + x_2) = 2(x_1 + x_2) = 2x_1 + 2x_2 = f(x_1) + f(x_2)$) but not an automorphism (since it is not a bijection). The map $x \mapsto -x$ is an automorphism of $(\mathbf{Z}, +)$.

We now discuss some simple properties of morphisms.

---

†Even though the term "*homomorphism*" is somewhat old-fashioned and less descriptive than its modern equivalent "epimorphism," we shall be tempted to use it almost exclusively. This "conservatism" conforms with the majority of books and papers in the physics literature. A word of caution: some authors use the term "homomorphism" to denote an *arbitrary* morphism.

‡There is no special name for the designation of endomorphisms that are only injective or only surjective.

THEOREM 3.1(1).   *The composite† of two isomorphisms (epimorphisms, monomorphisms)   is   again   an   isomorphism   (epimorphism, monomorphism).*

The *proof* follows simply from our previous consideration of composition of morphisms and from the composition property of injections and surjections (see Problem 2.1-1).

For *isomorphisms* we have, in addition, the following

THEOREM   3.1(2).   *The   inverse   of   an   isomorphism   is   also   an isomorphism.*

Proof. First we observe that since f is bijective, the inverse function f^{-1} exists. Now, if $f(a) = a'$ and $f(b) = b'$, then we have

$$f^{-1}(a' \ \Box' \ b') = f^{-1}(f(a) \ \Box' \ f(b)) = f^{-1}(f(a \ \Box \ b))$$
$$= a \ \Box \ b = f^{-1}(a') \ \Box \ f^{-1}(b'),$$

hence $f^{-1}(a' \ \Box' \ b') = f^{-1}(a') \ \Box \ f^{-1}(b')$, q.e.d.

We can now appreciate the importance of isomorphisms: *Isomorphism is an equivalence relation in the collection of all algebraic systems.* This follows clearly from Theorems 3.1(1) and 3.1(2) and from the trivial observation that the identity map 1_A furnishes an isomorphism of any algebraic system $(A, \Box)$ onto itself.‡

Isomorphism of algebraic systems will be denoted by the symbol $\approx$. Thus,

$$(A, \Box) \approx (A, \Box),$$
$$(A, \Box) \approx (B, \triangle) \quad \text{implies} \quad (B, \triangle) \approx (A, \Box),$$
$$(A, \Box) \approx (B, \triangle) \quad \text{and} \quad (B, \triangle) \approx (C, *)$$
$$\text{imply} \quad (A, \Box) \approx (C, *).$$

In conformity with the practice that, if no confusion can arise, the system $(A, \Box)$ is often denoted simply by A, we shall frequently use the notation $A \approx B$ to indicate that the two corresponding systems are isomorphic. If f furnishes the isomorphism from A to B, then f^{-1} gives the isomorphism from B to A. If, in addition, g provides an isomorphism from B to C, then the isomorphism from A to C is given by $g \circ f$.

†It is of course assumed that the composite is defined, i.e. that the domain of the second morphism contains the range of the first.

‡The map $1_A : (A, \Box) \rightarrow (A, \Box)$ is an automorphism.

Isomorphic algebraic systems are "equivalent" both in their set-theoretic as well as in their algebraic properties. Thus, they may be "identified" in these respects, or speaking more precisely, they are indistinguishable and differ only in the concrete nature of their elements and in the notation for the composition laws. Isomorphic algebraic systems belong to the same equivalence class (which is a subset of all algebraic structures).

It is customary to define an *algebraic property* as one which, if possessed by an algebraic system, is also possessed by every other isomorphic image of the system. For example, commutativity is an algebraic property. A deeper example is provided by the following:

THEOREM 3.1(3). *If $(A, \square)$ has a unit element e and if $f : (A, \square) \rightarrow (A', \square')$ is an isomorphism, then $(A', \square')$ also has a unit element e'. Furthermore, e' and e are images of each other.*

Proof. Since e is the unit element,

$$f(a \square e) = f(a).$$

On the other hand, $f(a \square e) = f(a) \square' f(e)$. Comparison shows that $f(e)$ is a right unit in $(A', \square')$. In a similar way one sees that it is also a left unit. Note that the proof tacitly utilized both the surjective and the injective nature of the mapping. (The "onto" property is necessary, otherwise $f(e)$ is not guaranteed to act as a unit for *all* elements of A'. Furthermore, to shows that $e = f^{-1}(e')$, we must be sure that the inverse *function* f^{-1} exists; hence f must be "one-to-one" as well.)

From the above theorem, there follows another statement on algebraic properties:

THEOREM 3.1(4). *Let $(A, \square)$ have a unit element and suppose some element $a \in A$ possesses an inverse a^{-1}. Let $(A', \square')$ be isomorphic to $(A, \square)$. Then $f(a) \in A'$ also has an inverse which is precisely $f(a^{-1})$.*

Proof. $f(a^{-1}) \square' f(a) = f(a^{-1} \square a) = f(e) = e'$, and similarly for the opposite order of factors. We summarize the theorem by saying that under an isomorphism, the image of an inverse element is the inverse element of the image.

PROBLEMS

3.1-1. Let $f: (A, \square) \to (A', \square')$ be a morphism. Show that then the diagram

$$
\begin{array}{ccc}
A \times A & \xrightarrow{\ \square\ } & A \\
{\scriptstyle f \times f} \big\downarrow & & \big\downarrow {\scriptstyle f} \\
A' \times A' & \xrightarrow{\ \square'\ } & A'
\end{array}
$$

commutes. Show, conversely that if for a map f the above diagram commutes, then f is a morphism. (*Note*: For the definition of $f \times f$, see Problem 2.1-5.)

3.1-2. Let $f: (A, \square) \to (A', \square')$ and $g: (A', \square') \to (A'', \square'')$ be two morphisms. Prove that if g and f are both injections, then $g \circ f$ is also an injection. Conversely, show that if $g \circ f$ is an injection, then f must be an injection.

3.1-3. Let

$$f: X \to Y \quad \text{and} \quad g: X \to Y$$

be two arbitrary morphisms between two algebraic systems. (In conformity with frequent usage, and also for practice, we use here and in the following the shorthand notation, disregarding the indication of the composition laws.) Let $m: Y \to Z$ be a monomorphism from Y to a third algebraic system Z. Suppose that

$$m \circ f = m \circ g,$$

and show that then $f = g$.

3.1-4. Let f and g be as stated in the preceding problem. Let $h: T \to X$ be a homomorphism (= epimorphism) from a third system T to X. Suppose that

$$f \circ h = g \circ h,$$

and show that then $f = g$.

3.1-5. Find all endomorphisms of the set of integers equipped with the usual addition as composition law. Which of these are automorphisms?

3.1-6. Find all endomorphisms of the system $(\mathbf{R}^+, \cdot)$. Which of these are automorphisms? (*Remark*: $\mathbf{R}^+$ is the set of *positive* reals.)

3.1-7. Show that the set of nonzero real numbers under multiplication cannot be isomorphic to the set of all real numbers under addition.

3.1-8. Let L be a lattice, L' another lattice and let $f: L \to L'$ be a morphism of lattices with respect to the binary composition laws $a \wedge b = \inf(\{a, b\})$ and $a \vee b = \sup(\{a, b\})$. Show that f is then automatically a morphism of order (i.e. $a \le b$ implies $f(a) \le f(b)$). (*Note*: The converse statement is not true in general, except if f is an *iso*morphism of order.)

4

Survey of Special Algebraic Systems

In this chapter we present a fairly detailed, yet by no means comprehensive, discussion of the most important algebraic systems.

All familiar algebraic systems of interest share one common property, namely associativity. *An algebraic system with a single internal associative composition law* is called a *semigroup.*

Example α. Let $A = \{a, b\}$, and define the composition law as specified by the following table:[†]

$\square$	a	b
a	a	b
b	a	b

The associativity is easily verified. Note that there is no unit element (a is a left unit but not a right unit, since $ba = a \neq b$).

Example β. The set $\mathbf{N}^+ = \{1, 2, 3, \ldots\}$ with ordinary addition. In contrast to the previous example, this is an *infinite* semigroup, which also happens to be commutative.

Example γ. Let X be an arbitrary set and consider the class $\mathscr{S}$ of all possible finite sequences with terms taken from X. Thus, typical elements of $\mathscr{S}$ are $s_n = (x_1, \ldots, x_n)$ or $t_m = (y_1, \ldots, y_m)$, with x_k or $y_k \in X$. Define in

---

[†]For small finite sets, it is customary and convenient to condense the law in a "composition table" (often called a multiplication table). The rows and columns are labeled by the elements. In the intersection of a row labeled x and a column labeled y stands the composite element $x \square y$.

$\mathscr{S}$ the composition law typified by

$$s_n \ \square \ t_m = (x_1, \ldots, x_n, y_1, \ldots, y_m).$$

This law, often called the concatenation of finite sequences, is obviously associative. The law has no other distinguishing features.

More interesting than semigroups are *associative systems which have a unit element*. Such a system is called a *monoid*.

Example δ.   The set $A = \{a, b\}$, with the composition table

$\square$	a	b
a	a	a
b	a	b

The (two-sided) unit element is b. Note that a has no inverse because the equation $a \ \square \ x = b$ has no solution.

Example ε.   The set $\mathbf{N} = \{0, 1, 2, \ldots\}$ with ordinary addition. The unit element is 0.

Example ζ.   The semigroup system of Example γ can be enlarged to become a monoid if we adjoin to the underlying set the ideal element "empty sequence" (which has no terms at all). This new element serves as a unit.

The real fun with algebraic systems starts if we go one step further and ensure that each element of the monoid possesses an inverse. If we modify the underlying set of Example ε to become the set $\mathbf{Z}$ of all integers (retaining addition as a composition law), then each element z will have an inverse, its opposite $(-z)$. Monoids where each element has an inverse are called groups and these extremely important structures will be the subject of the next section.

PROBLEMS

4-1. Let Q be the set of all rational numbers. Determine which of the following composition laws give rise to a semigroup:

(**α**) $a \ \square \ b = 0,$
(**β**) $a \ \square \ b = b,$
(**γ**) $a \ \square \ b = \frac{1}{2}(a + b),$
(**δ**) $a \ \square \ b = a + b - 1.$

Which of those are monoids?

4-2. Let $\mathcal{M}$ be the set $\mathbf{R} \times \mathbf{R}$, equipped with the composition law

$$(a, b) \,\square\, (c, d) = (a + c, b + d + 2bd).$$

Show that $\mathcal{M}$ is a commutative monoid.

4-3. Let X be a poset and Y any arbitrary set. Let $\mathcal{S}$ denote the set of all functions from X to Y. Define in $\mathcal{S}$ the composition law

$$(f \,\square\, g)(x) = \begin{cases} f(x) & \text{if } x < a \\ g(x) & \text{if } x \geqslant a, \end{cases}$$

where a is an arbitrary fixed element of X. Show that $(\mathcal{S}, \square)$ is a semigroup. Is it a monoid? (*Remark*: This system is a generalization of Example γ.)

4-4. Show by an example that a morphism of monoids is not necessarily a morphism of units (i.e. that $f(e)$ is not necessarily e').

4.1 GROUPS

By definition, *a group is an algebraic system* $(G, \square)$ *where the internal composition law is associative, there exists a unit element, and every element possesses an inverse.*

It is traditional not to use a symbol like $\square$, etc. for the group composition law but to rather write simply ab instead of $a \,\square\, b$. With this "multiplicative" notation, the *defining axioms of a group*† G are as follows:

(i) $a(bc) = (ab)c$ for all $a, b, c \in G$.
(ii) There exists an $e \in G$ such that $ea = ae = a$ for all $a \in G$.
(iii) For every $a \in G$ there exist an element $a^{-1} \in G$ such that $aa^{-1} = a^{-1}a = e$.

We recall that the uniqueness of e follows from the general theory of compositions (p. 58), whereas the uniqueness of inverse elements is a consequence of associativity (cf. Problem 3-4).

If G is a group with the additional property that, for any two elements a and b, we have $ab = ba$, we call G a *commutative group* or more frequently an *Abelian group*. Sometimes the term "additive group" is also used, and to emphasize the Abelian (i.e. "additive") nature of the composition law, the symbol $+$ is used explicitly.‡ Thus, for the composite of a and b we then write $a + b$. Similarly, the unit element of an additive group

---

†Without causing confusion, we can use the symbol G for the group as a *system*.

‡This is of course a reminder of the trivial fact that the best known Abelian group is $(\mathbf{Z}, +)$, where $+$ is elementary addition.

is written as o (rather than e), and the inverse of a is denoted by $-a$ (rather than by a^{-1}). Thus, $a + (-a) = o$.

One more comment on notation. Because of the associative law, in *any* group† the symbol a^k is perfectly well defined as a shorthand for

$$a^k \equiv aa \dots a \qquad \text{with } k \text{ "factors."}$$

It is advantageous to introduce the notation

$$a^0 = e,$$

and also set

$$a^{-k} \equiv a^{-1}a^{-1} \dots a^{-1} \qquad \text{with } k \text{ "factors."}$$

It should be clear that with these conventions

$$a^n a^m = a^m a^n = a^{n+m} \qquad \text{for any} \quad n, m = 0, \pm 1, \pm 2, \dots.$$

One also verifies that

$$(a^n)^m = (a^m)^n = a^{nm},$$

and that

$$a^{-n} = (a^n)^{-1}.$$

However, the reader should be aware that if $a \neq b$ and the group is not Abelian, then, in general, $(ab)^n \neq a^n b^n$.

We now give a few preliminary examples of groups.

Example α.   The set $G = \{a, b\}$ with the composition table

	a	b
a	a	b
b	b	a

Associativity is trivial; the unit element is a; the inverse of a is a itself as for any unit, and $b^{-1} = b$. It is natural to write e for a, and rename b, denoting it, say, by α. The group is denoted customarily by the symbol Z_2 with the table

	e	α
e	e	α
α	α	e

Example β.   Observe that in the previous example the element α has the property that $\alpha^2 = e$. We now define the group Z_n in the following

---

†Since we are considering *arbitrary* groups, we revert, naturally, to the multiplicative notation.

way. Let e be decreed to be a unit element and let α be another element. Consider the collection $\{e, \alpha, \alpha^2, \ldots, \alpha^{n-1}\}$ with the composition law

$$\alpha^k \alpha^l = \alpha^{k+l} \quad \text{if} \quad k + l < n$$

(as already implied by the notation) and

$$\alpha^k \alpha^l = \alpha^j \quad \text{with} \quad j = k + l - n \quad \text{if} \quad k + l \geq n.$$

This law gives rise to a group, called the *cyclic group Z_n of order n*. Note that this way of defining the group is merely a convenient way of simplifying notation. For example, for Z_3, we could write $\alpha^2 = \beta$ and the group table is then given by

	e	α	β
e	e	α	β
α	α	β	e
β	β	e	α

Note that, for any n, the cyclic group Z_n is Abelian.

Example γ.   Let G be the set of all pairs of real numbers with nonzero first terms, i.e.

$$G = \{(a, b)|a, b \in \mathbf{R}, \quad a \neq 0\}.$$

Define on G the composition law

$$(a, b)(c, d) = (ac, bc + d).$$

Associativity can be verified by direct calculation. The pair $(1, 0)$ is the unit element. The inverse of $(a, b) \in G$ is the pair $(1/a, -b/a)$. Note that this (infinite) group is not Abelian.

Example δ.   Let, for each real number α, the function $f_\alpha : \mathbf{R} \to \mathbf{R}$ be defined by $f_\alpha(x) = x + \alpha$. The set $G = \{\ldots f_\alpha \ldots\}$ all such functions is an infinite Abelian group, if the group composition law is defined to be the ordinary composition $f_\beta \circ f_\alpha$ of functions. Indeed, the composition law is given explicitly by $f_\beta \circ f_\alpha = f_{\beta+\alpha}$. It then follows clearly that the law is associative; the unit element is f_0; and the inverse of f_α is the element $f_{-\alpha}$. In addition, the group is Abelian.

Example ε.   Let S be some universal set and let $\mathscr{P}(S)$ be the power set, with elements $A, B, \ldots$. As follows from the discussion on p. 60, the formation of unions (or the formation of intersections) leads to a monoid, but not to a group, because (except for the unit elements $\emptyset$ and S, respectively) we do not have inverses. However, let us define on $\mathscr{P}(S)$ the new composition law, called *symmetric difference* and given by†

$$A \bigtriangleup B = (A - B) \cup (B - A).$$

The reader will easily verify that this composition law is associative. The unit element is the empty set, since $A \bigtriangleup \emptyset = A \cup \emptyset = A$, and also $\emptyset \bigtriangleup A = \emptyset \cup A = A$. Since we have $A \bigtriangleup A = \emptyset \cup \emptyset = \emptyset$, it is clear that every element A has an inverse which is itself. Thus, $(\mathscr{P}(S), \bigtriangleup)$ is a group, and in fact it is Abelian.

In the following we call attention to some *elementary theorems on groups.*

THEOREM 4.1(1).   *Every element of a group is a regular element. Hence the cancellation law applies and either one of the equations*

$$ac = bc \quad \text{or} \quad ca = cb$$

implies that $a = b$.

The proof follows from the existence of inverses and from associativity.

THEOREM 4.1(2).   *The inverse of a product‡ of any two group elements is the product of their inverses, in reverse order,* i.e.

$$(ab)^{-1} = b^{-1}a^{-1}.$$

The proof follows from the uniqueness of the inverse and from associativity.

Three additional simple theorems are given in Problems 4.1-3, 4.1-4, and 4.1-5.

We also take the opportunity to call the reader's attention to Problem 4.1-8 which gives the definition of the *direct product* of groups, a device which enables one to construct new groups from given ones.

Now we proceed to the discussion of *subgroups.*

Let G be a group and let Γ be a sub*system* of G as defined on p. 61. If Γ obeys the group axioms (i), (ii), and (iii) as specified at the beginning of

---

†The symbol $\bigtriangleup$ is traditional. The set $A \bigtriangleup B$ obviously contains all elements which are *either* in A *or* in B but not in both. It is worthwhile to observe that equivalently we may write

$$A \bigtriangleup B = (A \cup B) - (A \cap B).$$

‡Parenthetically we remark that it is customary to call the composite of two group elements a "product."

this section, then Γ is called a subgroup of G. It is important to realize that, by our definition of subsystems, Γ is certainly closed under the group composition law. This, however, does not yet guarantee in general the fulfillment of the second and third group axioms for Γ. (Associativity is trivially assured.) If G is a *finite* group, then closure *does* guarantee that Γ is a subgroup. (This follows from the observation that if a is an element of Γ, there must exist some integer k such that $a^k = e$, since we cannot go on generating an infinite number of elements. Then a^{k-1} is the inverse of a.) The *general criterion* is given by

THEOREM 4.1(3).   *Let Γ be a sub*SYSTEM *of G. Then Γ is a subgroup of G iff for every $\gamma \in \Gamma$, we have also $\gamma^{-1} \in \Gamma$.*

We leave the proof as an exercise.

A consequence of this theorem is the following, frequently used equivalent criterion:

THEOREM 4.1(4).   *A sub*SET† *Γ of G is a subgroup iff*

$$\gamma \in \Gamma \quad and \quad \delta \in \Gamma \quad imply \quad \gamma\delta^{-1} \in \Gamma.$$

In summary, we may say that a subset Γ of G which is itself a group in its own right for the same composition law that was defined on G, is a subgroup.

Some examples of subgroups follow.

Example ζ.   Every group G has two *trivial subgroups*, namely G itself and the one-element group $\{e\}$. All other subgroups are called *proper*.

Example η.   If G is the group of integers under addition, then the set of even integers is a subgroup. But the set of odd integers is not. The set of all natural numbers is also a subsystem, but only a monoid, not a subgroup.

Example ϑ.   Let G be a group of all nonzero rationals equipped with ordinary multiplication. The subset $\Gamma = \{2^n \,|\, n \in \mathbf{Z}\}$ is a subgroup. Indeed, together with $\gamma = 2^n$ and $\delta = 2^m$ also $\gamma\delta^{-1} = 2^{n-m}$ belongs to Γ, so that Theorem 4.1(4) applies.

---

†Note that this extremely powerful theorem refers to sub*sets*; we do not have to ascertain that Γ is closed.

Example κ.   Let G be the group of functions defined in Example δ. The subset $\{\ldots f_z \ldots\}$ with z an integer, is clearly a subgroup.

Example λ.   Let G be an arbitrary group and let $g \in G$ be an arbitrary fixed element. The subset $\Gamma_g = \{\ldots, g^{-2}, g^{-1}, e, g, g^2, \ldots\}$ is a subgroup, as is evident from Theorem 4.1(3). Γ_g is said to be a cyclic subgroup of G, generated by the element g. It is Abelian. Note that if G is an infinite group, then Γ_g may be either finite or infinite.

To conclude the discussion of subgroups, we consider briefly *the class $\mathscr{K}$ of all subgroups of a given group G*. Recalling from Subsection 1.2b that set inclusion is an order relation, it is clear that $\mathscr{K}$ is a poset. However, it is not a chain (if Γ_1 and Γ_2 are subgroups, we need not have either $\Gamma_1 \subset \Gamma_2$ or $\Gamma_2 \subset \Gamma_1$). On the other hand, the poset $\mathscr{K}$ has a first and a last element (the improper subgroups $\{e\}$ and G, respectively) and actually it is a lattice.

There is one more general topic in the introductory discussion of groups that deserves some comments, namely the question of *group morphisms*. Of course, we have a morphism between two groups G and G' if there exists a map $f: G \to G'$ such that $f(ab) = f(a)f(b)$. The novelty here is that, *for groups, any morphism* (not only an *iso*morphism) *is also a morphism of units and of inverses*, i.e.

$$f(e) = e', \qquad f(a^{-1}) = (f(a))^{-1} \equiv (a')^{-1}.$$

In other words, Theorems 3.1(3) and 3.1(4) hold for nonbijective morphisms as well. This follows easily from the unrestricted existence of inverses and from associativity.

Morphisms of groups are well visualized in some important aspects with the help of two concepts, the *image* Im (f) and the *kernel* Ker (f) of the morphism. If $f: G \to G'$ is a group morphism, then we define

$$\text{Im}\,(f) = \{f(g) | g \in G\} \tag{4.1}$$

and

$$\text{Ker}\,(f) = \{g | g \in G, \quad f(g) = e'\}. \tag{4.2}$$

Thus, Im (f) is simply the range of the morphism f (and may be denoted also by $f(G)$), whereas Ker (f) is the subset of elements of G which are mapped by the morphism onto the unit element e' of G' (and may be denoted also by $f^{-1}(\{e'\})$). Figure 4.1 visualizes these concepts. The importance of these concepts is shown by the following theorem, whose verification we leave as an easy exercise:

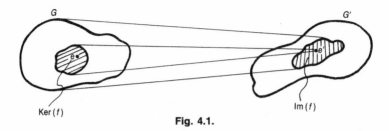

Fig. 4.1.

THEOREM 4.1(5). *Im* (f) *is a subgroup of* G' *and* Ker (f) *is a subgroup of* G.

To illustrate this theorem, let G be the additive group of all integers and let G' be the multiplicative group of all nonzero real numbers. The map $f: G \to G'$ given by

$$f(z) = \begin{cases} +1 & \text{if } z \text{ is even} \\ -1 & \text{if } z \text{ is odd} \end{cases}$$

is easily seen to be a morphism. Here Im $(f) = \{1, -1\}$ which is clearly a subgroup of G'. Furthermore, Ker $(f) = \mathbf{E}$, and this is obviously a subgroup of G.

Another useful theorem, giving a necessary and sufficient criterion for a group morphism to be a group *iso*morphism, is the following:

THEOREM 4.1(6). *A morphism is an isomorphism iff* Ker $(f) = \{e\}$ *and* Im $(f) = G'$.

The *proof* is almost trivial, since one realizes that bijectiveness is implied by the statements and vice versa.

Endo- and in particular automorphisms of groups play an important role in many applications. We give here a simple but very elegant theorem:

THEOREM 4.1(7). *If* G *is a group,*† *then the set of all its automorphisms is itself a group with respect to composition of morphisms.*

The *proof* is evident from the fact that the composition of two automorphisms is an automorphism (cf. Theorem 3.1(1)), 1_G is an

---

†As a matter of fact, this theorem holds not only for groups, but also for arbitrary algebraic systems.

automorphism, and the inverse of an automorphism is an automorphism (cf. Theorem 3.1(2)). The group of all automorphisms of G is usually denoted by Aut (G).

We illustrate this theorem on the following example. Let G be the cyclic group $Z_3 = \{e, \alpha, \alpha^2\}$. There are only two possible automorphisms; the first, g_1, is the identity map,

$$
\begin{array}{ccc}
e & \alpha & \alpha^2 \\
\downarrow & \downarrow & \downarrow \\
e & \alpha & \alpha^2
\end{array}
$$

and the second, g_2, is visualized by

The set $\{g_1, g_2\}$ is a group: the unit element is g_1, and since $g_2^2 \equiv g_2 \circ g_2 = g_1$, we have $g_2^{-1} = g_2$. Therefore, Aut $(Z_3) = \{g_1, g_2\}$. We observe that the composition law in $\{g_1, g_2\}$ is obviously that of Z_2, hence this group is isomorphic to Z_2, so we may say that Aut $(Z_3) \approx Z_2$.

PROBLEMS

4.1-1. Show that the following systems are groups:

(a) The set $\{z \,|\, z^n = 1\}$ with z a complex number, n positive integer, composition law ordinary multiplication.

(b) The unit circle $\{z \,|\, |z| = 1\}$ in the complex plane, with ordinary multiplication.

(c) The set $\mathbf{R}^n$ of all n-tuples of real numbers, with "componentwise addition," $(x_1, \ldots, x_n) + (y_1, \ldots, y_n) = (x_1 + y_1, \ldots, x_n + y_n)$.

(d) The set of all $n \times n$ matrices with nonzero determinant, where ordinary matrix multiplication is the composition law.

(e) The set of all surjective mappings of a finite set A onto itself, with the usual composition law of functions. Show that if A has more than two elements, this group cannot be Abelian.

4.1-2. Determine which of the systems listed below are groups:

(a) $G = \mathbf{N}^+$, $ab = \max (\{a, b\})$,

(b)† $G = \mathbf{R}$, $ab = a + b - ab$,

(c) $G = \mathbf{Z} \times \mathbf{Z}$, $(a, b)(c, d) = (a + c, b + d)$,

(d) $G = \mathbf{R} \times \mathbf{R}$, $(a, b)(c, d) = (ac + bd, ad + bd)$.

For systems that are not groups, indicate the axioms which are violated.

---

†In $a + b - ab$, the sum, difference, and product denote, of course, the usual operations in the set of reals.

4.1-3. Show that in any group, the equation $xx = x$ implies that $x = e$. (*Remark*: We may reexpress this theorem by saying that in a group, the only *idempotent element* is the unit.)

4.1-4. Show that in any group, the equations

$$ax = b \quad \text{and} \quad ya = b$$

have a unique solution x and y, respectively, for any given pair $a, b \in G$.

4.1-5. Prove that a group where for every element $xx = e$, is necessarily Abelian.

4.1-6. Let $a, b \in G$, and suppose $ab \neq ba$. Show that the elements e, a, b, ab, ba are all distinct. (*Hint*: You have to disprove each of the ten possible equalities separately.)

4.1-7. Using the result of the preceding problem, show that any noncommutative group has at least six elements. (*Hint*: Show that if G is noncommutative, then the above list of distinct elements may be enlarged by adjoining either aa or aba.)

4.1-8. Let G be a group with elements denoted by $a, b, \ldots$ and G' a group with elements $a', b', \ldots$. The *direct product* of G and G' is defined as the Cartesian product set $G \times G'$ with the composition law

$$(a, a')(b, b') = (ab, a'b').$$

Show that this system is indeed a group. What condition on G and G' has to be satisfied for $G \times G'$ to be Abelian?

4.1-9. Let G be the group of all nonzero rationals equipped with ordinary multiplication. Show that $\Gamma = \{(1 + 2n)/(1 + 2m) | n, m \in \mathbf{Z}\}$ is a subgroup.

4.1-10. Let G be a group and define its *center* as the set $C = \{a | a \in G, ax = xa$ for all $x \in G\}$. Prove that the center is a subgroup of G.

4.1-11. Let $\{\ldots \Gamma_\alpha \ldots\}$ an arbitrary indexed collection of subgroups of a group G. Show that $\cap_\alpha \Gamma_\alpha$ is also a subgroup of G. Is the same true for $\cup_\alpha \Gamma_\alpha$?

4.1-12. Show that every subgroup of a cyclic group is cyclic.

4.1-13. Let G be a group and let A, B be two *subsets*. Define the set

$$AB = \{ab | a \in A, \quad b \in B\}.$$

This is obviously a subset of G. Suppose now that A and B are actually sub*groups* of G, and that, furthermore, $AB = BA$. Show that then AB is a subgroup of G. (*Remark*: The condition $AB = BA$ does *not* imply that each element of A commutes with each element of B.)

4.1-14. Show that any two groups with three elements are isomorphic.

4.1-15. In each of the cases listed below, determine whether f is a morphism. If it is, decide whether it is a monomorphism, a homomorphism or an isomorphism. Find Im (f) and Ker (f):

(a) G = nonzero reals with multiplication,
   G' = positive reals with multiplication,
   $f(a) = |a|$.

(b) $G = G' =$ integers with addition,
  $f(a) = a + 1$.
(c) $G =$ integers with addition,
  $G' =$ rationals with addition,
  $f(a) = a/q$, where q is a given nonzero integer.
(d) $G = G' =$ integers with addition,
  $f(a) = na$, where n is a given integer.

4.1-16. Prove that every infinite cyclic group is isomorphic to the additive group of integers. (*Hint*: Recall that a group is said to be cyclic if it consists of all distinct (positive *and* negative, powers a^n of a single element a. Then consider the map $f(a^n) = n$, $n \in \mathbf{Z}$.)

4.1-17. Let $G \times G'$ be the direct product of two groups (cf. Problem 4.1-8). Show that the projections

$$p: G \times G' \to G \quad \text{and} \quad g: G \times G' \to G'$$

are morphisms of groups. Show also that $G \times G'$ and $G' \times G$ are isomorphic. Illustrate all these results for the cyclic groups $G = Z_2$ and $G' = Z_3$.

4.1a. Transformation Groups; G-Spaces; Orbits

Let X be an arbitrary set. Let $T = \{\ldots t_\alpha \ldots\}$ be some set of *bijective maps*

$$t_\alpha: X \to X,$$

subject to the following criteria:

(a) The identity map 1_X belongs to T,
(b) If $t_\alpha \in T$, then $t_\alpha^{-1} \in T$,
(c) If t_α and t_β belong to T, then the composite map $t_\beta \circ t_\alpha$ belongs to T.

The *set T is obviously a group under composition of maps*: (c) guarantees that it is closed; (a) assures the presence of a unit element; (b) assumes the existence of inverses; and the associativity is a basic property of composition of functions. The group T is called a *transformation group on the set X*. Its elements are bijective maps (transformations), and for the composite of two group elements we write simply $t_\beta t_\alpha$.

Example α.   Let X be the set of points which constitute the perimeter of some equilateral triangle (see Fig. 4.2). Let $T^{(1)} = \{1_X, R_{120}, R_{240}\}$, where R_θ denotes a counterclockwise rotation about the center by the angle θ. Thus, R_{120} maps the points of side 1–2 one-to-one and onto the points of side 2–3 and so on. Clearly, $R_{120}R_{120} = R_{240}$ and $R_{120}R_{240} = R_{240}R_{120} = 1_X$, so $R_{120}^{-1} = R_{240}$. Another transformation group is $T^{(2)} = \{1_X, D_1\}$, where D_1 is the reflection in the altitude through vertex 1. Here $D_1^{2} = 1_X$ so $D_1^{-1} = D_1$.

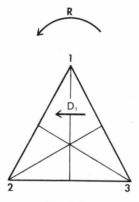

Fig. 4.2.

The largest set of bijections $X \to X$ which leaves distances between points unaltered is the set

$$T = \{1_X, R_{120}, R_{240}, D_1, D_2, D_3\}.$$

This is a group, usually called the *group of symmetries* of the equilateral triangle.† The reader is urged to compute the group composition table of T, by keeping track of where each vertex goes. The group T is not Abelian.

Example β.   The set of all maps $f_\alpha : \mathbf{R} \to \mathbf{R}$ given by $f_\alpha(x) = x + \alpha$ (see Example δ on p. 72) can be looked upon as a transformation group on the real line. It is called the group of translations in one dimension.

Example γ.   Let G be a group. The group Aut (G) may be considered as a transformation group *on the set* G, since each automorphism is a bijection $G \to G$ and the criteria (a), (b), and (c) listed above are satisfied.

As Examples α and β indicate, it is often profitable to give an alternative and somewhat less precise characterization of transformation groups. We may say that a transformation group T on X is a closed class of certain specified invertible bijective transformations of X onto itself. The group elements are the transformations themselves. Group composition is the successive application of transformations. The transformation which maps every point $x \in X$ onto itself is the unit element of the group T. The

---

†Quite generally, if X is some arbitrary set, the group consisting of all bijections $X \to X$ which leaves some specified property invariant, is called the group of symmetries, relative to that property. Such symmetry groups play a central role in modern physics.

transformation t^{-1} which reverses the effect of the transformation t is the inverse of the group element t.

We remark here that if X is a finite set (with N elements), then a transformation group T on X is usually called a *permutation group*. Each element of T is a permutation of elements of X. If T consists of n permutations, we denote it by P_n^N. In particular, the set of *all* possible permutations of elements of X is called the *symmetric group* S_N. It has $N!$ elements.

Since transformation groups are easily visualized and very familiar to every physicist and geometer, it is of great importance to realize that, actually, *any* group can be looked upon as a transformation group. The precise statement is given by *Cayley's theorem*:

THEOREM 4.1a(1).   *Any given group G is isomorphic to a group T of transformations.*

Proof.   Define for every $a \in G$ a function *on the set G*,

$$t_a : G \to G \quad \text{given by} \quad t_a(g) = ag \quad \text{for all} \quad g \in G. \qquad (4.3)$$

The map t_a is called *a left translation* on G by a. Each t_a is a *bijection* on G. Indeed, if $g_1 \neq g_2$, then $ag_1 \neq ag_2$ so that t_a (as defined by Eq. (4.3)) is one-to-one; furthermore, for any given $\gamma \in G$ and a fixed a, the equation $\gamma = ag$ always has a solution for g (namely $g = a^{-1}\gamma$), so that t_a is onto. Consider now the set T of all left translations t_a. Obviously 1_G belongs to T, since actually $1_G = t_e$, because $t_e(g) = eg = g$. Together with t_a, also t_a^{-1} belongs to T, since $(t_{a^{-1}} \circ t_a)(g) = t_{a^{-1}}(ag) = a^{-1}ag = g$, so that in fact $t_a^{-1} = t_{a^{-1}}$. Finally, the composite of two left translations is a left translation, because $(t_b \circ t_a)(g) = t_b(ag) = bag = t_{ba}(g)$, and $ba \in G$. Thus, all criteria (a), (b), and (c) given at the beginning of this subsection are satisfied so that *the set T of left translations is a transformation group on G*. For reference we note that, as was shown above

$$t_b t_a = t_{ba}, \qquad t_a^{-1} = t_{a^{-1}}, \qquad t_e = 1_G.$$

Now we proceed to exhibit an *isomorphism between G and the associated left translation group T*. Let

$$f : G \to T \quad \text{be defined by} \quad f(a) = t_a \quad \text{for each} \quad a \in G \qquad (4.4)$$

i.e. let us map each group element $a \in G$ on the left translation t_a which it defines. Since $f(ab) = t_{ab} = t_a t_b = f(a)f(b)$, we certainly have a morphism of groups. Since T consists of nothing else than all the possible t_a, we have a surjection. Since $a \neq b$ implies $t_a \neq t_b$ (because $ag \neq bg$ unless $a = b$), we have $f(a) \neq f(b)$ for distinct a and b, hence f is injective. Thus, f is bijective, so it is an isomorphism, and we have $G \approx T$, as claimed.

Remarks: (a) The above construction not only proved Cayley's assertion but actually identified the transformation group T as the left translation group on G.

(b) In place of the left translation group T, we could define the *right translation group* $\hat{T}$ with elements $\hat{t}_a$,

$$\hat{t}_a : G \to G \quad \text{given by} \quad \hat{t}_a(g) = ga \quad \text{for all} \quad g \in G.$$

All previous calculations work equally for $\hat{T}$. Thus, not only T but also $\hat{T}$ is isomorphic to G. It then follows that $\hat{T} \approx T$. (If, in particular, G is Abelian, then of course $\hat{T} = T$.)

A few simple examples may illuminate Cayley's theorem.

Example δ. If G is any finite group, then T consists obviously of some permutations of the elements of G. This implies that *every finite group is isomorphic to a group of permutations.*

Example ε. Let $G = Z_3 = \{e, \alpha, \alpha^2\}$. Then, by Cayley's construction, we find, for example, $t_\alpha(e) = \alpha e = \alpha$, $t_\alpha(\alpha) = \alpha\alpha = \alpha^2$, $t_\alpha(\alpha^2) = \alpha\alpha^2 = \alpha^3 = e$, and so on. The results are best visualized by using "matrix notation":

$$t_e = \begin{pmatrix} 1 & 0 & 0 \\ 0 & 1 & 0 \\ 0 & 0 & 1 \end{pmatrix}, \quad t_\alpha = \begin{pmatrix} 0 & 1 & 0 \\ 0 & 0 & 1 \\ 1 & 0 & 0 \end{pmatrix}, \quad t_{\alpha^2} = \begin{pmatrix} 0 & 0 & 1 \\ 1 & 0 & 0 \\ 0 & 1 & 0 \end{pmatrix}.$$

This convention is to be understood in the sense that, for example,

$$\begin{pmatrix} 0 & 1 & 0 \\ 0 & 0 & 1 \\ 1 & 0 & 0 \end{pmatrix} \begin{pmatrix} e \\ \alpha \\ \alpha^2 \end{pmatrix} = \begin{pmatrix} \alpha \\ \alpha^2 \\ e \end{pmatrix},$$

exhibiting the effect of t_α on the three group elements e, α, α^2. It is a simple matter of matrix multiplication to verify that the map $g \mapsto t_g$ is an isomorphism, i.e. that, for example, $t_\alpha t_{\alpha^2} = t_e$. Of course, in conformity with the statement of Example δ, each t_g is a permutation of the three "objects" e, α, α^2. For example, under t_α,

$$
\begin{array}{ccc}
e & \alpha & \alpha^2 \\
\downarrow & \downarrow & \downarrow \\
\alpha & \alpha^2 & e
\end{array}
$$

and so on. In other words, $Z_3 \approx P_3^3$. By resorting to the matrix notation, we actually exhibited a further isomorphism, namely one between the permutation group P_3^3 and the group of three matrices (as shown above) with matrix multiplication as the composition law.

Example ζ. Let G be the additive groups of reals, to be denoted by $\alpha, \beta, \gamma, \ldots$. Then $t_\alpha(\beta) = \alpha + \beta \equiv \beta + \alpha$ is a typical element of the left translation group,† and T is precisely the now familiar group of translations in one dimension, cf. Example β. We then have the map $f: (\mathbf{R}, +) \rightarrow T$ given by $f(\alpha) = t_\alpha$ for each real number α; so that the additive group of reals is seen to be isomorphic to the translation group in one dimension, the latter being considered as a transformation group on the reals.

We add here a few words on the *representation of groups*. If G is an arbitrary group and T some transformation group on a set X, then *a homomorphism from G to T is called a representation of G by T on the set X*. If, in particular, we have an *isomorphism* $G \approx T$, then we call the representation *faithful*. If the transformation group T on X consists of linear transformations,‡ we speak of a linear representation of G or a representation by linear operators. These linear representations are the most familiar ones. The importance of group representations lies primarily in the fact that transformation groups are much easier to handle than abstract groups. Unfortunately, we shall not have occasion to study the theory and practice of group representations in this book.§ However, at this point we note that the left *translation* (or the right translation) *group T associated with any given group G is*, clearly, *a faithful representation of G on the set G*. We call it the left (right) *regular representation* of G. Obviously, a given group can have many representations, apart from the regular, carried by the most diverse sets ("representation spaces") X. Actually, a fixed X may support several, *nonequivalent* representations of the same group G. We now leave this topic.

The well-explored and simple notion of transformation groups has an important and powerful generalization, which we shall study in the rest of this subsection.

Let X be some set and G some group. Suppose there is defined a fixed

---

†This particular case justifies the nomenclature "translation."

‡Linear maps will be introduced in Section 4.3. Linear operators (transformations) on topological spaces will be studied in detail in Chapter 12.

§As a matter of fact, the most important parts of representation theory (concerning the representations of infinite groups) can be attacked only in the possession of the full apparatus of Hilbert space theory, i.e. after having mastered all that will be said in this book.

map† $G \times X \to X$. It is then customary to say that the set X is a *group space* relative to G, or simply, X is a G-space. In other words, "X is a G-space" means that there is defined some external composition law for X, with "prefactors" taken from some group G. In general we shall denote the composite of a $g \in G$ and of an $x \in X$ by writing $g \cdot x$. Thus,

$$(g, x) \mapsto g \cdot x,$$

where, it should be remembered, $g \cdot x$ is an element of X (not of G). Now suppose the map $G \times X \to X$ obeys the following axioms:‡

(i) $e \cdot x = x$ for all $x \in X$,
(ii) $(g_1 g_2) \cdot x = g_1 \cdot (g_2 \cdot x)$ for all $x \in X$ and all $g_1, g_2 \in G$.

In that case we say that *the group G acts on the G-space X.* The "action" is given by the map $(g, x) \mapsto g \cdot x$.

Remark: Instead of acting on X from the left by G, we may also define a right action on X. This means, we have a map $X \times G \to X$, given by $(x, g) \mapsto x \cdot g$ and subject to the conditions $x \cdot e = x$, $x \cdot (g_1 g_2) = (x \cdot g_1) \cdot g_2$.

Example η. Transformation groups represent a special case of group action. In that instance, the acting group is the transformation group T, which consists of a certain class of bijective functions $t_\alpha : X \to X$. The action $T \times X \to X$ of T on X is given by the assignment $(t_\alpha, x) \mapsto t_\alpha \cdot x = t_\alpha(x)$.

Example ϑ. Suppose we have a representation of G by T on the set X, i.e. there exists a (homo)morphism $h: G \to T$. *This representation gives rise to an action of G on X,*

$$G \times X \to X \quad \text{with} \quad (g, x) \mapsto g \cdot x = (h(g))(x). \tag{4.5}$$

(To clarify the notation, note that $h(g)$ is image of $g \in G$ under the morphism h, hence $h(g)$ is an element of the transformation group T, i.e. a (bijective) map $X \to X$. The transform of $x \in X$ under the transformation $h(g) \in T$ is denoted by $(h(g))(x)$. Often, to simplify notation, one just writes $(hg)x$.) The fulfillment of the "action axioms" (i) and (ii) is easily verified.

Examples of a different and more sophisticated kind will appear in the

---

†Since we shall not need it, we do not introduce a function symbol f for the map.
‡In other words, the external composition law has a unit and is associative. We may call the system $G \times X$ an "external monoid."

following subsection. But first we wish to introduce an important concept connected with group actions.

Suppose G acts on X. Two points x and y of X are said to be *equivalent under G* if there exists some $g \in G$ such that $g \cdot x = y$. This relation E in X is indeed an equivalence relation. Since $e \cdot x = x$, we have $x E x$. Since $g \cdot x = y$ implies $g^{-1} \cdot y = x$ (because of property (ii) of actions), we see that $x E y$ implies $y E x$. Finally, if $g_1 \cdot x = y$ and $g_2 \cdot y = z$, then $(g_2 g_1) \cdot x = g_2 \cdot (g_1 \cdot x) = g_2 \cdot y = z$, so that $x E y$ and $y E z$ indeed imply $x E z$, q.e.d.

In consequence of the existence of the equivalence relation just defined, we can say that if G acts on X, then X will be partitioned into (disjoint) equivalence classes. An equivalence class with the representative element $x_0 \in X$ is the set

$$[x_0] = \{g \cdot x_0 | g \in G\}. \tag{4.6}$$

Each such equivalence class is called an *orbit of G on X*. The set $[x_0]$ is referred to as the orbit of the point x_0.

A preliminary example will both clarify the concept and justify the terminology. Let X be the set of points in the plane $\mathbf{R}^2$, let G be the group of rotations about the origin.† Each rotation is labeled by an angle variable θ. A point x is "carried along" a circle by the amount of θ when G acts on X. Consequently, the orbits are concentric circles, cf. Fig. 4.3. As is always the case, distinct orbits are disjoint sets and their union gives X.

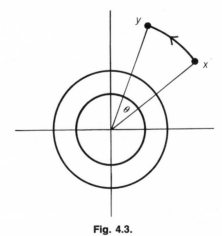

Fig. 4.3.

†In this simple case, G is actually a transformation group on X.

There is here some terminology worthwhile remembering. Suppose G acts on X in such a way that any pair x and y of elements X can be connected by some action, i.e. suppose that for each pair $x, y \in X$ there exists at least one $g \in G$ such that $y = g \cdot x$. We then say that G acts *transitively* on X. In view of Eq. (4.6) this means that we have only one orbit, i.e. the only orbit is the whole of X. A set X on which G acts transitively is called a *homogeneous space* † (under G). The above example tells us that $\mathbf{R}^2$ is not a homogeneous space under the rotation group of the plane. On the other hand, if $X = \mathbf{R}^2$ and T is the group of all translations of the plane, then $\mathbf{R}^2$ is a homogeneous space under T, because any pair of points can be connected by a suitable translation.

Another important concept relating to orbits will be introduced in Problems 4.1a-6 and 4.1a-7.

PROBLEMS

4.1a-1. Let X be the set of all nonzero reals. Let T be the set of the four functions $t_k \colon X \to X$, $k = 1, 2, 3, 4$, defined by

$$t_1(x) = x, \quad t_2(x) = \frac{1}{x}, \quad t_3(x) = -x, \quad t_4(x) = -\frac{1}{x}.$$

Is T a transformation group on X?

4.1a-2. Using the construction of Cayley's theorem, show that the group defined in Problem 4.1-1(b) is isomorphic to the group of rotations in the plane.

4.1a-3. Let G be the additive group of real numbers. Find the regular representation of the group $G \times G$ and characterize this representation as a transformation group on $\mathbf{R} \times \mathbf{R}$.

4.1a-4. Construct the left- and the right-regular representations of the symmetric group S_3. (*Remark*: Recall that S_3 consists of the set of all six permutations on three (abstract) objects.)

4.1a-5. Let $G = Z_3 = \{e, a, a^2\}$ and $X = \{0, 1, 2\}$. Consider the map $Z_3 \times X \to X$ given by

$$(e, 0) \mapsto 0, \quad (a, 0) \mapsto 1, \quad (a^2, 0) \mapsto 2.$$
$$(e, 1) \mapsto 1, \quad (a, 1) \mapsto 2, \quad (a^2, 1) \mapsto 0.$$
$$(e, 2) \mapsto 2, \quad (a, 2) \mapsto 0, \quad (a^2, 2) \mapsto 1.$$

Show that this defines an action of Z_3 on X. Determine the orbits of Z_3 on X. Is X a homogeneous space?

---

†A "homogeneous topological space" is a quite different notion and will be briefly mentioned in Section 5.6. However, in the theory of topological groups the two concepts "come together."

4.1a-6. Let a group G act on a set X. Let $x_0 \in X$ be an arbitrary given point of X. Show that the subset Γ_{x_0} of G, which consists of all elements γ such that $\gamma \cdot x_0 = x_0$, is actually a subgroup of G. This subgroup Γ_{x_0} is called the *stabilizer* of x_0 or the *little group* of x_0 (sometimes the term "isotropy group of x_0" is used). We say, the little group Γ_{x_0} *fixes* the point x_0.

4.1a-7. Let G act on X and let $[x_0]$ be an orbit. Show that the stabilizers of all points of the orbit are isomorphic subgroups of G (*Hint*: Show that if Γ_{x_0} and Γ_{x_1} are stabilizers of the points x_0 and x_1 of the same orbit, then $\Gamma_{x_1} = \{g\gamma g^{-1} | \gamma \in \Gamma_{x_0}$ and $g \in G$ such that $g \cdot x_0 = x_1\}$. In other words, Γ_{x_1} and Γ_{x_0} are conjugate groups. (*Remark*: This problem reveals that *the stabilizer is a characteristic of the orbit*.)

4.1a-8. Let G act on X. A subset V of X is said to be *invariant* under G if for all $v \in V$ and all $g \in G$, we have $g \cdot v \in V$ (i.e. if the action of G does not lead out of V). Show that every orbit $[x_0]$ is invariant under G (this is trivial), and that the smallest invariant subset V of X that contains a given point z is the orbit of z. Furthermore, show that every invariant subset of X is a union of some orbits.

4.1b. Conjugate Classes; Cosets

In this subsection we shall apply the powerful tools of the preceding subsection to derive, in a modern and unified manner, some important results of classical group theory.

Let G be a group and g a fixed element of G. Consider the map

$$i_g : G \to G \quad \text{given by} \quad x \mapsto gxg^{-1} \quad \text{for all} \quad x \in G. \tag{4.7}$$

This map is called a *conjugation by g*, and the reader will verify with ease that i_g is an *automorphism*. As a matter of fact, it is such an important type of automorphism that it deserves a special name. Any automorphism which can be expressed as a conjugation by some element of G is called an *inner automorphism* of G. All other automorphisms are called, accordingly, *outer automorphisms*.

From our present viewpoint, it is important to realize that *a conjugation may be considered as an action of the group G on itself*. Indeed, the map

$$G \times G \to G \quad \text{given by} \quad (g, x) \mapsto g \cdot x = gxg^{-1} \quad \text{for all} \quad g \in G, x \in G \tag{4.8}$$

is an action of the group G on the set G. The fulfillment of the action axioms (i) and (ii) on p. 84 can be checked directly without trouble.†

---

†The more sophisticated reader will observe that the explicit check can be avoided by noting that, as discussed above, for each g, $(g, x) \mapsto gxg^{-1}$ is an automorphism, and composition of two automorphisms is associative; furthermore, conjugation by $g = e$ is the identity automorphism.

The orbits under conjugation are easily found. If $x \in G$, then its orbit (4.6) consists of the subset of all elements y which have the form $y = gxg^{-1}$, where g is *some* element of G. In other words, each orbit contains all mutually conjugate elements. In the present case, an orbit is customarily called a *conjugate class*. Two elements g_1 and g_2 of G are conjugate iff they lie in the same conjugate class. Distinct conjugate classes are disjoint, and their union makes up the whole group. Note that the unit element e of G is a conjugate class by itself (and hence, cannot belong to any other conjugate class). Furthermore, if G is Abelian, then every element g is a class by itself.

An instructive example of conjugation is the following. Let G be the Euclidean group in two dimensions. That is, G is a transformation group of the Euclidean plane, consisting of all rotations,

$$\rho_\theta : \mathbf{x} \mapsto R(\theta)\mathbf{x},$$

all translations,

$$t_\mathbf{a} : \mathbf{x} \mapsto \mathbf{x} + \mathbf{a},$$

and all their combinations.† It is clear that all translations are in one single conjugation class. Indeed, $t_\mathbf{b} t_\mathbf{a} t_\mathbf{b}^{-1} = t_\mathbf{a}$ and $\rho_\theta t_\mathbf{a} \rho_\theta^{-1} = t_{R(\theta)\mathbf{a}}$. On the other hand, given a rotation ρ_θ, its conjugation class consists of all transformations that can be described as "ρ_θ followed by an arbitrary translation." That is, the conjugate class of ρ_θ (for $\theta \neq 0$) consists of all transformations $t_\mathbf{b}\rho_\theta$, with $\mathbf{b}$ arbitrary.‡

We now leave the topic of conjugation and consider another, equally important type of action.

Let Γ be a subgroup of G. We define the map§

$$\Gamma \times G \to G \quad \text{given by} \quad (\gamma, g) \to \gamma \cdot g = \gamma g \quad \text{for all } \gamma \in \Gamma, \quad g \in G. \quad (4.9)$$

Since $eg = g$ and $(\gamma_1 \gamma_2)g = \gamma_1(\gamma_2 g)$, this is obviously an *action of the subgroup Γ on the set G*. It is called *left multiplication by the subgroup Γ*. The corresponding orbits are called the *left cosets* of Γ in G. The standard

---

†We use here elementary vector notation. $R(\theta)$ is a "rotation matrix,"

$$\begin{pmatrix} \cos\theta & \sin\theta \\ -\sin\theta & \cos\theta \end{pmatrix}$$

carrying over the point $\mathbf{x} = (x_1, x_2)$ into its rotated image. The vector $\mathbf{a}$ describes a translation, carrying the point $\mathbf{x}$ into its translated image.

‡Indeed, $\rho_\varphi \rho_\theta \rho_\varphi^{-1} = \rho_\theta$ and $t_\mathbf{a} \rho_\theta t_\mathbf{a}^{-1} = t_{\mathbf{a} - R(\theta)\mathbf{a}} \rho_\theta$, as easily obtained from direct calculation with the above given formulae. Note that since $\mathbf{a}$ is arbitrary, so is $\mathbf{b} = \mathbf{a} - R(\theta)\mathbf{a}$.

§Here, naturally, γg denotes the standard group composite of the group elements γ and g.

notation for a left coset is Γg, so that

$$\Gamma g = \{\gamma g \mid \gamma \in \Gamma\}. \tag{4.10}$$

As the notation implies, this is the left coset associated with the (given) group element g, i.e. this subset of G is the orbit of g under the action of Γ.

Two elements x and y of G belong to the same left coset iff there exists an element $\gamma \in \Gamma$ such that $y = \gamma x$. This may be reexpressed as follows:

THEOREM 4.1b(1).   *Two elements x and y of G belong to the same left coset iff $yx^{-1} \in \Gamma$.*

This is the most convenient form of the equivalence relation on G generated by the action of Γ. Once again, we note that all distinct left cosets Γg are disjoint and the union $\cup_g \Gamma g$ of all cosets gives the entire group G. As we shall see later, this partitioning of G into cosets is very advantageous for studying the structure of G. In this respect, the quotient set corresponding to the coset partition will turn out to be particularly important. It is customary to call this quotient set the left *coset space* and to denote it by G/Γ_L. (Symbolically, here Γ_L refers to the equivalence relation generated by the left action of Γ on G.) Thus,

$$G/\Gamma_L = \{\Gamma g \mid \text{all distinct}\}.$$

We illustrate these concepts on the following example. Let G be the group of transformations $\{\ldots g_{ab} \ldots\}$, where each g_{ab} is a map $\mathbf{R} \to \mathbf{R}$ given by

$$x \mapsto ax + b, \qquad a, b \text{ real}, \quad a \neq 0.$$

Let Γ be the subgroup consisting of the group elements g_{1b}, i.e. the transformations

$$x \mapsto x + b.$$

Then the left coset Γg_{ab} consists of all elements $g_{1\beta} g_{ab}$, with β arbitrary real, i.e. it is the set of all transformations

$$x \mapsto ax + b + \beta \equiv ax + \gamma, \qquad \gamma \text{ arbitrary real}.$$

Thus, the elements of Γg_{ab} are those which consist of the (fixed) dilation $x \mapsto ax$ followed by an arbitrary translation. Each coset can be described by specifying a nonzero number a; no two such cosets have common elements; the union of all cosets is all of G.

In analogy to left cosets, one can define right cosets. The map

$$G \times \Gamma \to G \quad \text{given by} \quad (g, \gamma) \mapsto g \cdot \gamma = g\gamma \quad \text{for all } \gamma \in \Gamma, \ g \in G \quad (4.11)$$

is a right action of Γ on G, called *right multiplication by* Γ. The corresponding orbits are the *right cosets*†

$$g\Gamma = \{g\gamma | \gamma \in \Gamma\}. \quad (4.12)$$

The criterion for two elements x and y to belong to the same right coset is that $x^{-1}y \in \Gamma$. (This *differs* from the criterion expressed in Theorem 4.1b(1).) With the partition corresponding to right cosets, one can define the right coset space

$$G/\Gamma_R = \{g\Gamma | \text{all distinct}\}.$$

For illustration, we consider the same group G and subgroup Γ as above. The right coset $g_{ab}\Gamma$ is then the set of transformations which has the elements $g_{ab}g_{1\beta}$, i.e. the set of all transformations $x \mapsto a(x + \beta) + b = ax + a\beta + b \equiv ax + \gamma$, γ arbitrary real. Thus, in the present example $g_{ab}\Gamma = \Gamma g_{ab}$. The coincidence of left and right cosets is by no means a common phenomenon. For example, taking for G the same group as before but selecting the subgroup $\hat{\Gamma} = \{\ldots g_{\alpha 0} \ldots\}$ with α arbitrary real, the reader will easily verify that the left coset of g_{ab} is the set of transformations

$$x \mapsto \alpha a x + \alpha b,$$

while the right coset of g_{ab} is the set

$$x \mapsto \alpha a x + b.$$

Thus, now $g_{ab}\hat{\Gamma} \neq \hat{\Gamma}g_{ab}$.

At this point, it may be worthwhile to summarize, under a common viewpoint, our work done in the last two subsections. Our *guiding idea* was that of the *action of a group on a set*. Special cases considered in detail were transformation groups, conjugation, and left (right) multiplication by a subgroup. The salient features are given in the following table:

---

†Unfortunately, many authors use the terms left (right) cosets in exactly the opposite sense as defined here.

$G \times X \to X$	$g \cdot x$	Orbits
$T \times X \to X$	$t(x)$	geom. curves
$G \times G \to G$	gxg^{-1}	conjug. classes
$\Gamma \times G \to G$	γg	left cosets
$G \times \Gamma \to G$	$g\gamma$	right cosets

PROBLEMS

4.1b-1. Determine the conjugate classes of the group consisting of the numbers $\{1, -1, i, -i\}$, with composition being ordinary multiplication.

4.1b-2. Determine the conjugate classes of the group of symmetries of the equilateral triangle (p. 80).

4.1b-3. Let G be the group of transformations on the real numbers x, defined by

$$g_{ab} : x \mapsto ax + b, \qquad a \neq 0, \quad b \text{ real.}$$

Work out, in general terms, the composition law $g_{ab}g_{cd} = g_{ef}$. Using this, find all conjugation classes of G.

4.1b-4. Show that every left coset Γg of G has the same number of elements as every right coset $g\Gamma$.

4.1b-5. Let G be the group of symmetries of the equilateral triangle (p. 80). Let Γ be the subgroup $\Gamma = \{1_X, D_1\}$. Determine the elements of the left and right coset space. Do the same for $\hat{\Gamma} = \{1_X, R_{120}, R_{240}\}$.

4.1b-6. Let G be the Euclidean group of the plane (p. 88). Determine the left and right cosets of the translation subgroup. Do the same for the rotation subgroup.

4.1c. Normal Subgroups; Quotient Groups; Isomorphism Theorems

In this subsection the concepts introduced in the preceding subsection will be further developed and used.

Let Γ be a subgroup of G and suppose that, for every $g \in G$, the left coset Γg and the right coset $g\Gamma$ coincide. We then call Γ a *normal subgroup* of G.

Obviously, the trivial subgroups $\{e\}$ and G of any group are normal subgroups. A group G whose only normal subgroups are these two trivial ones, is called a *simple group*.† A group which has only non-Abelian proper normal subgroups is called *semisimple*.

†For Lie groups (which will not be discussed in this book), the terminology is slightly different: a Lie group is said to be simple if it has no proper invariant *Lie* subgroup.

Every subgroup of an Abelian group is normal: $g\Gamma = \{g\gamma | \gamma \in \Gamma\} = \{\gamma g | \gamma \in \Gamma\} = \Gamma g$. The center of any group (defined in Problem 4.1-10) is a normal subgroup: $gC = \{gc | c \in C\} = \{cg | c \in C\} = Cg$. (Thus, the center of a simple group can consist only of the single element e.)

A very useful characterization of normal subgroups is the following:

THEOREM 4.1c(1). *A subgroup N of G is normal iff together with any of its elements it also contains all conjugates of that element*, i.e. *N is normal iff $n \in N$ implies $gng^{-1} \in N$ for all $g \in G$.*

Proof. (a) Assume that $gng^{-1} \equiv h \in N$. This means that $gn = hg \in Ng$. But gn is the generic form of an element of gN. Hence, every element of gN belongs also to Ng. The opposite inclusion follows from observing that the hypothesis of our theorem also implies $g^{-1}ng \equiv l \in N$; hence $ng = gl \in gN$, where ng is a generic form for elements of Ng.

(b) Conversely, assume $gN = Ng$ for all $g \in G$. This means that for each $n \in N$ there must exist an $n' \in N$ such that $gn = n'g$. Therefore, $gng^{-1} = n' \in N$, as claimed.

It is customary to use the notation

$$gNg^{-1} = \{gng^{-1} | n \in N\}.$$

Then the theorem may be expressed by writing

$$gNg^{-1} = N \quad \text{for all} \quad g \in G. \tag{4.13}$$

We see that N is "invariant" under conjugation. For this reason, normal subgroups are frequently called *invariant subgroups*. Equation (4.13) is indeed often taken as the definition of a normal subgroup.

An alternative way to express these findings is to say that a normal subgroup contains the conjugate classes of all its elements and nothing else, i.e. any normal subgroup N is the union of conjugate classes from G.

Apart from checking the identity $gN = Ng$, or checking the conjugacy property $gNg^{-1} = N$ for all g, there is a third way to recognize normal subgroups:

THEOREM 4.1c(2). *The kernel $\mathrm{Ker}\,(f)$ of any morphism $f : G \to G'$ is a normal subgroup of G.*

Proof. We already know that $\mathrm{Ker}\,(f)$ is a subgroup of G. We have $f(k) = e'$ for every $k \in \mathrm{Ker}\,(f)$ by the very definition of the kernel. Then, for any $g \in G$, we see that $f(gkg^{-1}) = f(g)e'f(g^{-1}) = f(g)f(g^{-1}) =$

$f(g)(f(g))^{-1} = e'$. Hence gkg^{-1} also belongs to Ker (f), for any $k \in$ Ker (f), so that the previous theorem applies.

We illustrate our results on the following nontrivial example. Let $G = A \times B$ be the direct product of two groups.† Consider the morphism‡ $p : A \times B \to A$, given by $p(a, b) = a$. Here Ker $(p) = \{(e_A, \beta) | \beta \in B\}$, because $p(e_A, \beta) = e_A$ is the unit element of A. By our theorem this set is an invariant subgroup of $A \times B$. Indeed, we may check that

$$(a, b)(e_A, \beta) = (ae_A, b\beta) \equiv (a, b'),$$

$$(e_A, \beta)(a, b) = (e_A a, \beta b) \equiv (a, b'').$$

Letting β run through all of B, we see that g Ker(p) = Ker(p) g, for any $g \equiv (a, b)$, as claimed.

Theorem 4.1c(2) asserts that a necessary criterion for a subgroup Γ to be the kernel of some morphism is that Γ be normal. We shall see below that normality is also a sufficient criterion for Γ to be a kernel of some morphism. To prepare the proof, we must first study a remarkable and very important property of the coset space G/N associated with a normal subgroup.

Let N be a normal subgroup of G. The elements of G/N are the distinct cosets§ Ng_1, Ng_2, ... Let us define in this coset space a composition law, to be denoted by $*$, by the following rule:

$$Ng_1 * Ng_2 = Ng_1g_2 \tag{4.14}$$

This definition is stated in terms of coset representatives g_1 and g_2. To make sure that the definition is unambiguous,¶ we must show that actually it is independent of the arbitrary choice of representatives. Thus, we must show that if

$$N\bar{g}_1 = Ng_1 \tag{a}$$

and

$$N\bar{g}_2 = Ng_2, \tag{b}$$

then also

$$N\bar{g}_1 * N\bar{g}_2 = N\bar{g}_1\bar{g}_2 = Ng_1g_2. \tag{c}$$

But indeed, since N is a normal subgroup, Eq. (a) can be written as $\bar{g}_1 N = Ng_1$. This means that $\bar{g}_1 Ng_2 = Ng_1g_2$. Using then Eq. (b), we have

---

†The direct product of groups was defined in Problem 4.1-8.
‡Cf. Problem 4.1-17.
§Since N is normal, it does not matter whether we consider left or right cosets.
¶That is to say, that we really have a composition law, i.e. a map $G/N \times G/N \to G/N$.

$\bar{g}_1 N \bar{g}_2 \equiv \bar{g}_1 \bar{g}_2 N = N g_1 g_2$, and once again realizing that N is normal, we can write $\bar{g}_1 \bar{g}_2 N = N \bar{g}_1 \bar{g}_2$, so that Eq. (c) is proven.†

Next we observe that the composition law (4.14) on G/N obeys the axioms of a group. Indeed, the associativity follows from that of the group composition in G; the unit element is the coset $Ne \equiv N$; and the inverse of Ng is Ng^{-1}.

Thus, in summary, by the rule (4.14) *we have endowed the coset space G/N with a group structure.* This new group is called the *quotient group of G modulo the normal subgroup N* (or simply, the quotient group of G by N). Instead of $(G/N, *)$ we shall refer to the quotient group simply by the symbol G/N.

We illustrate quotient groups by the following example. Let G be the additive group of integers, $G = (\mathbf{Z}, +)$. It is easy to see that all subgroups are cyclic groups (hence Abelian) and normal. Let us take, for example, the subgroup that consists of the numbers $0, \pm n, \pm 2n, \ldots$, i.e. the group $N = \{kn | k = 0, \pm 1, \pm 2, \ldots ; n \text{ fixed nonnegative integer}\}$. The cosets are $Ng = \{g + kn | k \in \mathbf{Z}\}$ (where of course g is a given integer). Coset multiplication is given by

$$Ng_1 * Ng_2 = Ng_1 g_2 = \{g_1 + g_2 + kn | k \in \mathbf{Z}\}.$$

We observe that there are exactly n distinct cosets; they have the representative elements $g = 0, 1, 2, \ldots, n - 1$. Furthermore, the quotient group is isomorphic to the abstract cyclic group Z_n. Indeed, if α denotes the generating element of Z_n, the map

$$Ng \mapsto \alpha^g \quad (g = 0, 1, 2, \ldots, n - 1)$$

is obviously an isomorphism, $Ng_1 * Ng_2 \mapsto \alpha^{g_1 + g_2} = \alpha^{g_1 g_2}$. Thus, we may say that the quotient group $\mathbf{Z}/N$ is the cyclic group Z_n (up to isomorphism).

We are now prepared to state the promised converse of Theorem 4.1c(2):

THEOREM 4.1c(3).   *Let N be a normal subgroup of G. Then the canonical map $c : G \to G/N$ (defined by $g \mapsto Ng$) is a homomorphism from G to the quotient group G/N, with $\mathrm{Ker}(c) = N$.*

---

†It is obvious that the normality of N plays a crucial role. This is why no composition law is defined for arbitrary coset spaces G/Γ and hence why normal subgroups are so much more important than arbitrary subgroups.

Proof.   The map c is a morphism since $c(g_1 g_2) = N g_1 g_2 = N g_1 * N g_2 = c(g_1) * c(g_2)$. The morphism is surjective, because every canonical map is such. Furthermore, since the unit element of G/N is $Ne = N$, it is clear that all elements $n \in N$ are mapped by c onto $Nn = N$, and if $g \notin N$, we cannot have $Ng = N$.

As we see, this theorem gives us more than just the statement to the effect that every normal subgroup may serve as a kernel. It also tells us *what* exactly the morphism is for which the given N will act as a kernel.

The reader may have wondered for a while, why we use the term "kernel" in connection with group morphisms when in Section 2.2 we introduced the same expression in connection with equivalence relations induced by maps.† Our last two theorems justify the nomenclature. Even though the argument follows directly from the fact that cosets are equivalence classes for the action $N \times G \to G$, it is instructive to present a detailed analysis.

Let $f: G \to G'$ be a morphism. We know that Ker $(f) \equiv N$ is an invariant subgroup of G. Let us define on G an equivalence relation N by the following condition: "$a \, N \, b$ iff $f(a) = f(b)$." It is easy to see that $a \, N \, b$ iff a and b lie in the same coset modulo N. Indeed, if $a \in Ng$ and $b \in Ng$, then $ab^{-1} = n \in N$ (cf. Theorem 4.1b(1)). Hence $f(a)f(b^{-1}) = f(n) = e'$, or $f(a)(f(b))^{-1} = e'$, therefore $f(a) = f(b)$, as stated. Conversely, let $f(a) = f(b)$. Then $f(ab^{-1}) = f(a)(f(b))^{-1} = f(a)(f(a))^{-1} = e'$. Therefore, we must have $ab^{-1} \in N$, so that a and b are in the same coset, as claimed. Now, recalling the standard terminology, the relation N is the equivalence kernel of the map f, i.e. ker $(f) = N$. On the other hand, by the definition of the kernel of the morphism, Ker $(f) = N$. Thus, the symbol N, when interpreted once as an equivalence relation, once as an invariant subgroup, leads directly to both kernel notions.

As a matter of fact, the above discussion shows also an important feature of group morphisms. If N is the kernel of a morphism and $Ng_1, Ng_2, \ldots$ are the corresponding cosets, then each element of Ng_k is mapped on the image g_k' of g_k, i.e. $f(Ng_k) = g_k'$. This is so because all elements of Ng_k are in the same equivalence class. This result is schematically shown in Fig. 4.4.

In view of the insight that we gained concerning the role of the kernel in a morphism as the kernel of an equivalence relation, we can easily establish the following important *isomorphism theorem*:

---

†To avoid confusion, however, we used the symbols Ker and ker, respectively.

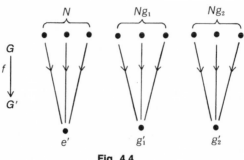

Fig. 4.4.

THEOREM 4.1c(4).   *If $f : G \to G'$ is an arbitrary morphism of groups, then* $G/\mathrm{Ker}\,(f)$ *is* ISOMORPHIC *to* $\mathrm{Im}\,(f)$.

Proof.   From the discussion following Theorem 2.2(3) on p. 40 and from the above comments, we know that the map f defines the chain

$$G \xrightarrow{\;c\;} G/\mathrm{Ker}\,(f) \xrightarrow{\;b\;} \mathrm{Im}\,(f) \xrightarrow{\;i\;} G'.$$

Here c is the canonical map, i is a simple insertion, and the *induced map b* is a *bijection*. All we have to show is that b is a *morphism*. We know that under b, $\mathrm{Ker}\,(f)g \mapsto f(g)$. (Recall that $\mathrm{Ker}\,(f) \equiv N$, so that Ng is an element of the quotient set.) So we must show that

$$b(Ng_1 * Ng_2) = b(Ng_1)\,b(Ng_2).$$

But indeed,

$$b(Ng_1 * Ng_2) = b(Ng_1g_2) = f(g_1g_2) = f(g_1)f(g_2) = b(Ng_1)\,b(Ng_2),$$

q.e.d.

Particularly important is the case when $\mathrm{Im}\,(f) = G'$, i.e. if we are dealing with a *homo*morphism to start with. Then, clearly, we have

THEOREM 4.1c(5).   *If G is homomorphic to G', then $G/\mathrm{Ker}\,(f)$ is isomorphic to G'.*

Since we know that $\mathrm{Ker}\,(f)$ is some normal subgroup we usually express the essential content of this theorem in the following manner: *If G is homomorphic to G', then there exists a normal subgroup N of G such that G/N is isomorphic to G'.* The theorem actually tells us how to find N: we must first locate $\mathrm{Ker}\,(f)$. Furthermore, from the proof of Theorem 4.1c(4) we know the map between G/N and G' which gives the desired isomorphism. It is the induced map, defined by $Ng \mapsto f(g)$, where $f(g)$ is the image of *any* element of the coset Ng under the homomorphism $G \to G'$.

The practical usefulness of our isomorphism theorem (and of other similar theorems) is the following. Suppose we have on our hand an unfamiliar group whose properties and structure we wish to determine. If we can show that our group is isomorphic to some well-known group, then our problem is completely solved, because isomorphic groups are essentially indistinguishable replicas of each other.

Because of the central importance of the isomorphism Theorem 4.1c(5), and because it is a prototype of similar theorems to be met in later sections, we give three different graphical illustrations. Figure 4.5 is simply

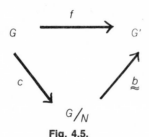

Fig. 4.5.

summarizing the situation with a commutative mapping diagram. Figure 4.6 shows the relevant sets and maps. Figure 4.7 illuminates details, under the assumption that we have a finite group. For simplicity, we wrote everywhere N in place of the more informative symbol Ker (f).

One more interesting remark is in order. From the foregoing considerations and from Theorem 4.1(6) it clearly follows that if a group G has no nontrivial normal subgroups, then any possible *homo*morphism to any group G' which has at least two elements is automatically an *iso*morphism.† This property may serve as an alternative definition of simple groups.

To conclude these elegant and important considerations, we give some illustrations.

Example α. Let $G = (\mathbf{Z}, +)$ be the additive group of integers and let $G' = \{1, -1\}$ with ordinary multiplication as composition law.‡ The map f

---

†Of course, arbitrary, not onto morphisms are possible, but not interesting. The point is that in the given case, a surjection is bound to be a bijection, if it is a morphism at all.
‡Note in passing, that G' is isomorphic to the cyclic group Z_2.

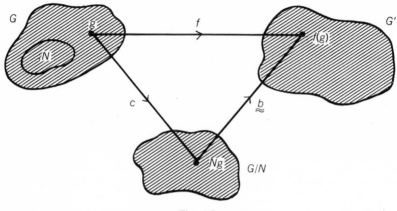

Fig. 4.6.

defined as

$$f(\text{even integer}) = +1$$
$$f(\text{odd integer}) = -1$$

is clearly a homomorphism from G to G'. We have $\text{Ker}(f) = \mathbf{E} \equiv \{0, \pm 2, \pm 4, \ldots\}$. The cosets of this normal subgroup are given by $\mathbf{E}g = \{g + 2k \,|\, k \in \mathbf{Z}\}$, where g is an integer. Obviously, there are two distinct cosets,† $\mathbf{E}0 = \mathbf{E}$ and $\mathbf{E}1 = \mathbf{O}$ (where $\mathbf{O}$ denotes the set of all odd numbers). Thus,

$$\mathbf{Z}/\mathbf{E} = \{\mathbf{E}, \mathbf{O}\}.$$

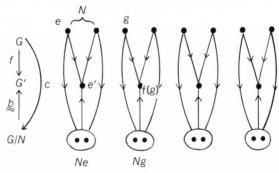

Fig. 4.7.

†We are dealing with a special case of the example discussed on p. 94, where now $n = 2$.

Since $\mathbf{E}g_1 * \mathbf{E}g_2 = \mathbf{E}g_1g_2$, it clearly follows that the coset composition law has the group table

	E	O
E	E	O
O	O	E

Theorem 4.1c(5) tells us that

$$\mathbf{Z}/\mathbf{E} \approx \{1, -1\},$$

where the isomorphism map is given by $\mathbf{E}g \mapsto f(g)$, i.e. by

$$\mathbf{E} \mapsto +1,$$
$$\mathbf{O} \mapsto -1.$$

A look at the above group table confirms this statement.†

Example β. Often it is desirable to determine a quotient group without explicitly calculating the cosets which are its members. In this (rather trivial) example we illustrate how Theorem 4.1c(5) can be used for this purpose. Let $\bar{Q}$ denote the multiplicative group of nonzero rationals. Let S be the invariant subgroup $\{1, -1\}$. Question: What is $\bar{Q}/S$? To answer this question, it is sufficient to exhibit a group K to which $\bar{Q}/S$ is isomorphic. In order to find K, we first must construct a homomorphism‡ $f\colon \bar{Q} \to K$ of which S is the kernel. If e_K is the unit of K, then surely we must have $f(1) = e_K$, and since S is to be the kernel, we also must have $f(-1) = e_K$. This suggests to try the map

$$f(r) = |r| \quad \text{for all} \quad r \in \bar{Q}.$$

It is easy to check that indeed this is a homomorphism of $\bar{Q}$ onto the multiplicative group Q^+ of all positive rationals. Thus, $K = Q^+$. Now we invoke Theorem 4.1c(5), which tells us that $\bar{Q}/S \approx Q^+$. This answers the original question. The reader is advised to explore the construction in detail. The important point to notice is that the cosets Sr are the sets $\{r, -r\}$.

At this point, we terminate our introduction to group theory.§ We only

---

†Part of this study may be summarized by saying that the quotient group of the additive group of integers by the additive group of even integers is isomorphic to the cyclic group Z_2.

‡Note that we are groping for two things: first, to guess a group K, and second, to show that $\bar{Q}$ is homomorphic to K with Ker $(f) = S$.

§Some further important topics are relegated to the problems, cf. in particular Problems 4.1c-10 through 4.1c-14.

scratched the surface and, apart from points of structural interest, we discussed only topics that will be needed in our future work.†

PROBLEMS

4.1c-1. Let G be the group of orthogonal matrices M with n rows and n columns, i.e. the infinite set of n by n matrices‡ such that $M\tilde{M} = I$. Consider the subgroup Γ of the "proper" elements, i.e. the subset of matrices for which det $M = +1$. Determine the left and right cosets of Γ. By comparison, show that Γ is an invariant subgroup. Verify this also directly by considering the conjugation classes of the elements of Γ. (*Remark*: The group G is of course isomorphic to the group of orthogonal transformations of the n-dimensional Euclidean space§ and Γ is isomorphic to the corresponding proper rotation group.)

4.1c-2. Determine all normal subgroups in the group of symmetries of the equilateral triangle.

4.1c-3. Let E be the Euclidean group in n dimensions, i.e. the set of all rotations and translations of the n-dimensional Euclidean space.† Consider the translation subgroup T and determine its left and right cosets. Do the same for the rotation subgroup R. Is T an invariant subgroup? Is R one?

4.1c-4. Let $G = \{1, -1, i, -i\}$ with ordinary multiplication, and let $N = \{1, -1\}$. Determine G/N.

4.1c-5. Using the results of Problem 4.1c-3, determine the quotient group of the Euclidean group by the translation subgroup.

4.1c-6. Show that G/N is Abelian if G is Abelian. More generally, let G be an arbitrary group and N a normal subgroup, and show that G/N is Abelian iff $aba^{-1}b^{-1} \in N$ for all $a, b \in G$.

4.1c-7. Prove the following theorem: Let $f: G \to G'$ be a homomorphism and let N be a normal subgroup of G such that Ker $(f) \subset N$. Then $f(N)$ is a normal subgroup of G', and moreover $G/N \approx G'/f(N)$.

4.1c-8. Let G be a group and define the group In (G) of inner automorphisms, as usual, as the set of maps $i_g: G \to G$ given by $i_g(x) = gxg^{-1}$. Verify that this set is really a group under composition. Prove that In (G) is a normal subgroup of the group of all automorphisms Aut (G). Now, let G be an arbitrary group. Let C be the center of G. Prove that G/C is isomorphic to the group In (G) of inner

---

†The interested reader is strongly urged to consult the extensive group theoretic textbook literature for further details. As a start, we suggest Kurosh[23], and concerning direct applications to simple problems in physics, Hamermesh[13] and Wigner[42]. Topological groups should be studied at a later stage, after having absorbed the rest of this book. For a first introduction, see McCarty[28]; the standard classic is Pontrjagin[32].

‡Here $\tilde{M}$ denotes the transpose of M and I represents the $n \times n$ unit matrix.

§By "n-dimensional Euclidean space" here we simply mean the set $\mathbf{R}^n$, whose elements are the n-tuples $x = (\xi_1, \ldots, \xi_n)$ of real numbers ξ_k, considered as "vectors" with n components.

automorphisms. (*Hint*: First construct a homomorphism $f: G \to \mathrm{In}\,(G)$. Then show that $\mathrm{Ker}\,(f) = C$.)

4.1c-9. Let $\bar{C}$ be the group of nonzero complex numbers equipped with ordinary multiplication. Represent complex numbers $z = x + iy$ by the pair (x, y) and work out the composition law in terms of this representation. Consider now the subgroup $\mathbf{R}^+ = \{(x, 0)|x > 0\}$ of positive real numbers and describe the coset space. Construct a morphism f of $\bar{C}$ into itself whose kernel is the positive real axis. (*Note*: The morphism will be neither surjective *nor* injective!) Determine and describe the quotient group $\bar{C}/\mathrm{Ker}\,(f)$.

4.1c-10. Let A, B be arbitrary groups and consider the direct product group $G = A \times B$. Show that B is isomorphic to a group $\hat{B}$ which is a normal subgroup of G, and similarly, A is isomorphic to a group $\hat{A}$ which is a normal subgroup of G. Show that $G/\hat{B} \approx \hat{A}$ and $G/\hat{A} \approx \hat{B}$. Show also that

(i) Every element of $\hat{A}$ commutes with every element of $\hat{B}$,

(ii) Every element of G can be uniquely written as $g = \hat{a}\hat{b}$, where $\hat{a} \in \hat{A}$, $\hat{b} \in \hat{B}$,

(iii) The set $\hat{A}\hat{B} = \{\hat{a}\hat{b}|\hat{a} \in \hat{A}, \hat{b} \in \hat{B}\}$ is equal to the set $G = A \times B$.

4.1c-11. This problem is essentially the converse of the preceding one. Let G be a given, arbitrary group and let A and B be subgroups of G with the following properties:

(a) Every element of A commutes with every element of B,

(b) Every element $g \in G$ can be uniquely written as $g = ab$, where $a \in A$, $b \in B$.

Prove that G is the direct product group $A \times B$, or more precisely, G is isomorphic to $A \times B$. (*Remark*: If subgroups A and B of G can be found which satisfy criteria (a) and (b), then we say that $G = A \times B$ is a direct product decomposition of G.)

4.1c-12. In this problem we relax the condition (a) of the previous one and define a new composite group which resembles direct products.

Let G be a given, arbitrary group and let M and N be subgroups of G, with the following properties:

(a) N is a normal subgroup.

(b) Every element $g \in G$ can be uniquely written as $g = mn$, where $m \in M$, $n \in N$ (in that order).

Consider now the Cartesian product set $M \times N$, with elements $\hat{g} = (m, n)$ and define the composition law

$$(m_1, n_1)(m_2, n_2) = (m_1 m_2, m_2^{-1} n_1 m_2 n_2).$$

Show that if conditions (a) and (b) hold, this law defines a group $\hat{G}$ which is isomorphic to the given group G. We call it the *semidirect product*† and write $\hat{G} = M \,\textcircled{s}\, N$.

†Even though the underlying sets are the same, $M \,\textcircled{s}\, N \neq M \times N$ because the composition laws are different, and coincide only if M and N commute. Note also that semidirect products cannot be defined for arbitrary pairs of groups: both must be subgroups of a given group and also the conditions (a) and (b) must hold.

Let now $\hat{G} = M \otimes N$. Show that the following statements are true:

(i) $M \cap N = \{e\}$,

(ii) The set $MN = \{mn \mid m \in M, n \in N\}$ is equal to the set G,

(iii) The isomorphism $G/N \approx M$ holds, and since $G \approx \hat{G}$, we may write $\hat{G}/\hat{N} \approx M$, where $\hat{N} = \{(e, n) \mid n \in N\}$.

4.1c-13. This will be essentially the converse of the preceding problem. Let G be a given, arbitrary group. Let M and N be subgroups of G with the following properties:

(a) N is a normal subgroup,

(b) $M \cap N = \{e\}$,

(c) The set $MN = \{mn \mid m \in M, n \in N\}$ is equal to the set G.

Prove that G is isomorphic to the semidirect product group $M \otimes N$. (*Remark*: If subgroups M and N of G can be found which satisfy criteria (a), (b), and (c), then we say that $G = M \otimes N$ is a semidirect product decomposition of G.)

4.1c-14. Give examples of direct product and semidirect product group decompositions. (*Suggestion*: Restudy some of the previous problems. Think of transformation groups of the Euclidean space.)

4.2 RINGS AND FIELDS

Groups are algebraic systems with one internal composition law. More complicated (and hence, richer) systems are obtained if we introduce a second internal composition law which is related† to the first. This is what we are going to do in this section.

By definition, *a ring is an algebraic system* $(\mathscr{R}, \square, *)$ *where the first internal composition law* $\square$ *is that of an Abelian group, the second law* $*$ *is that of a semigroup, and* $*$ *is distributive over* $\square$.

It is traditional to call the first law "addition" and denote it by $+$, and to call the second law "multiplication" and denote it by simple juxtaposition. Furthermore, we shall use the symbol o to denote the identity element of addition, which is often called the *zero element* of the ring. Finally, the "additive inverse" of x is denoted by $(-x)$, and called the negative of x. With these notations the *defining axioms of a ring*‡ $\mathscr{R}$ are as follows:

(ia) $a + (b + c) = (a + b) + c$ for all $a, b, c \in \mathscr{R}$,

(ib) There exists an element o $\in \mathscr{R}$ such that o $+ a = a + o = a$ for all $a \in \mathscr{R}$,

---

†It is obvious that nothing of interest is gained if the two composition laws are unrelated. Each could be then studied separately.

‡Without causing confusion, we can use the symbol $\mathscr{R}$ for the ring as a *system*.

(ic) For every $a \in \mathcal{R}$ there exists an element $-a \in \mathcal{R}$ such that
$a + (-a) = (-a) + a = o$,

(id) $a + b = b + a$ for all $a, b \in \mathcal{R}$.

(ii) $a(bc) = (ab)c$ for all $a, b, c \in \mathcal{R}$.

(iii) $a(b + c) = ab + ac$ and $(a + b)c = ac + bc$ for all $a, b, c \in \mathcal{R}$.

Remarks: 1. Because of the uniqueness of o and $(-a)$, one usually writes $a - b$ instead of $a + (-b)$.

2. From the uniqueness of o and from axioms (iii) and (id) it easily follows that
$$oa = ao = o \quad \text{for all} \quad a \in \mathcal{R}.$$

Thus, clearly, the zero element o is *not regular* for multiplication.

If in addition to the above axioms, we have $ab = ba$ for any pair of elements, $\mathcal{R}$ is called a *commutative ring*.

If there exists a unit element for multiplication (i.e. if the semigroup law is actually a monoid law), then we call $\mathcal{R}$ a *ring with unit*. We shall denote the unit by ϵ, so that $\epsilon a = a\epsilon = a$ for all $a \in \mathcal{R}$. We recall that if ϵ exists, it is unique. Note that a ring with unit need not be commutative and vice versa.

It is easy to convince oneself that, because of Remark 2 above, the equality $o = \epsilon$ is contradictory unless we have the trivial ring consisting of one element. Since we tacitly assume that we are considering only rings with more than one element, we then have the rule: $\epsilon \neq o$ for a ring with unit.

There is one more general remark, concerning notation. It is customary to use the shorthand†

$$ka \equiv a + a + \cdots + a \quad \text{with } k \text{ "terms."}$$

Similarly, we write

$$-ka \equiv -a - a - \cdots - a \quad \text{with } k \text{ "terms."}$$

One then easily verifies the following formal rules, for any integers n, m and any $a, b \in \mathcal{R}$:
$$(n + m)a = na + ma,$$
$$(nm)a = n(ma),$$
$$n(a + b) = na + nb,$$
$$n(ab) = (na)b = a(nb),$$

†One should emphasize that the *symbol ka* does *not* represent a ring product: the integer k is not a member of $\mathcal{R}$ (unless, of course, $\mathcal{R}$ consists of ordinary numbers).

$$(na)(mb) = (nm)ab,$$
$$na = (n\epsilon)a \qquad \text{(if } \epsilon \text{ exists).}$$

We now give a few preliminary examples of rings.

Example α.   Let $R = \{o, \alpha, 2\alpha, 3\alpha\}$ with the addition and multiplication tables

+	o	α	2α	3α
o	o	α	2α	3α
α	α	2α	3α	o
2α	2α	3α	o	α
3α	3α	o	α	2α

·	o	α	2α	3α
o	o	o	o	o
α	o	α	2α	3α
2α	o	2α	o	2α
3α	o	3α	2α	α

With respect to $+$, we have here the cyclic group Z_4, written additively. For multiplication, we clearly have a monoid. The only thing to check is distributivity. We easily see that, for example, $\alpha(\alpha + \alpha) = \alpha\alpha + \alpha\alpha = \alpha + \alpha = 2\alpha$, and on the other hand, $\alpha(\alpha + \alpha) = \alpha(2\alpha) = 2\alpha$, so it checks. We see that we have a commutative ring (the multiplication table is symmetric), and there is a unit element, $\epsilon = \alpha$.

Example β.   The set of reals with ordinary addition and multiplication is a commutative ring with unit ($\epsilon = 1$, $o = 0$). The same holds for $(\mathbf{Q}, +, \cdot)$ and $(\mathbf{Z}, +, \cdot)$. The corresponding system of even integers, $(\mathbf{E}, +, \cdot)$ is also a commutative ring, but has no unit. $(\mathbf{O}, +, \cdot)$ is not a ring.

Example γ.   Let $\mathscr{R}$ be the set of all ordered pairs of real numbers and define

$$(a, b) + (c, d) = (a + c, b + d),$$
$$(a, b)(c, d) = (ac, bc + d).$$

An elementary check shows that all ring postulates are fulfilled. The zero element is $(0, 0)$. The ring is not commutative but has the unit element $(1, 0)$.

Example δ.   Let $\mathscr{R}$ be the set of all n by n matrices (with real or even with complex numbers as entries), equipped with standard matrix addition

$$(A + B)_{\alpha\beta} = A_{\alpha\beta} + B_{\alpha\beta} \qquad (\alpha, \beta = 1, \ldots, n)$$

and matrix multiplication

$$(AB)_{\alpha\beta} = \sum_{\gamma=1}^{n} A_{\alpha\gamma}B_{\gamma\beta} \qquad (\alpha, \beta = 1, \ldots, n).$$

This is a ring, with o being the null matrix $A_{\alpha\beta} = 0$. The element $(-A)$ is the matrix with elements $(-A)_{\alpha\beta} = -A_{\alpha\beta}$. There is a unit element ϵ, namely the unit matrix $A_{\alpha\alpha} = 1$, $A_{\alpha\beta} = 0$, $(\alpha \neq \beta)$. The ring is not commutative.

***Example* ϵ.**   The ring of the preceding example has many subsets which provide rings with interesting properties. For example, take the set of all n by n matrices where the entries in all but the last line vanish, i.e. matrices of the form

$$\begin{pmatrix} 0 & 0 & 0 & \cdots & 0 \\ 0 & 0 & 0 & \cdots & 0 \\ \cdot & & & & \cdot \\ \cdot & & & & \cdot \\ \alpha & \beta & \gamma & \cdots & \nu \end{pmatrix}$$

They obviously constitute a ring which is remarkable inasmuch as it is neither commutative nor has it a unit element.

In most rings, there are, besides o, many elements which are not regular. Their existence is closely related to the existence of so-called divisors of zero. Consider the equation

$$ab = o.$$

It is perfectly possible that, even though $a \neq o$ and $b \neq o$, the equation holds true. We then say that both a and b are divisors of zero. More formally, *an element†* d *is called a divisor of zero if there exist some* $x \neq o$ *such that* $dx = o$, *or if there exist some* $y \neq o$ *such that* $yd = o$. We now establish the following.

THEOREM 4.2(1).   *A divisor of zero cannot be a regular element and conversely, a regular element cannot be a divisor of zero.* Thus, for a ring, the nonregular elements are precisely the divisors of zero.

Proof.   Suppose d is a divisor of zero. Consider $db = do$. Since $do = o$, we cannot conclude that $b = o$, i.e. the cancellation law does not hold, d is not regular. Conversely, suppose $r \neq o$ is regular, and suppose $rx = o$. We may write this as $rx = ro$, and then, by cancellation, we must have $x = o$.

The reader is urged to check Examples α–ϵ for divisors of zero, i.e. for

†In this definition it is tacitly assumed that $d \neq o$. Thus, the null element is *not* called a divisor of zero, even though $oa = ao = o$ for all a.

nonregular elements. For example, in the set of n by n matrices, it is very easy to find nonzero matrices such that $AB = o$. A more interesting example is the following. Let $\mathcal{R}$ and $\mathcal{S}$ be rings, and define the *direct product ring* $\mathcal{R} \times \mathcal{S}$ as the Cartesian product set equipped with the composition laws

$$(r, s) + (r', s') = (r + r', s + s'),$$
$$(r, s)(r', s') = (rr', ss').$$

It is easy to verify that this is a ring,† with zero element $(o_{\mathcal{R}}, o_{\mathcal{S}})$. Here all elements of the form $(o_{\mathcal{R}}, s)$ or $(r, o_{\mathcal{S}})$ are obviously divisors of zero, i.e. nonregular. Indeed, $(o_{\mathcal{R}}, s)(r, o_{\mathcal{S}}) = (o_{\mathcal{R}}, o_{\mathcal{S}})$.

Let us now focus our attention to rings with unit. Then we may talk about the *inverse of an element*. As for all systems, a is said to have an inverse a^{-1} iff $aa^{-1} = a^{-1}a = \epsilon$. Elements that have an inverse‡ are called *invertible* (or *nonsingular*) elements. We note that *invertible elements are always regular* (this follows from associativity) hence an invertible element cannot be a divisor of zero. Conversely, a *divisor of zero* (i.e. a nonregular element) *cannot be invertible*. However, the opposite is *not* true! In other words, *an element which does not have an inverse, may be a regular element* (i.e. need not be a divisor of zero). A good example is furnished by the set of integers equipped with ordinary addition and multiplication. The unit element is the number 1. The only two elements that are invertible are $+1$ and -1; nevertheless, all elements are regular (except of course the number 0). This is obvious since the equation $xy = 0$ can be satisfied only if either x or y is zero. Thus, for any integer the cancellation laws apply.§

We now consider a type of rings which is particularly important in later applications. *A ring with unit where every element* (*except the zero element* o) *is invertible, is called a field*. We may express this by saying that the *nonzero* elements of the field $\mathcal{F}$ form a group under ring multiplication.¶ It will be worthwhile to explicitly write down the defining axioms of a field. They are as follows:

---

†Incidentally, the ring $\mathcal{R} \times \mathcal{S}$ is commutative if both $\mathcal{R}$ and $\mathcal{S}$ are, and has a unit element $(\epsilon_{\mathcal{R}}, \epsilon_{\mathcal{S}})$ provided both component rings have one.

‡We recall here that, for any associative system, if a has an inverse, it is unique. Furthermore, ϵ is always invertible and $\epsilon^{-1} = \epsilon$.

§It is customary to call a commutative ring with unit an *integral domain* if it has no divisors of zero. The terminology derives from our example.

¶One would be now tempted to consider rings where *all* elements form a group under the multiplication. It is easy to see that this is impossible, because $\epsilon \neq o$.

(ia) $a + (b + c) = (a + b) + c$ for all $a, b, c \in \mathscr{F}$.

(ib) There exists an element $o \in \mathscr{F}$ such that $o + a = a + o = a$ for all $a \in \mathscr{F}$.

(ic) For every $a \in \mathscr{F}$ there exists an element $-a \in \mathscr{F}$, such that $a + (-a) = (-a) + a = o$.

(id) $a + b = b + a$ for all $a, b \in \mathscr{F}$.

(iia) $a(bc) = (ab)c$ for all $a, b, c \in \mathscr{F}$.

(iib) There exists an element $\epsilon \in \mathscr{F}$ such that $\epsilon a = a\epsilon = a$ for all $a \in \mathscr{F}$.

(iic) For every $a \neq o$ there exists an element $a^{-1} \in \mathscr{F}$ such that $aa^{-1} = a^{-1}a = \epsilon$.

(iii) $a(b + c) = ab + ac$ and $(a + b)c = ac + bc$ for all $a, b, c \in \mathscr{F}$.

We ought to add here an important remark on terminology. Most mathematicians call the system defined above a *skew field* or a *division ring*, and reserve the term "field" for the case when, in addition, multiplication is commutative, i.e. when $ab = ba$ for all $a, b \in \mathscr{F}$. We, however, shall not demand this restriction and speak of fields irrespective of whether they are commutative or not. If necessary, we shall speak specifically of a "commutative field."

Since, in a field, every element (except zero) has an inverse, it is clear that *a field has no divisors of zero.*† Actually, for fields the stronger statement holds: *The equations $xa = b$ and $ay = b$ have, for $a \neq o$, $b \neq o$, a unique solution* (namely $x = ba^{-1}$ and $y = a^{-1}b$, respectively).

We give a few examples of fields.

Example ζ.   Let $\mathscr{F} = \{o, \epsilon\}$ with the composition tables

+	o	ϵ		·	o	ϵ
o	o	ϵ		o	o	o
ϵ	ϵ	o		ϵ	o	ϵ

As a matter of fact, this is the simplest nontrivial ring with identity, and it is actually a commutative field. The only nontrivial point to check is the distributivity. We have $\epsilon(\epsilon + \epsilon) = \epsilon\epsilon + \epsilon\epsilon = \epsilon + \epsilon = o$ and also $\epsilon(\epsilon + \epsilon) = \epsilon o = o$, so that indeed $\epsilon(\epsilon + \epsilon) = \epsilon\epsilon + \epsilon\epsilon$. It can be shown that every finite field is commutative. Finite fields are usually called *Galois fields*. Note that not every finite commutative ring with identity is a field: For example, the ring of Example α is not a field, because 2α is a divisor of zero.

---

†Thus, every commutative field is an integral domain.

Example η.   The set of real numbers with ordinary addition and multiplication is a commutative field. So is the set of complex numbers. But the set of integers or the set of n by n matrices are not fields, since not all nonzero elements are invertible.

Example ϑ.   The set of all rational functions of a single variable is easily seen to be a commutative field.

Example κ.   Consider the subset of all 2×2 matrices that have the form

$$q = \begin{pmatrix} \alpha + i\delta & -\gamma + i\beta \\ \gamma + i\beta & \alpha + i\delta \end{pmatrix},$$

where α, β, γ, δ are arbitrary real numbers. The ring axioms are obviously satisfied with respect to matrix addition and multiplication. The unit element is the 2×2 unit matrix, with $\alpha = 1$, $\beta = \gamma = \delta = 0$. Moreover, every q (except the null matrix) has an inverse, since

$$\det q = \alpha^2 + \beta^2 + \gamma^2 + \delta^2 \neq 0$$

unless $\alpha = \beta = \gamma = \delta = 0$ and actually, we have

$$q^{-1} = \frac{1}{\alpha^2 + \beta^2 + \gamma^2 + \delta^2} \begin{pmatrix} \alpha - i\delta & \gamma - i\beta \\ -\gamma - i\beta & \alpha + i\delta \end{pmatrix}.$$

Thus, the above set is a field. However, it is not commutative, because, for example,

$$\begin{pmatrix} 0 & -1 \\ 1 & 0 \end{pmatrix} \begin{pmatrix} i & 0 \\ 0 & -i \end{pmatrix} \neq \begin{pmatrix} i & 0 \\ 0 & -i \end{pmatrix} \begin{pmatrix} 0 & -1 \\ 1 & 0 \end{pmatrix}.$$

This very important noncomutative field is called the field of *quaternions*. It is useful to consider the four simplest elements of this field. They are, with conventional notation,†

$$\epsilon = \begin{pmatrix} 1 & 0 \\ 0 & 1 \end{pmatrix}, \qquad j = \begin{pmatrix} 0 & i \\ i & 0 \end{pmatrix}, \qquad k = \begin{pmatrix} 0 & -1 \\ 1 & 0 \end{pmatrix}, \qquad l = \begin{pmatrix} i & 0 \\ 0 & -i \end{pmatrix}.$$

We easily verify that

$$\epsilon^2 = \epsilon, \qquad j^2 = k^2 = l^2 = -\epsilon, \tag{4.15a}$$

†In terms of the familiar Pauli matrices,

$$\sigma_0 = \begin{pmatrix} 1 & 0 \\ 0 & 1 \end{pmatrix}, \qquad \sigma_1 = \begin{pmatrix} 0 & 1 \\ 1 & 0 \end{pmatrix}, \qquad \sigma_2 = \begin{pmatrix} 0 & -i \\ i & 0 \end{pmatrix}, \qquad \sigma = \begin{pmatrix} 1 & 0 \\ 0 & -1 \end{pmatrix}$$

we have

$$\epsilon = \sigma_0, \qquad j = i\sigma_1, \qquad k = -i\sigma_2, \qquad l = i\sigma_3.$$

$$jk = -kj = l,$$
$$kl = -lk = j, \qquad (4.15b)$$
$$lj = -jl = k.$$

We also observe that the complex numbers can be considered as a subset, in fact a subfield, of the quaternion field. Indeed, the set of matrices of form

$$z = \begin{pmatrix} \alpha & i\beta \\ i\beta & \alpha \end{pmatrix}$$

can be written as $z = \alpha\epsilon + \beta j$, and α and β may be identified with the "real" and "imaginary" part of z. Thus, the quaternions are a generalization of complex numbers. At this point we note that the field of complex numbers is best defined as the set of pairs (α, β) of real numbers with the composition laws

$$(\alpha, \beta) + (\alpha', \beta') = (\alpha + \alpha', \ \beta + \beta'),$$
$$(\alpha, \beta)(\alpha', \beta') = (\alpha\alpha' - \beta\beta', \ \alpha\beta' + \beta\alpha').$$

The pair $(0, 0)$ is the zero element, the pair $(1, 0)$ the unit, the pair $(0, 1)$ corresponds to the "imaginary unit" i. Now, in a similar way, *the field of quaternions can be defined as the set of quadruples $(\alpha, \beta, \gamma, \delta)$ of real numbers* with the composition laws†

$$(\alpha, \beta, \gamma, \delta) + (\alpha', \beta', \gamma', \delta') = (\alpha + \alpha', \beta + \beta', \gamma + \gamma', \delta + \delta'),$$
$$(\alpha, \beta, \gamma, \delta)(\alpha', \beta', \gamma', \delta')$$
$$= (\alpha\alpha' - \beta\beta' - \gamma\gamma' - \delta\delta', \beta\alpha' + \alpha\beta' + \delta\gamma' - \gamma\delta',$$
$$\gamma\alpha' - \delta\beta' + \alpha\gamma' + \beta\delta', \delta\alpha' - \beta\gamma' + \gamma\beta' + \alpha\delta').$$

We easily recognize that

$$o = (0, 0, 0, 0),$$
$$\epsilon = (1, 0, 0, 0),$$
$$j = (0, 1, 0, 0),$$
$$k = (0, 0, 0, 1),$$
$$l = (0, 0, 1, 0).$$

One final remark about quaternions. If we permitted $\alpha, \beta, \gamma, \delta$ to be *complex* numbers, we still had a ring with unit, but no longer a field. This

†This definition is superior to the one we gave to start with, because it does not necessitate reference to matrices. Our original definition may be considered as a realization of the quaternions in terms of 2×2 matrices.

is most easily seen from our original matrix-definition, since for complex numbers, $\alpha^2 + \beta^2 + \gamma^2 + \delta^2 = 0$ is possible even if $\alpha, \beta, \gamma, \delta$ are not all zero. (For example, $\alpha = \beta = 1$, $\gamma = \delta = i$.) Consequently, infinitely many elements would lack an inverse.

We now leave the special case of fields and return to the study of some simple topics relating to rings in general.

Let $\mathscr{R}$ be a ring and $\mathscr{S}$ a sub*system* which obeys the ring axioms. Then $\mathscr{S}$ is called a *subring* of $\mathscr{R}$. Thus, a subring $\mathscr{S}$ is a sub*set* of $\mathscr{R}$ which is closed under the two composition laws of $\mathscr{R}$ and is a ring in its own right. Noting that $\mathscr{R}$ is a group for $+$ and recalling the criterion of Theorem 4.1(4) for subsets being subgroups, as well as adopting the notation for additive groups, we easily see the validity of the following criterion:

THEOREM 4.2(2).   *A subset $\mathscr{S}$ of a ring $\mathscr{R}$ is a subring iff $\gamma \in \mathscr{S}$ and $\delta \in \mathscr{S}$ imply $\gamma - \delta \in \mathscr{S}$ and $\gamma\delta \in \mathscr{S}$.*

It is important to realize that, since for multiplication a ring, in general, is only an associative structure, the following situations may occur for rings with unit:

1. A subring need not have a unit, even if the parent ring has one.
2. Both the ring and its subring have units, but they are different.
3. A certain subring possesses a unit, even though the parent ring has no unit.

We note here that Cases 2 and 3 can occur only if the unit of the subring happens to be a divisor of zero in the parent ring. Now, a few examples.

Example λ.   Every ring has two *trivial subrings*, namely $\mathscr{R}$ itself and $\{o\}$. All other subrings are called *proper*.

Example μ.   If $\mathscr{R} = (\mathbf{Z}, +, \cdot)$, then $\mathscr{S} = (\mathbf{E}, +, \cdot)$ is a subring. Note that $\mathscr{S}$ has no unit while $\mathscr{R}$ has.

Example ν.   The set $\mathbf{R} \times \mathbf{R}$ with

$$(a, b) + (c, d) = (a + c, b + d),$$
$$(a, b)(c, d) = (ac, bd)$$

is a ring (the product ring of $(\mathbf{R}, +, \cdot)$ with itself) and has the unit $(1, 1)$. The subset $\mathscr{S}$ consisting of elements of form $(a, 0)$ is easily seen to be a subring. It also has a unit, namely $(1, 0)$ which, however, differs from the unit of the parent ring.

Example ξ.   The subset $\mathscr{S}$ of the ring $\mathscr{R}$ defined in Example **ε**, consisting of all $n \times n$ matrices of the form

$$\begin{pmatrix} 0 & 0 & \cdots & 0 \\ 0 & 0 & \cdots & 0 \\ \cdot & & & \cdot \\ \cdot & & & \cdot \\ \cdot & & & \cdot \\ 0 & 0 & \cdots & \alpha \end{pmatrix}$$

is clearly a subring of $\mathscr{R}$ (use Theorem 4.2(2)). In contrast to its parent ring, $\mathscr{S}$ has a unit, namely the matrix with $\alpha = 1$; moreover, $\mathscr{S}$ is commutative, whereas $\mathscr{R}$ was not.

In connection with subrings we briefly mention the concept of *ring cosets*.† Since $\mathscr{S}$ is a subgroup of $\mathscr{R}$ for addition, we can act with $\mathscr{S}$ on the set $\mathscr{R}$, defining the action

$$(s, r) \mapsto s \cdot r = s + r \quad \text{for all} \quad s \in \mathscr{S}, r \in \mathscr{R}.$$

The corresponding orbits will be (two-sided) additive cosets for the addition. It is customary to use additive notation and denote the coset of $\mathscr{S}$ with the representative element $r \in \mathscr{R}$ by the symbol $\mathscr{S} + r$. Thus,

$$\mathscr{S} + r = \{s + r | s \in \mathscr{S}\}. \tag{4.16}$$

We shall use and illustrate this construction later, for the only case of importance when $\mathscr{S}$ is a very special type of subring. Here we note that, obviously, coset formation for rings would make sense even if $\mathscr{S}$ were not a subring, but only an additive subgroup of $\mathscr{R}$ (considered as a group).

We terminate the general discussion of rings with some comments on *ring morphisms*. In accord with the terminology adopted in Section 3.1, we of course say that a map $f: \mathscr{R} \to \mathscr{R}'$ between two rings is a morphism if it preserves both composition laws, i.e. if

$$f(a + b) = f(a) + f(b) \quad \text{and} \quad f(ab) = f(a)f(b).$$

Since, for addition, a ring is a group, we can take over the results we learned for groups concerning the preservation of the (additive) identity element and of the (additive) inverse. Thus, if f is a morphism of rings, we always have

$$f(o) = o' \quad \text{and} \quad f(-a) = -f(a) \equiv -a'.$$

---

†Also called *residue classes.*

If, in particular, $\mathcal{R}$ and $\mathcal{R}'$ are rings with unit, *and f is onto* (i.e. a homomorphism), then also $f(\epsilon) = \epsilon'$. However, if f is not surjective, this does not necessarily hold.† To avoid this unpleasantness, it is standard practice to *define* an arbitrary morphism to be one with the additional *requirement* that $f(\epsilon) = \epsilon'$. This will be understood in the following. Now, if $f(\epsilon) = \epsilon'$, then it can be easily seen that, for each invertible element $a \in \mathcal{R}$, we have $f(a^{-1}) = (f(a))^{-1} \equiv (a')^{-1}$, as expected.

Similarly as we saw for groups, it is found that

$$\text{Im}(f) \equiv \{f(r)|r \in \mathcal{R}\} = f(\mathcal{R})$$

and

$$\text{Ker}(f) \equiv \{r|r \in \mathcal{R}, \quad f(r) = 0'\} = f^{-1}(\{0'\})$$

are subrings of $\mathcal{R}'$ and $\mathcal{R}$, respectively. It should be remembered that if $\mathcal{R}$ is a ring with unit, there is no reason for ϵ to belong to Ker (f).

Theorem 4.1(6) of group morphisms also has its counterpart for rings: *A morphism of rings is an isomorphism of rings iff* Ker $(f) = \{0\}$ *and* Im $(f) = \mathcal{R}'$.

We conclude with some examples. Let $\mathcal{R}$ be an arbitrary ring with unit, and let r be a fixed invertible element of $\mathcal{R}$. The map

$$f_r: \mathcal{R} \to \mathcal{R} \quad \text{given by} \quad x \mapsto rxr^{-1} \quad \text{for all} \quad x \in \mathcal{R}$$

is a morphism from $\mathcal{R}$ into itself,‡ i.e. an endomorphism.

A counterexample for ring morphisms is the following. Let $\mathcal{R} = (\mathbf{Z}, +, \cdot)$ and let $\mathcal{R}' = (\mathbf{E}, +, \cdot)$. The map $a \mapsto 2a$ is *not* a morphism of rings, because even though $f(a + b) = f(a) + f(b)$, we have $f(ab) = 2ab \neq 2a2b = f(a)f(b)$.

PROBLEMS

4.2-1. Let $\mathcal{R}$ be the set of ordered pairs of reals. Determine which of the composition laws listed below yield a commutative ring with unit:
(a) $(a, b) + (c, d) = (ac, bc + d)$, $(a, b)(c, d) = (ac, bd)$,
(b) $(a, b) + (c, d) = (a + c, b + d)$, $(a, b)(c, d) = (ac + bd, ad + bd)$,
(c) $(a, b) + (c, d) = (a + c, b + d)$, $(a, b)(c, d) = (ac, ad + bc)$.
For the systems not answering the requirement, indicate the axioms which are violated.

4.2-2. Let $(G, +)$ be an Abelian group. Define multiplication by

†It turns out that $f(\epsilon)$ is only guaranteed to be the unit for the subring $f(\mathcal{R})$.
‡The reader is urged to compare this morphism with conjugation for groups. In particular, he should study various cases to determine Ker (f_r).

$$(\boldsymbol{\alpha}) \ ab = a \quad \text{and} \quad (\boldsymbol{\beta}) \ ab = o.$$

In which case do you get a ring $(G, +, \cdot)$?

4.2-3. Let X be some universal set and let $\mathscr{R}$ be a class of subsets such that if A and B belong to $\mathscr{R}$, so does $A \cup B$ and $A - B$. Such a class $\mathscr{R}$ of sets is called a *ring of sets*. (*Remarks*: One can show that from the definition it follows that $A \cap B$ also belongs to $\mathscr{R}$. Recall also that $A - B = A \cap B^c$.) Define the two† composition laws on $\mathscr{R}$:

$$A + B = (A - B) \cup (B - A),$$

$$AB = A \cap B.$$

Show that with respect to these operations, $\mathscr{R}$ is indeed a commutative ring. What is its zero element? Show that $\mathscr{R}$ has an identity iff it contains a set I which contains all other sets. (In particular, if X belongs to $\mathscr{R}$, then X is the identity.)

4.2-4. Let X be some universal set and consider the power set $\mathscr{P}(X)$. Is this a ring of sets? Which elements have an inverse? What are the regular elements? What are the divisors of zero? Under what conditions is this ring a field?

4.2-5. Let A be an arbitrary set and consider the set of all functions $\{f, g \ldots\}$ which map A into the real numbers (i.e. consider all "real valued functions" on A). Denoting an arbitrary element of A by a, define the "pointwise" sum and product of functions by

$$(f + g)(a) = f(a) + g(a),$$

$$fg(a) = f(a)g(a).$$

Show that the set of all real valued functions on A is a commutative ring with identity. (*Remark*: Instead of the set of functions into the reals, you could consider the set of functions into an arbitrary commutative field with identity.)

4.2-6. Take the set A of the preceding problem to be the interval $0 \le x \le 1$ on the real line and consider the ring $\mathscr{R}$ of all real valued functions defined on A. What are the divisors of zero, the regular elements, the noninvertible elements?

4.2-7. Let G be an arbitrary *additive* group. Consider the set of all endomorphisms f_α of G. This set will be denoted by End (G). Define the sum and product of two endomorphisms by setting, for any $g \in G$,

$$(f_\alpha + f_\beta)(g) = f_\alpha(g) + f_\beta(g),$$

$$f_\alpha f_\beta(g) = (f_\alpha \circ f_\beta)(g) \equiv f_\alpha(f_\beta(g)).$$

Show that End (G), equipped with these two composition laws, is a ring which has an identity but which, in general, is not commutative. What new feature emerges if instead of End (G), we consider Aut (G)?

---

†The first composition law is the symmetric difference, which was discussed in Section 4.1, p. 73, but we use a different notation.

4.2-8. Let $\mathcal{R}$ be a ring with unit. Show that the subset of invertible elements forms a group with respect to multiplication.

4.2-9. Let $\mathcal{R}$ be a ring with unit and suppose there are no divisors of zero. Show that then $xx = x$ implies either $x = o$ or $x = \epsilon$.

4.2-10. Let $\mathcal{R}$ be a ring with unit. Show that if x is an invertible element, then $(-x)$ is also invertible and in fact $(-x)^{-1} = -x^{-1}$.

4.2-11. Prove that if in a ring for every element one has $xx = x$, then $x + x = o$. Show also that the ring is necessarily commutative.

4.2-12. Let $\mathcal{R}$ be a commutative field, or only an integral domain (cf. footnote on p. 106). Show that the equation $x^2 = \epsilon$ implies either $x = \epsilon$ or $x = -\epsilon$.

4.2-13. Let $\mathcal{R}$ be a commutative field or only an integral domain. Show that the only element x for which $x^n = o$ for some positive integer n, is the zero element. (*Remark*: We say that, under the condition stated, o is the only *nilpotent element*.)

4.2-14. Let $\mathcal{R}$ be an arbitrary ring. If there exists a positive integer n such that† $nx = o$ for all $x \in \mathcal{R}$, we call the smallest such n the *characteristic* of the ring. If no such $n > 0$ exists (i.e. if $nx = o$ for all $x \in \mathcal{R}$ implies $n = 0$), then we say that $\mathcal{R}$ has characteristic zero. Give examples of rings with characteristic zero. Show that the ring of Problem 4.2-4 has characteristic two.

4.2-15. Let $\mathcal{R}$ be a ring with unit. Show that $\mathcal{R}$ has characteristic $n > 0$ iff n is the smallest positive integer for which $n\epsilon = o$.

4.2-16. Consider the sets
(a) $F = \{a - b\sqrt{2}\,|a, b \in \mathbf{Z}\}$,
(b) $F = \{a + b\sqrt[3]{2}\,|a, b \in \mathbf{Q}\}$.
Which of these is a field with respect to ordinary addition and multiplication?

4.2-17. Let $\mathcal{R}$ be a ring. The *center* of $\mathcal{R}$ is the set

$$c = \{c \in \mathcal{R} \,|\, cx = xc \quad \text{for all} \quad x \in \mathcal{R}\}.$$

Show that the center is a subring of $\mathcal{R}$. Let now $\mathcal{R}$ be a (skew)-field and show that its center is a commutative field.

4.2-18. Let $\mathcal{S}$ and $\mathcal{T}$ be subrings of $\mathcal{R}$. Show that $\mathcal{S} \cap \mathcal{T}$ is also a subring.

4.2-19. If $\mathcal{F}$ is a field, a subring Φ that is itself a field is called, naturally, a subfield. Show that a sub*set* Φ of $\mathcal{F}$ is a subfield iff
(a) Φ has at least one nonzero element,
(b) $a \in \Phi$ and $b \in \Phi$ imply $a - b \in \Phi$,
(c) $a \in \Phi$ and $b \in \Phi$, with $b \neq o$, imply $ab^{-1} \in \Phi$.

4.2-20. Let C be the field of complex numbers. Consider the map $f: \mathbf{C} \to \mathbf{C}$ given by $x + iy \mapsto x - iy$ (i.e. $z \mapsto z^*$) and determine whether this is an isomorphism.

4.2-21. Let $(\mathbf{Z}, +, \cdot)$ be the ring of integers and let $\mathcal{R}$ be an arbitrary ring with unit ϵ. Show that the map $f: \mathbf{Z} \to \mathcal{R}$ given by $n \mapsto n\epsilon$ for all $n \in \mathbf{Z}$ is a morphism. Deter-

---

†Recall the notation: $nx \equiv x + x + \cdots + x$, n terms.

mine the kernel of the morphism. (*Note*: Do *not* assume that the characteristic of $\mathcal{R}$ (cf. Problem 4.2-14) is zero.)

4.2-22. Let $\mathcal{R}$ and $\mathcal{R}'$ be arbitrary rings. Show that the map $f: \mathcal{R} \to \mathcal{R}'$ given by $f(x) = o'$ for all x is a morphism, called the trivial morphism.

4.2-23. Let $(\mathbf{Z}, +, \cdot)$ be the ring of integers. Show that there are only two endomorphisms: the trivial one (cf. preceding problem) and the identity map 1_Z. (*Hint*: Consider the images of $n = 1 + \cdots + 1$, of $(-n)$, and of 0.)

4.2-24. Let a be a fixed element of a ring $\mathcal{R}$. Define left multiplication by a as the map $t_a: \mathcal{R} \to \mathcal{R}$ given by $x \mapsto ax$ for all $x \in \mathcal{R}$. Show that the set $\mathcal{T}$ of all such functions is a ring with respect to pointwise addition of functions and composition of functions. Show that $\mathcal{R}$ is homomorphic to $\mathcal{T}$. (*Hint*: Consider the map $f: \mathcal{R} \to \mathcal{T}$ given by $f(a) = t_a$. *Remark*: This problem is the ring-analog of Cayley's theorem.) What condition on the ring will guarantee that f is actually an *iso*morphism?

4.2-25. Let f be an endomorphism of the field $\mathcal{F}$. Let $\mathcal{K}$ be the set of elements which are left fixed under f, i.e. let $\mathcal{K} = \{a \in \mathcal{F} | f(a) = a\}$. Assuming that $\mathcal{K} \neq \{o\}$, show that $\mathcal{K}$ is a subfield of $\mathcal{F}$. (*Hint*: You may use the result of Problem 4.2-19.)

4.2a. Ideals; Quotient Rings; Isomorphism Theorems

In this subsection we shall consider the ring-theoretic analog of normal subgroups and connected subjects. We start with the following

DEFINITION 4.2a(1).   *An ideal $\mathcal{I}$ of a ring $\mathcal{R}$ is a subring of $\mathcal{R}$ which has the additional property that together with any element $l \in \mathcal{I}$, it contains also all elements of the form lr and of the form rl, where r is an arbitrary element of $\mathcal{R}$.*

Informally, we may say that an ideal is a subring which "captures" all products of its elements with arbitrary elements of the parent ring. Symbolically, we may say that if $\mathcal{I}$ is an ideal, then it is a subring such that $\mathcal{I}r \subset \mathcal{I}$ and $r\mathcal{I} \subset \mathcal{I}$, for all $r \in \mathcal{R}$.

If $\mathcal{I}_L$ is a subring such that $l \in \mathcal{I}_L$ implies only $lr \in \mathcal{I}_L$ (but not $rl \in \mathcal{I}_L$), then we call $\mathcal{I}_L$ a *left ideal*. Right ideals are defined analogously. Thus, in our usage, "ideal" means a "two-sided ideal." It is clear that for commutative rings any left (right) ideal is automatically a (two-sided) ideal.

Obviously, the trivial subrings $\{o\}$ and $\mathcal{R}$ of any ring are ideals. A ring $\mathcal{R}$ whose only ideals are these two trivial ones, is called a *simple ring*.

A very useful criterion for identifying ideals is the following

THEOREM 4.2a(1). *A* subSET $\mathscr{I}$ *of* $\mathscr{R}$ *is an ideal iff* (a) $l_1 \in \mathscr{I}$ *and* $l_2 \in \mathscr{I}$ *imply* $l_1 - l_2 \in \mathscr{I}$ *and* (b) $l \in \mathscr{I}$ *implies* $lr \in \mathscr{I}$ *and* $rl \in \mathscr{I}$ *for all* $r \in \mathscr{R}$.

Condition (a) is one part of the condition that $\mathscr{I}$ be a subring; condition (b), when r is restricted to belong to $\mathscr{I}$, completes the subring requirement, but at the same time, for arbitrary $r \in \mathscr{R}$, guarantees the definition of an ideal to be fulfilled.

Example α. Let $(\mathbf{Z}, +, \cdot)$ be the ring of integers. The subring $\mathscr{I} = \mathbf{E}$ is clearly an ideal, since the product of any even number with any integer is even. More generally, the subset $\mathscr{I} = \{na \,|\, n \in \mathbf{Z}\}$, where a is a fixed, but arbitrary integer, is an ideal, since $na - ma = (n - m)a \equiv ka \in \mathscr{I}$, and $(na)x = (nx)a \equiv ma \in \mathscr{I}$ for any integer x, as well as $x(na) = (xn)a \equiv ma \in \mathscr{I}$.

Example β. Let $\mathscr{R}$ be the ring of all n by n matrices, cf. Example δ on p. 104. Let $\mathscr{I}_L$ be the set of matrices of the form

$$L = \begin{pmatrix} 0 & 0 & \cdots & 0 \\ 0 & 0 & \cdots & 0 \\ \cdot & & & \cdot \\ \cdot & & & \cdot \\ \cdot & & & \cdot \\ \alpha & \beta & \cdots & \nu \end{pmatrix}.$$

Clearly, $L_1 - L_2 \in \mathscr{I}_L$ and also, for any matrix M, $LM \in \mathscr{I}_L$. However, in general, ML will have nonzero elements also in rows preceding the last, so that $ML \notin \mathscr{I}_L$. Hence $\mathscr{I}_L$ is only a *left* ideal.†

We now present two interesting results on ideals.

THEOREM 4.2a(2). *If* $\mathscr{R}$ *is a ring with unit* ϵ, *then no proper ideal* $\mathscr{I}$ *of* $\mathscr{R}$ *can contain* ϵ.

Proof. Suppose $\epsilon \in \mathscr{I}$. Then $r\epsilon = r \in \mathscr{I}$ for all $r \in \mathscr{R}$, hence $\mathscr{I} = \mathscr{R}$.

COROLLARY. *If* $\mathscr{R}$ *is a ring with unit and* $\mathscr{I}$ *is a proper ideal, then no element of* $\mathscr{I}$ *is invertible.*

Proof. If $l \in \mathscr{I}$ and if l had an inverse l^{-1} in $\mathscr{R}$, then $ll^{-1} = \epsilon$ should belong to $\mathscr{I}$, which is impossible.

---

†Actually, it can be shown that the ring of *all* n by n matrices has no two-sided proper ideal.

Since we know that, in a field, every nonzero element is invertible, we immediately have the following, rather surprising result:

THEOREM 4.2a(3). *Every field is a simple ring, i.e. it has no nontrivial ideals.*

We now proceed to show that (similarly to the role of normal subgroups in group theory) the importance of ideals rests in the close connection between ideals and kernels of ring morphisms. We have, first of all,

THEOREM 4.2a(4). *The kernel* Ker *(f) of any morphism f: $\mathcal{R} \to \mathcal{R}'$ is an ideal of $\mathcal{R}$.*

We leave the simple proof of this and of the subsequent theorems as an exercise, and give only an example. Let $\mathcal{R}$ be the ring of integers and let $\mathcal{R}'$ be the Galois field $\{o, \epsilon\}$ of Example ζ on p. 107. The map which assigns to every even integer o and to every odd integer ϵ is a (homo)morphism. Its kernel is the ring of even numbers, which, as we know, is indeed an ideal of the integers.

In analogy to what we learned in group theory, we expect that if $\mathcal{I}$ is an ideal, then it will be the kernel of some morphism. This is indeed the case, and we now sketch the details.

Let $\mathcal{I}$ be an ideal of the ring $\mathcal{R}$, whose elements we shall denote by $x, y, z, \ldots$. Let us form the ring cosets

$$\mathcal{I} + x = \{l + x \,|\, l \in \mathcal{I}\}.$$

We know that the class of all distinct cosets is a partition† of $\mathcal{R}$. This partition is precisely the coset space $\mathcal{R}/\mathcal{I}$. We introduce in this set an algebraic structure by defining the two composition laws $\boxplus$ and $*$ as follows:

$$(\mathcal{I} + x) \boxplus (\mathcal{I} + y) = \mathcal{I} + (x + y), \tag{4.17a}$$

$$(\mathcal{I} + x) * (\mathcal{I} + y) = \mathcal{I} + xy. \tag{4.17b}$$

It is not difficult to show that, *because $\mathcal{I}$ is an ideal*, these definitions are in fact independent of the choice of representatives, and hence, they yield well-defined composition laws on $\mathcal{R}/\mathcal{I}$. Furthermore, it is almost obvious that *the coset space $\mathcal{R}/\mathcal{I}$ equipped with the addition $\boxplus$ and multiplication $*$ is,* in fact, *a ring.* We call this ring the *quotient ring of $\mathcal{R}$*

---

†We remark here that x and y are in the same coset iff $x - y \in \mathcal{I}$.

by $\mathcal{I}$. We see that the zero element of $\mathcal{R}/\mathcal{I}$ is the coset $\mathcal{I} + o$. The negative of $\mathcal{I} + x$ is the coset $\mathcal{I} - x$. In addition we note that, if $\mathcal{R}$ is commutative, then $\mathcal{R}/\mathcal{I}$ is also commutative. If $\mathcal{R}$ has a unit ϵ, then $\mathcal{R}/\mathcal{I}$ has the unit $\mathcal{I} + \epsilon$.

We illustrate the construction of quotient rings on a very simple example. Let $\mathcal{R} = (\mathbf{Z}, +, \cdot)$. Take the ideal $\mathbf{E}$ consisting of the even integers (cf. Example α). The cosets are of the form $\mathbf{E} + x = \{2n + x \,|\, n \in \mathbf{Z}\}$. Clearly, there are only two distinct cosets, namely $\mathbf{E} + 0 \equiv \mathbf{E}$ and $\mathbf{E} + 1 \equiv \mathbf{O}$. Thus, $\mathbf{Z}/\mathbf{E} = \{\mathbf{E}, \mathbf{O}\}$. The ring operations in the quotient ring $\mathbf{Z}/\mathbf{E}$ are given by the tables

$\boxplus$	$\mathbf{E}$	$\mathbf{O}$
$\mathbf{E}$	$\mathbf{E}$	$\mathbf{O}$
$\mathbf{O}$	$\mathbf{O}$	$\mathbf{E}$

$*$	$\mathbf{E}$	$\mathbf{O}$
$\mathbf{E}$	$\mathbf{E}$	$\mathbf{E}$
$\mathbf{O}$	$\mathbf{E}$	$\mathbf{O}$

which follow directly from Eqs. (4.17a,b). We observe that $\mathbf{Z}/\mathbf{E}$ is essentially the two-element Galois field[†] $\{o, \epsilon\}$.

We can now easily establish the analog of Theorem 4.1c(3):

THEOREM 4.2a(5).   *Let $\mathcal{I}$ be an ideal of $\mathcal{R}$. Then the canonical map $c: \mathcal{R} \to \mathcal{R}/\mathcal{I}$ (defined by $r \mapsto \mathcal{I} + r$) is a homomorphism from $\mathcal{R}$ to the quotient ring $\mathcal{R}/\mathcal{I}$, with $\operatorname{Ker}(f) = \mathcal{I}$.*

As a consequence, we have the isomorphism theorem:

THEOREM 4.2a(6).   *If $f: \mathcal{R} \to \mathcal{R}'$ is an arbitrary morphism of rings, then $\mathcal{R}/\operatorname{Ker}(f)$ is isomorphic to $\operatorname{Im}(f)$.*

As a special case, we obtain

THEOREM 4.2a(7).   *If $\mathcal{R}$ is homomorphic to $\mathcal{R}'$, then $\mathcal{R}/\operatorname{Ker}(f)$ is isomorphic to $\mathcal{R}'$.*

Thus, if $\mathcal{R}$ and $\mathcal{R}'$ are homomorphic, then there exists an ideal $\mathcal{I}$ of $\mathcal{R}$ (namely $\mathcal{I} \equiv \operatorname{Ker}(f)$) such that $\mathcal{R}/\mathcal{I} \approx \mathcal{R}'$. The isomorphism is actually furnished by the induced map $\mathcal{I} + r \mapsto f(r)$. The reader is urged to visualize the isomorphism theorem of rings with the use of a mapping diagram and other, more detailed diagrams. Of course, these diagrams will be analogous to those for groups, i.e. to the diagrams of Figs. 4.5, 4.6, and 4.7.

---

†The map $\mathbf{E} \mapsto o$, $\mathbf{O} \mapsto \epsilon$ is an isomorphism.

In conclusion, we observe that, if a ring is simple (i.e. has no nontrivial ideals), then any possible *homo*morphism to any ring which has at least two elements is automatically an isomorphism. Thus, in particular, any nontrivial field homomorphism is a field isomorphism.

At this point we terminate our brief introduction to rings. It was not our aim to enter into the very rich theory of rings for its own sake. We only collected those basic features which will be needed in our future work.

PROBLEMS

4.2a-1. Consider the ring $\mathcal{R}$ defined in Problem 4.2-6. Let X be a subset of $0 \le x \le 1$, and consider the set of functions

$$\mathcal{I} = \{f | f(x) = 0 \quad \text{for every} \quad x \in X\}.$$

Show that $\mathcal{I}$ is an ideal of $\mathcal{R}$. Determine and analyze $\mathcal{R}/\mathcal{I}$.

4.2a-2. Let $\mathcal{R}$ be the ring of all functions $f: \mathbf{R} \to \mathbf{R}$, equipped with pointwise sum and pointwise product (as in Problem 4.2-5). Show that the subset

$$\mathcal{I} = \{f | f(1) = 0\}$$

is an ideal. Determine and study $\mathcal{R}/\mathcal{I}$.

4.2a-3. Consider the ring of sets defined in Problem 4.2-4, and show that the class of all finite subsets of X is an ideal in this ring.

4.2a-4. Let $\{\ldots \mathcal{I}_\alpha \ldots\}$ be an indexed collection of ideals of a ring $\mathcal{R}$. Show that $\cap_\alpha \mathcal{I}_\alpha$ is also an ideal.

4.2a-5. Let $\mathcal{I}$ be an ideal of $\mathcal{R}$. Define the *annihilator* of $\mathcal{I}$ as the set

$$\text{ann } \mathcal{I} = \{r \in \mathcal{R} | rl = o \quad \text{for all} \quad l \in \mathcal{I}\}.$$

Show that this set (equipped with the same sum and product as $\mathcal{R}$ itself) is an ideal of $\mathcal{R}$.

4.2a-6. Let $\mathcal{I}_1$ and $\mathcal{I}_2$ be two ideals of $\mathcal{R}$ such that $\mathcal{I}_1 \cap \mathcal{I}_2 = \{o\}$. Show that $l_1 l_2 = o$ for every pair $l_1 \in \mathcal{I}_1$, $l_2 \in \mathcal{I}_2$.

4.2a-7. Consider the ring of sets defined in Problem 4.2-4. Let F be a fixed subset of X. Show that the map

$$f: \mathcal{P}(X) \to \mathcal{P}(X) \quad \text{given by} \quad A \mapsto A \cap F \quad \text{for all} \quad A \in \mathcal{P}(X),$$

is an endomorphism and determine its kernel. Verify that this kernel is an ideal of the ring.

4.2a-8. Let $\mathcal{I}$ be an ideal of a ring $\mathcal{R}$. Suppose that $\mathcal{R}$ has no divisors of zero. Show that, nevertheless, the quotient ring $\mathcal{R}/\mathcal{I}$ may have divisors of zero.

4.2a-9. Let $\mathcal{R}$ be a ring, $\mathcal{S}$ an arbitrary subring, and $\mathcal{I}$ an ideal. Suppose $\mathcal{S} \cap \mathcal{I} = \{o\}$. Prove that $\mathcal{S}$ is isomorphic to a subring of the quotient ring $\mathcal{R}/\mathcal{I}$. (*Hint*: Consider the map $f: \mathcal{S} \to \mathcal{R}/\mathcal{I}$ given by $f(s) = \mathcal{I} + s$.)

4.3 LINEAR SPACES

Groups and rings are systems with only internal composition laws. Rings, in particular, have two internal composition laws. In this section we shall study systems which have one internal and one external composition law.

Let $\mathscr{L}$ be an additive group and let K be a commutative field. Let us introduce an external composition law $K \times \mathscr{L} \to \mathscr{L}$ by defining the multiplicative action[†] of K on $\mathscr{L}$. This means that to every pair (α, x) with $\alpha \in K$ and $x \in \mathscr{L}$ we assign a composite $\alpha \cdot x$ (to be denoted in the sequel simply by the juxtaposition αx) such that $(\alpha\beta)x = \alpha(\beta x)$ for every $\alpha, \beta \in K$ and all $x \in \mathscr{L}$, and further such that $\epsilon x = x$ for all $x \in \mathscr{L}$ (where ϵ is the unit of K). This external composition law will be called *scalar multiplication*. We relate scalar multiplication to the internal additive group composition law (*addition*) in $\mathscr{L}$ by demanding the mixed distributive laws $\alpha(x + y) = \alpha x + \alpha y$ and $(\alpha + \beta)x = \alpha x + \beta x$, for all $\alpha, \beta \in K$ and $x, y \in \mathscr{L}$. The algebraic system so defined is called a linear space. In summary, a *linear space is an algebraic system* $(\mathscr{L}, K, \cdot)$, *where $\mathscr{L}$ is an additive group, K is a commutative field, the external composition law $\cdot$ is an action of K on $\mathscr{L}$, and the internal law $+$ in $\mathscr{L}$ is related to $\cdot$ by mixed distributive laws.* This somewhat terse definition is made clearer by writing down explicitly the defining axioms of a linear space[‡] $\mathscr{L}$ as follows:

(ia) $x + (y + z) = (x + y) + z$ for all $x, y, z \in \mathscr{L}$.
(ib) There exists an element $o \in \mathscr{L}$ such that $o + x = x + o = x$ for all $x \in \mathscr{L}$.
(ic) For every $x \in \mathscr{L}$ there exists an element $-x \in \mathscr{L}$ such that $x + (-x) = o$.
(id) $x + y = y + x$ for all $x, y \in \mathscr{L}$.
(iia) $\alpha(\beta x) = (\alpha\beta)x$ for all $\alpha, \beta \in K$ and all $x \in \mathscr{L}$.
(iib) $\epsilon x = x$ for all $x \in \mathscr{L}$ (ϵ is unit of K).
(iiia) $\alpha(x + y) = \alpha x + \alpha y$ for all $\alpha \in K$ and all $x, y \in \mathscr{L}$.
(iiib) $(\alpha + \beta)x = \alpha x + \beta x$ for all $\alpha, \beta \in K$ and all $x \in \mathscr{L}$.

It is customary to call the elements $x, y, z \ldots$ of the underlying set $\mathscr{L}$ *vectors*. Correspondingly, a linear space is often referred to as a *vector*

[†]We recall that, disregarding the null element of K, we have a multiplicative group in K. The action of the null element of K onto an $x \in \mathscr{L}$ may be defined separately to yield the zero element o of $\mathscr{L}$. This is consistent with the other requisites of composition.

[‡]Without causing confusion, we may use the symbol $\mathscr{L}$ to denote the linear space as a *system*. Thus, strictly speaking, $\mathscr{L}$ has three different connotations: it stands for the underlying set, for the additive group structure, and for the linear space structure as a whole.

space. The internal composition law + is called vector addition. The element o has the name *null vector*.

The elements α, β, γ, ... of the commutative field K are called *scalars*, and K is referred to as the field of scalars. Linear spaces with the same set $\mathscr{L}$ but different fields of scalars K are, in general, different systems. It is, therefore, advisable to refer to a given linear space as "a linear space $\mathscr{L}$ *over* the scalars K." We also note in passing that the sum in K is denoted by the same symbol (namely +) as is the vector sum in $\mathscr{L}$. Furthermore, the product in K is denoted in the same way (namely by juxtaposition $\alpha\beta$) as is the scalar product αx between $\alpha \in K$ and $x \in \mathscr{L}$. These notational conventions should not cause misunderstandings.

We also alert the reader that the zero element of K should not be confused with the null vector o of $\mathscr{L}$.

At this point we make an important specialization. *In all of our future work we shall deal exclusively with linear spaces where the field of scalars is either the field of real numbers or the field of complex numbers.* Correspondingly, we shall speak of a *real linear space* or of a *complex linear space* according to whether the field of scalars is **R** or **C**. Some of the theorems we shall derive later are not true for vector spaces over arbitrary scalars. However, in most applications to physics, only our real or complex linear spaces play a sufficiently important role.†

Our restriction to real or complex linear spaces simplifies somewhat the notation. We shall denote the zero element of our field of scalars explicitly by the number 0 (which, of course, has to be distinguished in principle from the null vector o), and the multiplicative unit of the scalar field will be denoted explicitly by the number 1. Thus, axiom (iib) will be written as $1x = x$.

Before giving examples of linear spaces, we wish to point out some elementary, but important relations which follow easily from the defining axioms of linear spaces. We have

$$\alpha o = o \quad \text{for all scalars } \alpha,$$

$$0x = o \quad \text{for all vectors‡ } x,$$

$$(-1)x = 1(-x) = -x \quad \text{for all vectors } x.$$

†As a matter of fact, there exists a generalization of linear spaces inasmuch as one may replace the commutative field K by an arbitrary ring with unit. These systems are called *modules* and will not be discussed in this book, even though they have a well-developed mathematical theory and play some role in certain applications. For example, spaces with K being the *skew* field of quaternions have been occasionally used for physical theories. In the novel theory of automata, even more exotic modules appear to have a role.

‡For this reason, it is a common, but deplorable practice to use the symbol 0 for indicating the null vector as well.

Furthermore,

$$\alpha x = o \quad \text{implies that either} \quad x = o \quad \text{or} \quad \alpha = 0.$$

Now, some simple examples.

Example α.   The set consisting of the single element o, with $o + o = o$ and $\alpha o = o$ for all scalars, is a linear space, called *zero space*, and is denoted usually by $\{o\}$. Even though this is a trivial space, it does play a role (as a subspace) which must not be overlooked.

Example β.   The set $\mathscr{L} = \mathbf{R}$ of real numbers, with ordinary addition as vector addition and with $K = \mathbf{R}$ as the field of scalars, is a real vector space, if scalar multiplication αx is defined as ordinary product of real numbers. We shall call this space the *linear space of the reals*. Similarly, the set $\mathscr{L} = \mathbf{C}$ of complex numbers, with ordinary addition, and with $K = \mathbf{C}$ as scalars, is a *complex* vector space, if α is defined as the usual product of two complex numbers. We call this the *complex linear space of the complex numbers*. We may take also the set $\mathbf{C}$ for the underlying set (again with ordinary complex number addition) but use the field $K = \mathbf{R}$ as scalars, where now αz means the usual product of a real and a complex number. Then we have a *real* linear space, to be denoted by $\check{\mathbf{C}}$ and called the *real linear space of the complex numbers*. $\check{\mathbf{C}}$ must be distinguished from $\mathbf{C}$. Note that there is no "complex linear space of the real numbers," since αx with $\alpha \in \mathbf{C}$ and $x \in \mathbf{R}$ does not belong to $\mathbf{R}$.

Example γ.   The vector spaces of Example β have a simple and familiar generalization. Let as usual, $\mathbf{R}^n$ denote the Cartesian product set $\mathbf{R} \times \cdots \times \mathbf{R}$ whose elements are the ordered n-tuples $(\xi_1, \xi_2, \ldots, \xi_n)$ of real numbers. Let $K = \mathbf{R}$ and define the "componentwise" sum and scalar product

$$(\xi_1, \ldots, \xi_n) + (\eta_1, \ldots, \eta_n) = (\xi_1 + \eta_1, \ldots, \xi_n + \eta_n),$$
$$\alpha(\xi_1, \ldots, \xi_n) = (\alpha \xi_1, \ldots, \alpha \xi_n).$$

It is trivial to check out that all linear space axioms hold. This space, $\mathbf{R}^n$, is the *real n-dimensional† vector space*. Next, let $\mathbf{C}^n \equiv \mathbf{C} \times \cdots \times \mathbf{C}$ be the set whose elements are the ordered n-tuples $(\zeta_1, \ldots, \zeta_n)$ of complex numbers. Let now $K = \mathbf{C}$ and define sum and scalar product as before, i.e. componentwise. In this way the *complex n-dimensional† vector space* $\mathbf{C}^n$ is obtained. Alternatively, we may take for K the field of real numbers and we then have the *real vector space of the complex n-tuples*, which we

†The term "dimension" will be explained and justified later.

shall denote by $\check{C}^n$. It is also customary to call C^n the complex linear n-space over the complex scalars, and to call $\check{C}^n$ the complex linear n-space over the real scalars. The latter, by the way, does not play an important role.

Example δ.   The spaces introduced in Example γ have a simple, but important generalization. We may consider the set of real (infinite) *sequences*, $(\xi_1, \ldots, \xi_k, \ldots)$ and using $K = R$, define componentwise sum and scalar product. This will be the space R^∞, called the *real infinite dimensional vector space of sequences.* The corresponding space of all complex sequences $(\zeta_1, \ldots, \zeta_k, \ldots)$ with $K = C$ is the space C^∞ which we shall refer to as the *complex infinite dimensional vector space of sequences.* (The space $\check{C}^\infty$ may be also constructed, but plays no role in applications.)

Example ε.   Let X be some arbitrary set and let $\mathscr{L}$ be the collection of all functions from X to the real (or complex) numbers R (or C). Let K be R (or C). When we define the "pointwise sum"

$$(f + g)(x) = f(x) + g(x)$$

and "pointwise scalar product"

$$(\alpha f)(x) = \alpha(f(x)),$$

we have a real (complex) vector space. The null vector of this space is the constant function o which is defined by $o(x) = 0$ for all $x \in X$.

Example ζ.   The set of all n by n real (complex) matrices with $K = R(C)$ as scalars, if we define vector addition as the usual sum $(M + N)_{ab} = M_{ab} + N_{ab}$ of matrices and set $(\alpha M)_{ab} = \alpha M_{ab}$, is a linear space.

Following the same pattern that we used in the study of other algebraic systems, we now turn our interest to the relevant subsystems.

Let $\mathscr{L}$ be a linear space and let $\mathscr{M}$ be a subsystem which obeys the axioms of linear spaces in itself. Then $\mathscr{M}$ is called a *linear manifold*† of $\mathscr{L}$. Thus, a linear manifold $\mathscr{M}$ is a subset of $\mathscr{L}$ which is closed under vector addition and scalar multiplication that was defined in $\mathscr{L}$ and which is a linear space in its own right.

A little consideration shows‡ that the sufficient and necessary criterion

---

†Many texts on algebra use the term *subspace* instead of manifold. However, it is advisable to reserve this expression for certain subsystems of topological linear spaces. This topic will be discussed in Chapter 9.

‡Closure under scalar multiplication implies that, together with x, also $-x = (-1)x$ belongs to $\mathscr{M}$. This, together with closure under vector addition, will guarantee that $x - y \in \mathscr{M}$ if $x, y \in \mathscr{M}$.

for a sub*set* to be a linear manifold is given by the following

THEOREM 4.3(1). *A subset $\mathcal{M}$ of a linear space $\mathcal{L}$ is a linear manifold iff*

(a) $x \in \mathcal{M}$ *and* $y \in \mathcal{M}$ *imply* $x + y \in \mathcal{M}$

and

(b) $x \in \mathcal{M}$ *and* $\alpha \in K$ *imply* $\alpha x \in \mathcal{M}$.

Thus, the novelty is that closure alone guarantees the fulfillment of the vector space axioms.

The reader will easily verify that our criterion may be reformulated in the following elegant manner:

THEOREM 4.3(2). *A subset $\mathcal{M}$ of a linear space $\mathcal{L}$ is a linear manifold iff*

$$x, y \in \mathcal{M}, \qquad \alpha, \beta \in K \quad imply \quad \alpha x + \beta y \in \mathcal{M}.$$

We proceed to give some examples.

Example η. Every linear space has two *trivial linear manifolds*, namely $\mathcal{L}$ itself and the zero space $\{o\}$. All other manifolds are called *proper*.

Example ϑ. Let $\mathcal{L}$ be the two-dimensional linear space $\mathbf{R}^2$ (cf. Example γ). Using the explicit forms $x = (\xi_1, \xi_2)$ and $y = (\eta_1, \eta_2)$ of two vectors, the reader will easily convince himself† that all proper linear manifolds of $\mathbf{R}^2$ are precisely the straight lines that pass through the origin. Similarly, in $\mathbf{R}^3$, the proper manifolds are the straight lines and the planes through the origin. The reader will generalize these statements to $\mathbf{R}^n$ and $\mathbf{C}^n$ without difficulty. He is also advised to practice on $\mathbf{R}^2$ (and $\mathbf{R}^3$) the two different traditional ways of visualizing vectors. In our approach, we identify the vector x with the *point* (ξ_1, ξ_2). It is, however, equally possible to associate (in our mind!) an "arrow" with x, whose foot is at the origin of the "coordinate system" and whose head is at the point with the coordinates (ξ_1, ξ_2).

Example κ. In the function space of Example ϵ, the subset of all constant functions is a manifold. (If $f(x) = c_1$ and $g(x) = c_2$ for all $x \in X$, then $\alpha f + \beta g$ is also a constant function.)

---

†Note that $\alpha x + \beta y = (\alpha \xi_1 + \beta \eta_1, \alpha \xi_2 + \beta \eta_2)$, and x lies on the same line through the origin as y iff $\eta_2 \xi_1 = \xi_2 \eta_1$.

Example λ.   In the linear space of 2 by 2 matrices (cf. Example ζ), the subset of matrices of the form

$$\begin{pmatrix} \mu & \nu \\ -\nu & \mu \end{pmatrix}$$

is easily seen to be a linear manifold.

Example μ.   Consider the linear spaces $\mathbf{R}^n$ and $\mathbf{C}^n$, as defined in Example γ. Here $\mathbf{R}^n$ is *not* a linear manifold of $\mathbf{C}^n$. Indeed, even though $\mathbf{R}^n \subset \mathbf{C}^n$ and $x + y \in \mathbf{R}^n$ if $x, y \in \mathbf{R}^n$, the scalar product αx, for $x \in \mathbf{R}^n$ and some *nonreal* α, does *not* belong to $\mathbf{R}^n$.

We now turn our interest to a very important problem, which is this: when can a linear space be thought of as uniquely being "built up" from some of its own manifolds?

To start with, let $\mathscr{L}$ be a linear space and let $\mathscr{M}_1$ and $\mathscr{M}_2$ be two linear manifolds. Suppose that every vector $x \in \mathscr{L}$ can be written in the form $x = x_1 + x_2$, where $x_1 \in \mathscr{M}_1$ and $x_2 \in \mathscr{M}_2$. (This means that the *set* $\mathscr{M}_1 + \mathscr{M}_2 \equiv \{x_1 + x_2 | x_1 \in \mathscr{M}_1, x_2 \in \mathscr{M}_2\}$ is equal to the set $\mathscr{L}$.) Suppose that, in addition, the only common vector of $\mathscr{M}_1$ and $\mathscr{M}_2$ is the null vector,† i.e. $\mathscr{M}_1 \cap \mathscr{M}_2 = \{o\}$. Then the decomposition $x = x_1 + x_2$ is *unique*. Indeed, suppose $x = x_1 + x_2$ and also $x = x_1' + x_2'$. Then $x_1 + x_2 = x_1' + x_2'$, so that $x_1 - x_1' = x_2' - x_2$. But here the l.h.s. is a vector from $\mathscr{M}_1$ and the r.h.s. is a vector from $\mathscr{M}_2$. By our assumption $\mathscr{M}_1 \cap \mathscr{M}_2 = \{o\}$, this implies that $x_1 - x_1' = o$ and $x_2' - x_2 = o$, so that $x_1' = x_1$ and $x_2' = x_2$, i.e. the decomposition was actually unique.

Let us now consider the converse situation. Suppose every vector $x \in \mathscr{L}$ has a unique decomposition $x = x_1 + x_2$ with $x_1 \in \mathscr{M}_1$ and $x_2 \in \mathscr{M}_2$, where $\mathscr{M}_1$ and $\mathscr{M}_2$ are manifolds. We wish to show that then $\mathscr{M}_1 \cap \mathscr{M}_2 = \{o\}$. Suppose the opposite holds, i.e. suppose that $\mathscr{M}_1 \cap \mathscr{M}_2$ contains a nonzero vector z. Then we could write $z = z_1 + z_2$ with $z_1 = z \in \mathscr{M}_1$ and $z_2 = o \in \mathscr{M}_2$ *or also* $z = z_1' + z_2'$ with $z_1' = o \in \mathscr{M}_1$ and $z_2' = z \in \mathscr{M}_2$, contrary to the assumption of uniqueness.

At this point we introduce the following

DEFINITION 4.3(1).   *A linear space $\mathscr{L}$ is said to be the direct sum of two of its linear manifolds $\mathscr{M}_1$ and $\mathscr{M}_2$ if every $x \in \mathscr{L}$ has a unique decomposition $x = x_1 + x_2$ with $x_1 \in \mathscr{M}_1$ and $x_2 \in \mathscr{M}_2$. We then write $\mathscr{L} = \mathscr{M}_1 \oplus \mathscr{M}_2$.*

With this definition, we may formalize our above findings by the

†Since $\mathscr{M}_1$ and $\mathscr{M}_2$ are linear spaces, both *must* contain o.

following

THEOREM 4.3(2).   *Let every vector of the linear space $\mathcal{L}$ be expressible as the sum of a vector in the linear manifold $\mathcal{M}_1$ and a vector in the linear manifold $\mathcal{M}_2$. Then $\mathcal{L} = \mathcal{M}_1 \oplus \mathcal{M}_2$ iff $\mathcal{M}_1 \cap \mathcal{M}_2 = \{o\}$.*

We give some elementary examples of direct sum decompositions.

Example ν.   In   the   real   vector   space $\mathbf{R}^3$,   the   "axis" $\mathcal{M}_1 = \{(\xi_1, 0, 0)|\xi_1 \in \mathbf{R}\}$ and the "coordinate plane" $\mathcal{M}_{23} = \{(0, \xi_2, \xi_3)|\xi_2, \xi_3 \in \mathbf{R}\}$ are manifolds. Obviously, $\mathbf{R}^3 = \mathcal{M}_1 \oplus \mathcal{M}_{23}$. This example also warns us that, naturally, a direct sum decomposition is by no means unique. For example, we could take here as direct summands the axis $\mathcal{M}_2$ and the plane $\mathcal{M}_{13}$. As a matter of fact, to any line through the origin we can take any plane through the origin which does not contain it, and $\mathbf{R}^3$ will be then the direct sum of these manifolds.

Example ξ.   Let $\mathcal{L}$ be the function space (cf. Example **ε**) consisting of functions $f: \mathbf{R} \to \mathbf{R}$. We may decompose this space into the direct sum of the manifold of even functions, $\mathcal{M}_1 = \{h \,|\, h(-x) = h(x)\}$ and the manifold of odd functions, $\mathcal{M}_2 = \{g \,|\, g(-x) = -g(x)\}$. Indeed, for any f we can write $f = h + g$, where $h(x) = \frac{1}{2}(f(x) + f(-x))$ and $g(x) = \frac{1}{2}(f(x) - f(-x))$; furthermore, the only function which is both symmetric and antisymmetric, is the null function. Thus, Theorem 4.3(2) applies and $\mathcal{L} = \mathcal{M}_1 \oplus \mathcal{M}_2$.

The direct sum decomposition of $\mathcal{L}$ can be generalized for more than two summands.† We say that $\mathcal{L}$ is the direct sum of the manifolds $\mathcal{M}_1, \mathcal{M}_2, \ldots, \mathcal{M}_r$ if any $x \in \mathcal{L}$ can be *uniquely* represented in the form $x = x_1 + x_2 + \cdots + x_r$ with $x_k \in \mathcal{M}_k, k = 1, 2, \ldots, r$. It is easy to see that the corresponding generalization of Theorem 4.3(2) then requires the criterion $\mathcal{M}_i \cap \mathcal{M}_k = \{o\}$ for each pair $i, k = 1, 2, \ldots, r$. An example is given by the decomposition $\mathbf{R}^3 = \mathcal{M}_1 \oplus \mathcal{M}_2 \oplus \mathcal{M}_3$, where the $\mathcal{M}_k$ are the three axes, or, for that matter, any triple of not coplanar straight lines through the origin.

We now consider the converse of direct sum decomposition and study briefly an important method for *constructing a new vector space from given spaces*. Let $\mathcal{L}_1$ and $\mathcal{L}_2$ be two given, arbitrary linear spaces over the same field. Let us form the Cartesian product set $\mathcal{L}_1 \times \mathcal{L}_2$ (which, naturally consists of all pairs (u_1, u_2) with $u_1 \in \mathcal{L}_1$ and $u_2 \in \mathcal{L}_2$) and equip it with

---

†The meaningful generalization to infinite many summands requires topological methods and will be discussed only in Chapter 10.

the following composition laws:

$$(x_1, x_2) + (y_1, y_2) = (x_1 + y_1, x_2 + y_2),$$
$$\alpha(x_1, x_2) = (\alpha x_1, \alpha x_2) \quad \text{(for all} \quad \alpha \in K\text{)}.$$

It is evident that we now have a new linear space. *We call this $\mathscr{L}$ the direct sum of $\mathscr{L}_1$ and $\mathscr{L}_2$ and write $\mathscr{L} = \mathscr{L}_1 \oplus \mathscr{L}_2$.* This terminology is justified as follows. The subset $\hat{\mathscr{L}}_1$ of $\mathscr{L}$ consisting of pairs of the form (x_1, o_2) is a manifold of $\mathscr{L}$, and the subset $\hat{\mathscr{L}}_2$ of $\mathscr{L}$ consisting of pairs of the form (o_1, x_2) is another manifold. Any $x \in \mathscr{L}$ can be written as the sum of one element from each of these manifolds, and clearly, $\hat{\mathscr{L}}_1$ and $\hat{\mathscr{L}}_2$ have only the null vector (o_1, o_2) of $\mathscr{L}$ in common. Hence, $\mathscr{L} = \hat{\mathscr{L}}_1 \oplus \hat{\mathscr{L}}_2$. As a matter of fact $\hat{\mathscr{L}}_1$ and $\hat{\mathscr{L}}_2$ are isomorphic† to $\mathscr{L}_1$ and $\mathscr{L}_2$, respectively, as is easily seen by considering the projection map $\hat{\mathscr{L}}_1 \to \mathscr{L}_1$ given by $(x_1, o_2) \mapsto x_1$ and similarly for $\hat{\mathscr{L}}_2 \to \mathscr{L}_2$. Thus, it is perfectly in order to write $\mathscr{L} = \mathscr{L}_1 \oplus \mathscr{L}_2$.

The procedure of constructing direct sums of linear spaces carries over without difficulty to several terms. Then the underlying set is a Cartesian product $\mathscr{L}_1 \times \mathscr{L}_2 \times \cdots \times \mathscr{L}_r$.

In conclusion we call the reader's attention to the rather obvious fact that if $\mathscr{L} = \mathscr{M}_1 \oplus \mathscr{M}_2$, then the additive *group* of $\mathscr{L}$ has the direct product decomposition‡ into the additive groups associated with $\mathscr{M}_1$ and $\mathscr{M}_2$. Conversely, if $\mathscr{L} = \mathscr{L}_1 \oplus \mathscr{L}_2$, then the additive group structure of $\mathscr{L}$ arises as the direct product§ of the additive groups associated with $\mathscr{L}_1$ and $\mathscr{L}_2$.

PROBLEMS

4.3-1. Let K be a commutative field. Show that, using its own composition laws, K can be considered as a linear space over K itself.

4.3-2. Let K be a commutative field. Consider the Cartesian product set $K^n = K \times K \times \cdots \times K$ with elements $(k_1, k_2, \ldots, k_n)$ where $k_i \in K$. Define

$$(k_1, \ldots, k_m) + (k_1', \ldots, k_n') = (k_1 + k_1', \ldots, k_n + k_n'),$$
$$k(k_1, \ldots, k_n) = (kk_1, \ldots, kk_n), \quad (k \in K).$$

Show that K^n is a linear space over the field of scalars K.

4.3-3. Show that the set P of all polynomials (with real coefficients) defined, say, on the closed interval $0 \leqslant x \leqslant 1$ form a real linear space, if the addition of two polynomials, as well as the multiplication of a polynomial by a real number, are defined in the usual elementary manner. (*Remark*: The polynomial which is

---

†Isomorphisms of linear spaces will be discussed later on, but the reader no doubt understands our anticipated argument.

‡Concerning the direct product decomposition of groups, we refer back to Problem 4.1c-11.

§Compare with Problem 4.1c-10.

identically zero and all nonzero constant polynomials are assumed to belong to P. The former has no degree at all; the latter have degree zero.) Now let P_n be the subset of P which consists of the identically zero polynomial and all polynomials which have degree less than n. Show that P_n is a manifold of P.

4.3-4. Show that in any linear space the following two cancellation laws hold:
(a) If $x \neq o$, then $\alpha x = \beta x$ implies $\alpha = \beta$,
(b) If $x \neq o$, $y \neq o$, and $\gamma \neq 0$, then $\gamma x = \gamma y$ implies $x = y$.

4.3-5. For each of the following subsets of $\mathbf{R}^n$, determine whether it is a linear manifold of the vector space $\mathbf{R}^n$:
(a) $A = \{(\xi_1, \ldots, \xi_n) | \xi_1 + \cdots + \xi_n \neq 0\}$,
(b) $B = \{(\xi_1, \ldots, \xi_n) | \xi_1 = \xi_2 = \cdots = \xi_n\}$,
(c) $C = \{(\xi_1, \ldots, \xi_n) | \xi_1 \xi_2 = 0\}$,
(d) $D = \{x \in \mathbf{R}^n | Mx = o, M \text{ is a fixed } n \text{ by } n \text{ matrix}\}$.

4.3-6. Show that the subset of $\mathbf{R}^\infty$ (or $\mathbf{C}^\infty$) consisting of all sequences where all but a finite number of terms are zero, is a linear manifold.

4.3-7. Each of the following conditions determines a subset of the real linear space $\mathscr{F}[-1, 1]$ which is defined as the set of all continuous real functions f on the interval $-1 \leq x \leq 1$:
(a) f is differentiable,
(b) f is a polynomial of degree 3,
(c) f is an even function,
(d) f is an odd function,
(e) $f(0) = 0$,
(f) $f(0) = 1$,
(g) $f(x) \geq 0$ for all x.
Which of these subsets are manifolds of $\mathscr{F}[-1, 1]$?

4.3-8. Let $\{\ldots \mathscr{M}_\alpha \ldots\}$ be an arbitrary indexed collection of linear manifolds of $\mathscr{L}$. Show that $\cap_\alpha \mathscr{M}_\alpha$ is a linear manifold of $\mathscr{L}$.

4.3-9. Let A be the subset of $\mathbf{R}^3$ consisting of all triplets of the form $(a, b, a + b)$ and let B be the subset of all triplets of the form (c, c, c), where a, b, c are real numbers. Show that $\mathbf{R}^3 = A \oplus B$.

4.3-10. Let $\mathscr{L}$ be the linear space of n by n matrices. Let $\mathscr{D}$ be the linear manifold consisting of all multiples of the unit matrix. Find another manifold $\mathscr{N}$ such that $\mathscr{L} = \mathscr{D} \oplus \mathscr{N}$. Is $\mathscr{N}$ unique?

4.3a. Linear Independence, Bases, and Dimension

In this subsection we discuss those special features of linear spaces which, on one hand, render their study relatively simple and, on the other, are closely tied up with the appearance and importance of vector spaces in many branches of physics.

A *finite collection* $S = \{x_1, x_2, \ldots, x_k\}$ of vectors in a linear space $\mathscr{L}$ is said to be *linearly independent* iff the equation

$$\alpha_1 x_1 + \alpha_2 x_2 + \cdots + \alpha_k x_k = o$$

implies $\alpha_1 = \alpha_2 = \cdots = \alpha_k = 0$.

An *infinite collection*† $S = \{\ldots x_\alpha \ldots\}$ of vectors in $\mathscr{L}$ is called linearly independent if *every* finite subcollection of S is linearly independent in the above sense.

Remarks:‡ (1) A linearly independent set S cannot contain the null vector o.

(2) The primary importance of linear independence resides in the fact that it makes *linear combinations* of vectors unique. Indeed, if $x = \Sigma_k \alpha_k x_k$ with all x_k linearly independent, then $x = \Sigma_k \beta_k x_k$ implies $\beta_k = \alpha_k$ for all k. Conversely, if $x = \Sigma_k \alpha_k x_k$ and $x = \Sigma_k \beta_k x_k$, but β_k is not equal to α_k for all k, then the set $\{\ldots x_k \ldots\}$ is not linearly independent.

(3) If a vector y is a linear combination $y = \Sigma_k \alpha_k x_k$ of vectors from some collection, then the collection containing the original vectors x_k *and* y cannot be linearly independent. Conversely, if S is a linearly independent collection, then a vector in S cannot be expressed as a linear combination of the other members of S.

We give now some examples.

Example α. Let $\mathscr{L} = \mathbf{R}^3$ and consider the set $T = \{(1, 1, 0), (1, 0, 1), (0, 1, 0), (1, 1, 1)\}$. To check linear independence, we must ascertain whether the equation

$$\alpha_1(1, 1, 0) + \alpha_2(1, 0, 1) + \alpha_3(0, 1, 0) + \alpha_4(1, 1, 1) = (0, 0, 0)$$

has or has not a solution for α_1, α_2, α_3, α_4 (not all zero). The conditions of a solution are

$$\alpha_1 + \alpha_2 + \alpha_4 = 0, \qquad \alpha_1 + \alpha_3 + \alpha_4 = 0, \qquad \alpha_2 + \alpha_4 = 0.$$

Clearly, one has infinitely many nontrivial solutions (e.g., $\alpha_1 = 0$, $\alpha_2 = -1$, $\alpha_3 = -1$, $\alpha_4 = 1$), so that T is not a linearly independent set. We observe that $(1, 1, 1)$ can be written as a linear combination of the others:

$$(1, 1, 1) = (1, 0, 1) + (0, 1, 0).$$

Hence, removing $(1, 1, 1)$ from T, we get a linearly independent set.

---

†The collection S need not be countable.

‡We leave the proof of the following simple and familiar results to the reader. The same applies to some of the rest of the material in this subsection.

Example β.   In $\mathbf{R}^n$, the set S consisting of the vectors

$$e_1 = (1, 0, \ldots, 0)$$
$$e_2 = (0, 1, \ldots, 0)$$
$$\cdot$$
$$\cdot$$
$$\cdot$$
$$e_n = (0, 0, \ldots, 1)$$

is linearly independent. Indeed, the equation

$$\sum_{k=1}^{n} \alpha_k e_k = \mathrm{o}$$

means $(\alpha_1, \alpha_2, \ldots, \alpha_n) = (0, 0, \ldots, 0)$ which implies $\alpha_1 = \alpha_2 = \cdots = \alpha_n = 0$.

Example γ.   In $\mathbf{R}^\infty$, the countably infinite set S consisting of the vectors

$$e_1 = (1, 0, \ldots, 0, \ldots)$$
$$e_2 = (0, 1, \ldots, 0, \ldots)$$
$$\cdot$$
$$\cdot$$
$$\cdot$$
$$e_k = (0, 0, \ldots, 1, \ldots) \qquad \text{(1 stands in the } k\text{th position)}$$
$$\cdot$$
$$\cdot$$
$$\cdot$$

is linearly independent since, in view of Example β, every finite subcollection is such.

Our first use of the concept of linear independence is the rigorous formalization of the intuitive concept of dimension:

DEFINITION 4.3(1).   *A linear space $\mathscr{L}$ is said to be n-dimensional†if it contains a collection of n linearly independent vectors but does not contain any collection of n + 1 linearly independent vectors. A space $\mathscr{L}$ is called infinite dimensional if it is not n-dimensional for any finite n, i.e. if, for any positive integer n, we can find in $\mathscr{L}$ a collection of n linearly independent vectors.*

Since the zero space $\mathscr{L} = \{\mathrm{o}\}$ has no linearly independent sets at all, we must make a special convention. We call this space *zero dimensional*.

For infinite dimensional spaces, it is customary to define the "algebraic

---

†We also say that $\mathscr{L}$ has n dimensions or that the dimension of $\mathscr{L}$ is n. We write dim $\mathscr{L} = n$.

dimension" as the cardinal number of a maximal linearly independent collection of vectors. This is then a natural generalization of the concept of the dimensional number n.

The above Example β shows that $\mathbf{R}^n$ is n-dimensional.† Example γ clearly indicates that $\mathbf{R}^\infty$ is not finite dimensional. However, it would be a mistake to conclude that $\mathbf{R}^\infty$ has the algebraic dimension $\aleph_0$. The reader will easily convince himself that one can find linearly independent collections of vectors in $\mathbf{R}^\infty$ which have at least the cardinal number of the continuum.‡ Further, less trivial, examples will follow later.

We now proceed to use linear combinations for the study of spaces and manifolds. Suppose $\mathcal{L}$ is an arbitrary linear space and let $V = \{x_1, \ldots, x_k\}$ be a *finite* arbitrary collection of vectors. Consider the set of *all* vectors that have the form

$$x = \sum_{l=1}^{k} \alpha_l x_l.$$

It is almost trivial to convince oneself that *this set* $\{\ldots x \ldots\}$ *is a linear manifold* of $\mathcal{L}$. More important, however, is the converse situation. Suppose $\mathcal{M}$ is a given linear manifold of $\mathcal{L}$. Suppose that *every* vector $m \in \mathcal{M}$ can be written as a linear combination

$$m = \sum_{l=1}^{k} \alpha_l x_l$$

of vectors x_l taken from a fixed *finite* collection $V = \{x_1, \ldots, x_k\}$ of vectors from $\mathcal{L}$. We then say that $\mathcal{M}$ *is spanned*§ *by the set V*. In particular, we may have a collection V such that any vector of $\mathcal{L}$ itself can be written as a linear combination of vectors from V. We then say that V spans the whole space $\mathcal{L}$.

It is quite easy to see that the linear manifold spanned by a set V is the smallest manifold that contains V. For this reason, $\mathcal{M}$ spanned by V is often called the *linear hull* of the set V. Reformulating these observations in more precise language, we find that the linear hull of a collection V of vectors is the intersection of all manifolds that contain the set V.

---

†Any vector $(\xi_1, \ldots, \xi_n)$ can be written as a linear combination of the collection $\{e_1, \ldots, e_n\}$, hence, because of Remark 3 above, we cannot have $n + 1$ linearly independent vectors.

‡Take the set of all vectors of the form $(\alpha, \alpha^2, \alpha^3, \ldots, \alpha^k, \ldots)$, with α an arbitrary real number.

§Sometimes one says "$\mathcal{M}$ is *generated* by the set V."

Before proceeding, we must comment on the restriction we made on V, namely that it consists of *finitely* many vectors. The reason for this is that, in general, infinite linear combinations cannot be defined in purely algebraic terms. The concept "sum" of infinitely many terms of whatever kind always involves some notion of convergence. Convergence, on the other hand, is a typical concept of topology. Thus, in general, "infinite linear combinations" of vectors can be meaningfully defined only when the underlying set possesses not only an algebraic, but also some topological structure. We shall consider this question in detail† only in Chapter 9. However, we proceed now to formulate our subsequent theorems in a language which applies for the general case, i.e. irrespective of whether the "spanning" requires either finitely many or infinitely many vectors. We also note here that, in a special case, infinite linear combinations may be defined in purely algebraic terms. Suppose that $\{\ldots x_\nu \ldots\}$ is some infinite (not even necessarily countable) collection of vectors. The linear combination $\Sigma_\nu \alpha_\nu x_\nu$ still makes sense, *provided all but finitely many coefficients α_ν are zero*. The reader may use this case to illustrate statements about infinite linear combinations. For example, consider the infinite collection $V = \{e_1, \ldots, e_k, \ldots\}$ of vectors from $\mathbf{R}^\infty$, defined in Example γ. The set of vectors

$$\mathcal{M} = \left\{ \sum_{k=1}^{n} \alpha_k e_k \,\middle|\, \alpha_k \in \mathbf{R}, \quad n \text{ some positive integer} \right\}$$

is easily seen to be a (proper) linear manifold of $\mathbf{R}^\infty$. (It consists of all vectors which have only finitely many (but arbitrarily many) nonvanishing "components.") Thus, V spans $\mathcal{M}$ (but $\mathcal{M} \neq \mathbf{R}^\infty$).

We now return to our general study of spanning. Suppose $\mathcal{M}$ is spanned by V. Even though, by definition, this means that every $x \in \mathcal{M}$ can be represented as a linear combination of vectors from V, surely this representation is not unique. For example, let $\mathcal{L}$ be $\mathbf{R}^3$ and let $V = \{(1, 0, 0), (1, 1, 0), (2, 1, 0)\}$. The manifold of the coordinate plane $\mathcal{M}_{12}$ is spanned by V. But a given vector $x = (\xi_1, \xi_2, 0) \in \mathcal{M}_{12}$ can be written as $x = \alpha_1(1, 0, 0) + \alpha_2(1, 1, 0) + \alpha_3(2, 1, 0)$ whenever $\alpha_1 + \alpha_2 + 2\alpha_3 = \xi_1$ and $\alpha_2 + \alpha_3 = \xi_2$, i.e. with infinitely many choices of coefficients $\alpha_1, \alpha_2, \alpha_3$.

According to Remark 2 above, the representation of members $x \in \mathcal{M}$ will be *unique* if the spanning set V is a *linearly independent set of vectors*. This important observation gives rise to the following

---

†We shall be able to define linear combinations of not only countable, but even uncountably infinite collections of vectors.

DEFINITION 4.3a(2). *A linearly independent collection B of vectors which spans a manifold $\mathcal{M}$ is called a basis† for $\mathcal{M}$. If, in particular, B spans the whole space $\mathcal{L}$, we call it a basis of the linear space $\mathcal{L}$.*

In view of our previous discussions, this definition can be paraphrased as follows. *A collection‡ $B = \{\ldots e_k \ldots\}$ of vectors is a basis for $\mathcal{M}$ iff any vector $x \in \mathcal{M}$ can be written as a unique linear combination $x = \Sigma_k \alpha_k e_k$. Any basis is a linearly independent set.* We note here that, in the above, the manifold $\mathcal{M}$ may be the whole space $\mathcal{L}$.

We introduce one more technical term. If $B = \{\ldots e_k \ldots\}$ is a basis and $x = \Sigma_k \alpha_k e_k$, then we call the set of scalars $\{\ldots \alpha_k \ldots\}$ *the components of the vector x* relative to the basis B. For a given basis, the components are, of course, unique. But, clearly, a space admits, in general, many different bases, and a given vector x will have different components in the different bases.

We illuminate our new concepts on some examples.

Example δ. The linearly independent set $\{\ldots e_k \ldots\}$ defined in Example β is obviously a basis for $\mathbf{R}^n$. In this basis, the vector $x = (\xi_1, \ldots, \xi_n)$ has precisely the components $\xi_1, \ldots, \xi_n$. For this reason, the basis in question is usually called the *standard basis* of $\mathbf{R}^n$. In condensed notation, we often designate the standard basis by writing

$$e_k = (\delta_{1k}, \delta_{2k}, \ldots, \delta_{nk}), \qquad k = 1, 2, \ldots, n,$$

where δ_{lk}, the *Kronecker symbol*, is $+1$ if $l = k$, and zero if $l \neq k$. Of course, $\mathbf{R}^n$ has infinitely many other bases. For example, $e'_k = (-\delta_{1k}, \delta_{2k}, \ldots, \delta_{nk})$, $k = 1, 2, \ldots, n$ is also a basis.

Example ε. The same set $\{\ldots e_k \ldots\}$ with $e_k = (\delta_{1k}, \delta_{2k}, \ldots, \delta_{nk})$ is also a basis for $\mathbf{C}^n$. Indeed, any complex n-tuple $z = (\zeta_1, \ldots, \zeta_n)$ can be written as $z = \Sigma_{k=1}^n \alpha_k e_k$, with suitably chosen *complex* coefficients α_k (namely taking $\alpha_k = \zeta_k$), and any linear combination $\Sigma_{k=1}^n \alpha_k e_k = 0$ implies that all $\alpha_k = 0$.

Example ζ. In the vector space of all n by n matrices (cf. Example ζ on p. 123) a possible basis consists of the set of matrices $E^{(ij)}$ $(i, j = 1, 2, \ldots, n)$ which have the entry 1 in the ith row and jth column and the entry 0 in all other spaces, i.e. the set $\{\ldots E^{(ij)} \ldots\}$ with $(E^{(ij)})_{kl} =$

†The term "base" is also used.

‡It will be our practice to denote members of a basis by the specific symbols e_k. Members of arbitrary spanning sets will be denoted by noncommittal symbols like x_k.

$\delta_{ik}\delta_{jl}$, where $k, l = 1, 2, \ldots, n$ for each $i, j = 1, 2, \ldots, n$. Clearly, any given matrix M (with matrix elements M_{kl}) can be written as

$$M = \sum_{i,j=1}^{n} M_{ij}E^{(ij)}.$$

The matrices $E^{(ij)}$ are linearly independent, since $\Sigma_{i,j=1}^{n} \alpha_{ij}E^{(ij)}$ is the zero matrix iff all coefficients α_{ij} vanish. Note that the basis $\{\ldots E^{(ij)} \ldots\}$ contains n^2 elements.

In the following we list a few theorems about bases. We omit the rather technical proofs.[†]

THEOREM 4.3a(1).   *If S is an arbitrary linearly independent set of vectors in $\mathcal{L}$, then there exists a basis B of $\mathcal{L}$ which contains S.*

In other words, any linearly independent set can be enlarged to a basis. In particular, we see that a basis of a manifold can be extended to a basis of $\mathcal{L}$.

Since, in any space $\mathcal{L}$, the set $\{x\}$ consisting of any nonzero vector, is obviously a linearly independent set, the previous theorem immediately leads us to the following

THEOREM 4.3a(2).   *Every[‡] linear space possesses a basis.*

This theorem is very pleasing because it reassures us that we can always rely on the convenient method of uniquely representing vectors in terms of suitably chosen basis vectors. In other words, any linear space can be *coordinatized*, where by "coordinates of a point" we simply mean the components of the vector x relative to some basis. On the other hand, we already know that a space admits, in general, many different bases. However, different bases for a given $\mathcal{L}$ "do not differ too much," as shown by the following

THEOREM 4.3a(3).   *Let $\mathcal{L}$ be a finite n-dimensional linear space. Then every basis of $\mathcal{L}$ has exactly n members.*

The importance of this theorem is reflected in the fact that it enables us to introduce an alternative definition of dimensionality: *The dimension of a finite dimensional linear space equals the number of elements in an (arbitrary) basis of the space.*

[†]See, for example, Jacobson[17], Simmons[36], or Burton[2].
[‡]This clearly does *not* apply to the trivial zero space $\mathcal{L} = \{o\}$, which we disregard in this respect.

We illustrate this on the following example. Let $\mathscr{L} = \check{\mathbf{C}}^n$ be the vector space of complex n-tuples over the field of *real* numbers (cf. Example γ on p. 122). Contrary to what we saw above for $\mathbf{C}^n$, the set $e_1 = (1, 0, \ldots, 0), \ldots, e_n = (0, \ldots, 1)$ is *not* a basis for $\check{\mathbf{C}}^n$, since, for example, the vector $(i, 0, \ldots, 0)$ cannot be expressed, with *real* coefficients, as a linear combination of the above vectors. However, if we take the larger set

$$e_1 = (1, 0, \ldots, 0)$$
$$e_2 = (0, 1, \ldots, 0)$$
$$\cdot$$
$$\cdot$$
$$\cdot$$
$$e_n = (0, 0, \ldots, 1)$$
$$e_{n+1} = (i, 0, \ldots, 0)$$
$$e_{n+2} = (0, i, \ldots, 0)$$
$$\cdot$$
$$\cdot$$
$$\cdot$$
$$e_{2n} = (0, 0, \ldots, i),$$

then we obviously have a basis. It then follows that $\check{\mathbf{C}}^n$ has dimension $2n$.

Another simple application of our dimension theorem is given by the space of n by n matrices. Since we observed that a basis $E^{(ij)}$ can be constructed which has n^2 elements, it follows that our space is n^2-dimensional.

Theorem 4.3a(3) can be generalized so as to include also the case of infinite dimensional spaces. We have

THEOREM 4.3a(4). *If $\mathscr{L}$ is an arbitrary linear space, any two bases of $\mathscr{L}$ have the same cardinal number.*

The reader is cautioned that this theorem is, in a sense, weaker than the preceding one and does *not* imply that, for infinite dimensional spaces the algebraic dimension of $\mathscr{L}$ equals the cardinal number of base.†

For finite dimensional spaces, Theorem 4.3a(3) has something like a converse:

THEOREM 4.3a(5). *If $\mathscr{L}$ has the (finite) dimension n, then every set of n linearly independent vectors is a basis, i.e. it spans $\mathscr{L}$.*

---

†The cardinal number of the bases can be less than that of the maximal linearly independent sets. These are intricate questions which we shall briefly illustrate in Chapter 10.

The important thing here to keep in mind is that, in contrast to finite dimensional spaces, in infinite dimensional spaces linear independence of a set, even if it has sufficiently high cardinal number, does not necessarily imply that the set spans the entire space.

In addition to the topics discussed in this subsection, there are other interesting and important features related to bases. These topics involve the notion of orthogonality and obtain their full importance in connection with topological linear spaces. For this reason, we relegate these questions to Chapter 10.

PROBLEMS

4.3a-1. Let $S = \{x_1, \ldots, x_k\}$ be a linearly independent set in an arbitrary space $\mathscr{L}$. Show that

$$S' = \{x_1 - \alpha_1 x_n, x_2 - \alpha_2 x_n, \ldots, x_{n-1} - \alpha_{n-1} x_n\},$$

with α_k arbitrary scalars, is also linearly independent.

4.3a-2. Let $B = \{l_1, l_2, l_3\}$ be a basis for $\mathbf{R}^3$. Show that $\{l_1 + l_2, l_1 + l_3, l_2 + l_3\}$ is also a basis.

4.3a-3. Show that the set of vectors

$$(3 - i, 2 + 2i, 4), \qquad (2, 2 + 4i, 3), \qquad (1 - i, -2i, 1)$$

is a basis for $\mathbf{C}^3$. Take now the standard basis of $\mathbf{C}^3$ and determine the components of each member of this set relative to the above given basis.

4.3a-4. Show that the collection $\{1, x, x^2, \ldots, x^n, \ldots\}$ is a basis of the linear space P of polynomials defined in Problem 4.3-3. What is the dimension of P?

4.3a-5. Let $\mathscr{L}$ be the vector space of all n by n matrices. Show that the set $\mathscr{M}$ of all diagonal matrices is a manifold of $\mathscr{L}$. Determine the dimension of $\mathscr{M}$.

4.3a-6. Let $\mathscr{M}$ be a proper linear manifold of a finite dimensional vector space $\mathscr{L}$. Show that $\dim \mathscr{M} < \dim \mathscr{L}$.

4.3a-7. Let $\mathscr{M}$ and $\mathscr{N}$ be manifolds of $\mathbf{R}^n$, such that $\dim \mathscr{M} > n/2$ and $\dim \mathscr{N} > n/2$. Show that $\mathscr{M} \cap \mathscr{N} \neq \{o\}$.

4.3a-8. Let $\mathscr{L}$ be a finite dimensional linear space, with some basis $\{e_1, \ldots, e_n\}$. Let $\mathscr{M}_k$ be the manifold generated by the vector $e_k (k = 1, 2, \ldots, n)$. Show that $\mathscr{L} = \mathscr{M}_1 \oplus \mathscr{M}_2 \oplus \cdots \oplus \mathscr{M}_n$.

4.3a-9. Let $\{e_1, \ldots, e_n, d_1, \ldots, d_m\}$ be a basis for an $n + m$ dimensional linear space $\mathscr{L}$. Suppose $\{e_1, \ldots, e_n\}$ and $\{d_1, \ldots, d_m\}$ are bases for the manifolds $\mathscr{M}$ and $\mathscr{N}$ of $\mathscr{L}$, respectively. Show that $\mathscr{L} = \mathscr{M} \oplus \mathscr{N}$.

4.3a-10. Let $\mathscr{L}$ be a finite dimensional linear space, and let $\mathscr{M}$ and $\mathscr{N}$ be manifolds. Show that

$$\dim \mathscr{M} + \dim \mathscr{N} = \dim (\mathscr{M} + \mathscr{N}) + \dim (\mathscr{M} \cap \mathscr{N}),$$

where, as usual,

$$\mathcal{M} + \mathcal{N} = \{m + n \,|\, m \in \mathcal{M}, n \in \mathcal{N}\}.$$

(*Remark*: From Problem 4.3-8 you already know that $\mathcal{M} \cap \mathcal{N}$ is a manifold. You will easily convince yourself that $\mathcal{M} + \mathcal{N}$ is also a manifold.) Use the above result to prove that

$$\dim(\mathcal{M} \oplus \mathcal{N}) = \dim \mathcal{M} + \dim \mathcal{N}.$$

4.3b. Morphisms (Linear Transformations); Quotient Spaces

Let $\mathcal{L}$ and $\mathcal{L}'$ be two linear spaces *over the same field of scalars*. In conformity with the general terminology of Section 3.1, the map $f: \mathcal{L} \to \mathcal{L}'$ is called a morphism if it preserves both composition laws, i.e. if

$$f(x + y) = f(x) + f(y) \quad \text{and} \quad f(\alpha x) = \alpha f(x).$$

It is clear that these two properties of f are equivalent to the following:

$$f(\alpha x + \beta y) = \alpha f(x) + \beta f(y) \quad \text{for all} \quad x, y \in \mathcal{L} \quad \text{and all} \quad \alpha, \beta \in K.$$

$$(4.18)$$

A map from a linear space to another linear space which has the property (4.18), is usually called a linear function or a *linear transformation*. Thus, *the terms "linear transformation" and "morphism between linear spaces" are synonymous*. We shall usually use the term "linear transformation." Since linear functions occur so frequently in many branches of applied mathematics, we have here one reason why the theory of linear spaces is so important in physics.

We recall that, because of the additive group structure of a linear space, we have, for any morphism (i.e. for any linear transformation)

$$f(o) = o' \quad \text{and} \quad f(-x) = -f(x) = -x'.$$

It should not come as a surprise that

$$\text{Im}(f) \equiv \{f(x) \,|\, x \in \mathcal{L}\} = f(\mathcal{L})$$

is a linear manifold of $\mathcal{L}'$ and

$$\text{Ker}(f) \equiv \{x \,|\, x \in \mathcal{L}, \quad f(x) = o'\} = f^{-1}(\{o'\})$$

is a linear manifold of $\mathcal{L}$. Since morphisms of linear spaces are linear transformations, it is customary to refer to $\text{Im}(f)$ simply as *the range of the linear transformation* and to call $\text{Ker}(f)$ *the null space of the linear transformation*.

Recalling our previous result for groups we easily see that *a morphism of linear spaces is an isomorphism iff* Ker $(f) = \{o\}$ *and* Im $(f) = \mathcal{L}'$ (i.e. iff the null space of the corresponding linear transformation is the zero vector space and the range is all of the codomain).

We give some simple examples.

Example α.   Let $\mathcal{L} = \mathbf{R}^n$ and $\mathcal{L}' = \mathbf{R}$. The projection $(\xi_1, \ldots, \xi_n) \mapsto \xi_1$ (as well as all other projections) is clearly a homomorphism, with the kernel being the "coordinate plane" consisting of points $(0, \xi_2, \ldots, \xi_n)$.

Example β.   The map $\mathbf{R}^n \to \mathbf{R}^n$ given by $(\xi_1, \ldots, \xi_n) \mapsto (\xi_1, 0, \ldots, 0)$ is an endomorphism, with the range being the "coordinate axis" $\mathcal{M}_1$ and the kernel being again the coordinate plane as in the preceding example. Note that this linear transformation is neither injective nor surjective.

Example γ.   Let $\mathcal{L} = \mathbf{R}^3$. Any rotation about an arbitrary axis through the origin is easily seen to be a linear transformation,† and hence, an endomorphism of $\mathbf{R}^3$. Actually, the map is injective and surjective, so that we have an automorphism of the linear space $\mathbf{R}^3$. A reflexion through the origin given by $(\xi_1, \xi_2, \xi_3) \mapsto (-\xi_1, -\xi_2, -\xi_3)$ is also an automorphism. So is a translation in any direction, given by $(\xi_1, \xi_2, \xi_3) \mapsto (\xi_1 + a_1, \xi_2 + a_2, \xi_3 + a_3)$, where a_1, a_2, a_3 are arbitrary real numbers. All these observations generalize easily to $\mathbf{R}^n$.

Example δ.   Let $\mathcal{L} = \mathbf{R}^\infty$ (or $\mathbf{C}^\infty$). The *shift-map* given by $(\xi_1, \xi_2, \ldots, \xi_k, \ldots) \mapsto (\xi_2, \xi_3, \ldots, \xi_k, \ldots)$ is an endomorphism. It is surjective but not injective. The kernel is the one-dimensional manifold $\{\ldots (\xi_1, 0, \ldots, 0, \ldots) \ldots\}$.

There is an important relation between the dimensions of domain, range, and null space of a linear transformation, given by *Sylvester's law*:‡

THEOREM 4.3b(1).   *If* $f: \mathcal{L} \to \mathcal{L}'$ *is a morphism, then* $\dim \mathcal{L} = \dim \text{Ker} (f) + \dim \text{Im} (f)$.

Sketch of proof.   Let $\{e_1, \ldots, e_k\}$ be a basis for the manifold§ Ker (f). Extend it to a basis $\{e_1, \ldots, e_k, d_1, \ldots, d_m\}$ of $\mathcal{L}$ which is, say, n-

---

†The reader should try to establish this first by geometrical intuition and then by giving the analytic form (say, by using a rotation matrix) of a rotation.

‡As stated, the theorem applies for finite dimensional spaces. For the infinite dimensional case, the word "dimension" must be replaced by "cardinal number of a base."

§We assume, for simplicity, that Ker $(f) \neq \{o\}$.

dimensional (i.e. $k + m = n$). If $x = \Sigma_{l=1}^{k} \alpha_l e_l + \Sigma_{l=1}^{m} \beta_l d_l \in \mathscr{L}$, then $f(x) = \Sigma_{l=1}^{m} \beta_l f(d_l)$, because $f(e_l) = o'$. Hence, Im (f) is spanned by $V = \{f(d_1), \cdots, f(d_m)\}$. Furthermore, V is a linearly independent set of Im (f). Indeed, suppose $\Sigma_{l=1}^{m} \gamma_l f(d_l) = o'$; then $f(\Sigma_{l=1}^{m} \gamma_l d_l) = o'$, so that $y \equiv \Sigma_l \gamma_l d_l \in$ Ker (f). Since $\{\ldots e_l \ldots\}$ is a basis for Ker (f), we then can write $\Sigma_{l=1}^{m} \gamma_l d_l = \Sigma_{l=1}^{n} \rho_l e_l$. Since the e_l and d_l together are a basis of $\mathscr{L}$ and hence, are linearly independent, all γ_l (and ρ_l) in this equation must vanish. Thus, $\Sigma_{l=1}^{m} \gamma_l f(d_l) = o'$ implies that all γ_l must be zero, so that the set V is linearly independent as claimed, and is thus a basis for Im (f). Consequently, dim Im $(f) = m$. Therefore, dim Im $(f) +$ dim Ker $(f) = m + k = n =$ dim $\mathscr{L}$, q.e.d.

It is customary to call the dimension of the range of a linear transformation its *rank*, and to call the dimension of the null space its *nullity*. Thus, Sylvester's law is often expressed by saying that *the sum of the rank and nullity of a linear transformation equals the dimension of its domain.* The above examples of morphisms give elementary illustrations of Sylvester's law.

An extremely important consequence of Sylvester's law is the following

THEOREM 4.3b(2).   *If $\mathscr{L}$ and $\mathscr{L}'$ are linear spaces of the same finite dimension n, then any monomorphism or any epimorphism is automatically an isomorphism. In other words, for linear transformations between finite dimensional spaces, injectiveness implies surjectiveness and surjectiveness implies injectiveness.*

Proof.   (a) Let dim $\mathscr{L} =$ dim $\mathscr{L}' = n$. Suppose f is an epimorphism (surjective). Then dim Im $(f) =$ dim $\mathscr{L}' = n$. Therefore, by Sylvester's law, dim Ker $(f) = 0$, hence Ker $(f) = \{o\}$. Consequently, we have both criteria for an isomorphism satisfied.

(b) Suppose f is a monomorphism (injective). Then we must have Ker $(f) = \{o\}$, hence dim Ker $(f) = 0$. By Sylvester's law, then, dim $\mathscr{L} =$ dim Im (f), i.e. dim Im $(f) = n$. But, if Im (f) were a *proper* manifold of $\mathscr{L}'$, its dimension ought to be less than n (cf. Problem 4.3a-6). Thus, we have Im $(f) = \mathscr{L}'$, so the map is surjective.

This theorem is very useful inasmuch as it simplifies the proof of isomorphisms: it is sufficient to demonstrate only injectiveness or only surjectiveness. A simple illustration is given by the study of the geometrical automorphisms of Example β above. On the other hand, the reader should be aware that Theorem 4.3b(2) fails miserably for infinite dimensional spaces! Example δ is a good illustration of this statement.

Our next aim is to establish a remarkable *universality theorem* which tells us that for each finite dimension n, there exists essentially *only one* linear space with a given field of scalars:

THEOREM 4.3b(3). *If $\mathscr{L}$ and $\mathscr{L}'$ are linear spaces over the same field of scalars and have the same finite dimension, then they are isomorphic.*

Proof. Let $B = \{e_1, \ldots, e_n\}$ be a basis of $\mathscr{L}$ and let $D = \{d_1, \ldots, d_n\}$ be a basis of $\mathscr{L}'$. Define the map $f: B \to D$ by setting

$$f(e_k) = d_k \quad \text{for} \quad k = 1, 2, \ldots, n.$$

Extend this map by linearity to all of $\mathscr{L}$, i.e. define, for every $x \in \mathscr{L}$,

$$f(x) \equiv f\left(\sum_{k=1}^{n} \alpha_k e_k\right) = \sum_{k=1}^{n} \alpha_k f(e_k) = \sum_{k=1}^{n} \alpha_k d_k.$$

By construction, this map $f: \mathscr{L} \to \mathscr{L}'$ is linear. Because of the uniqueness of expansions in a given basis, f is clearly injective. Theorem 4.3b(2) then guarantees that it is also surjective. Hence, f is an isomorphism, q.e.d.

The somewhat less important converse statement also holds:

THEOREM 4.3b(4). *If two finite dimensional linear spaces are isomorphic, then they have the same dimension.*

Proof. Isomorphism demands $\text{Ker}(f) = \{o\}$, (hence $\dim \text{Ker}(f) = 0$) and $\text{Im}(f) = \mathscr{L}'$ (hence $\dim \text{Im}(f) = \dim \mathscr{L}'$). Sylvester's law then yields $\dim \mathscr{L} = \dim \mathscr{L}'$.

The reader is again warned that the last two isomorphism theorems are *not valid*, in general, for infinite dimensional spaces. Some special cases will be discussed in detail in Chapters 10 and 11.

Returning to Theorem 4.3b(3), we observe that, since $\mathbf{R}^n$ is a linear space with dimension n, *every "abstract" n-dimensional real linear space is isomorphic to $\mathbf{R}^n$*. Similarly, *every n dimensional complex linear space is isomorphic to $\mathbf{C}^n$*. Thus, $\mathbf{R}^n$ (or $\mathbf{C}^n$) is the *universal model* or *prototype* of any real (or complex) n-dimensional linear space. The correspondence of vectors $x \in \mathscr{L}$ with the n-tuples of $\mathbf{R}^n (\mathbf{C}^n)$ is exhibited by choosing some basis $\{\ldots e_k \ldots\}$ in $\mathscr{L}$, writing $x = \sum_{k=1}^{n} \alpha_k e_k$, and making the assignment

$$x \mapsto (\alpha_1, \alpha_2, \ldots, \alpha_n). \tag{4.19}$$

This follows immediately from the method of the proof of Theorem 4.3b(3), but the reader is urged to check it out directly.

Even though we can now visualize or represent *any* n-dimensional linear space by $\mathbf{R}^n (\mathbf{C}^n)$, this correspondence is less useful as it appears. Indeed, the correspondence is *not unique*, inasmuch as *the assignment* (4.19) *depends on the basis that we chose in* $\mathcal{L}$. Most properties of linear spaces are formulated and clarified in the best way by avoiding any specific basis. Hence, it is general policy to treat linear spaces in a basis-free manner, whenever possible. This approach, eminently suitable for the discovery of structural features (and for generalizations to infinite dimensional spaces) reduces considerably the usefulness of the isomorphism (4.19).

So far we studied individual morphisms of linear spaces. In the following, we focus our attention to the *totality* of all morphisms between given spaces.

Let $\mathcal{L}_1$ and $\mathcal{L}_2$ be two arbitrary linear spaces over the same field, each of which may have any, not necessarily finite, dimension. Consider the class $\mathscr{C}(\mathcal{L}_1, \mathcal{L}_2)$ of all possible morphisms from $\mathcal{L}_1$ to $\mathcal{L}_2$. Denote the elements of this set, i.e. an arbitrary linear transformation from $\mathcal{L}_1$ to $\mathcal{L}_2$, by the generic symbol t. Define in this set $\{\ldots t \ldots\}$ the usual pointwise sum and scalar product of functions,†

$$\begin{aligned}(t_1 + t_2)(x) &= t_1(x) + t_2(x), \\ (\alpha t)(x) &= \alpha t(x). \end{aligned} \tag{4.20}$$

The class $\{\ldots t \ldots\}$ is closed under these two composition laws. Indeed, if t_1 and t_2 are linear transformations, then

$$\begin{aligned}(t_1 + t_2)(\alpha x + \beta y) &= t_1(\alpha x + \beta y) + t_2(\alpha x + \beta y) \\ &= \alpha t_1(x) + \beta t_1(y) + \alpha t_2(x) + \beta t_2(y) \\ &= \alpha(t_1 + t_2)(x) + \beta(t_1 + t_2)(y),\end{aligned}$$

so that $t_1 + t_2$ is also a linear transformation. Similarly,

$$\begin{aligned}(\alpha t)(\gamma x + \delta y) &= \alpha t(\gamma x + \delta y) = \alpha \gamma t(x) + \alpha \delta t(y) \\ &= \gamma(\alpha t)(x) + \delta(\alpha t)(y),\end{aligned}$$

so that, together with t, αt is also a linear transformation. Thus, $\mathscr{C}(\mathcal{L}_1, \mathcal{L}_2)$ is an algebraic system, and it is trivially seen that the composition laws

†Observe that $x \in \mathcal{L}_1$, whereas $t(x) \in \mathcal{L}_2$.

(4.20) satisfy the axioms of a linear space. Thus, we established the following *structure theorem* of eminent significance:

THEOREM 4.3b(5). *The class* $\mathscr{C}(\mathscr{L}_1, \mathscr{L}_2)$ *of all linear transformations from a linear space* $\mathscr{L}_1$ *to a linear space* $\mathscr{L}_2$ *is itself a linear space, with respect to pointwise sum and scalar product.*†

We note that the null vector of $\mathscr{C}(\mathscr{L}_1, \mathscr{L}_2)$ is the linear transformation t_0 which sends every $x \in \mathscr{L}_1$ to the null vector of $\mathscr{L}_2$. Similarly, the negative of t is the map $(-1)t$.

The fact that morphisms of linear spaces lead to a new linear space is one of the reasons why linear spaces play a central role in the study of structures.

If both $\mathscr{L}_1$ and $\mathscr{L}_2$ are finite dimensional, we have, in addition to the above, the following:

THEOREM 4.3b(6). *If* $\mathscr{L}_1$ *and* $\mathscr{L}_2$ *are finite dimensional, then* $\dim \mathscr{C}(\mathscr{L}_1, \mathscr{L}_2) = \dim \mathscr{L}_1 \cdot \dim \mathscr{L}_2.$

We omit the lengthy and rather uninspiring proof,‡ as well as the construction of examples. Instead, we briefly discuss two particularly important special cases.

First, suppose $\mathscr{L}_1 = \mathscr{L}_2 \equiv \mathscr{L}$. Then the linear space $\mathscr{C}(\mathscr{L}, \mathscr{L})$ is written as $\mathscr{C}(\mathscr{L})$ and it consists of all endomorphisms of $\mathscr{L}$. The space $\mathscr{C}(\mathscr{L})$ is usually called the *linear space of all linear operators*§ *on* $\mathscr{L}$. In passing we note that $\mathscr{C}(\mathscr{L})$ arises in a somewhat similar manner from $\mathscr{L}$ as Aut G (the group of all automorphisms of G) arises in group theory. However, the existence of the linear space $\mathscr{C}(\mathscr{L})$ is a stronger structural statement than the existence of Aut G, because $\mathscr{C}(\mathscr{L})$ contains *all* morphisms of $\mathscr{L}$, whereas Aut G contains only the *iso*morphisms of G. Furthermore, as we shall demonstrate in Section 4.4, the system $\mathscr{C}(\mathscr{L})$ has *more* algebraic structure to it than just that of being a linear space.

In order to illustrate the importance of $\mathscr{C}(\mathscr{L})$ in a rather familiar setting, we now consider the case when $\mathscr{L}$ happens to be *finite dimensional*. Let us take a basis $\{e_1, \ldots, e_n\}$ in $\mathscr{L}$. Let $t: \mathscr{L} \to \mathscr{L}$ be a linear operator, i.e. a

---

†It is, of course, implicit in the whole argument that $\mathscr{L}_1$ and $\mathscr{L}_2$ have the same field of scalars, and then $\mathscr{C}(\mathscr{L}_1, \mathscr{L}_2)$ has also the same field of scalars.

‡See, for example, Burton[2], p. 285.

§It is customary, though not a uniform practice, to call linear transformations which map a linear space $\mathscr{L}$ into itself, linear operators on $\mathscr{L}$.

member of $\mathscr{C}(\mathscr{L})$. If e_k is an arbitrary basis vector of $\mathscr{L}$, its image $t(e_k)$ will be some vector of $\mathscr{L}$, which we can expand in terms of the basis. Thus, we may write

$$t(e_k) = \sum_{l=1}^{n} A_{lk} e_l. \tag{4.21}$$

We can do this for all e_k $(k = 1, 2, \ldots, n)$, and in this way we associate with the given linear operator t the n^2 scalars A_{lk}. Now we define an n by n matrix A by placing into its lth row and kth column precisely the number A_{lk}. In other words, we have a correspondence $t \mapsto A$ between the linear operator t and the matrix A, constructed by the prescription (4.21). The matrix A completely represents the action of t on any $x \in \mathscr{L}$, since, expanding x in terms of the selected basis, we have

$$t(x) = t\left(\sum_{k=1}^{n} \alpha_k e_k\right) = \sum_{k=1}^{n} \alpha_k t(e_k)$$

$$= \sum_{k=1}^{n} \sum_{l=1}^{n} \alpha_k A_{lk} e_l.$$

We can associate with *any* $t \in \mathscr{C}(\mathscr{L})$ a matrix A in the above manner. Thus, we have a map f from the linear space $\mathscr{C}(\mathscr{L})$ to the linear space M^{n^2} of all n by n matrices. We now proceed to show that this map is actually an *isomorphism* between $\mathscr{C}(\mathscr{L})$ and M^{n^2}.

To start with, suppose that

$$t_1 \mapsto A^{(1)} \quad \text{and} \quad t_2 \mapsto A^{(2)}.$$

Then

$$(\alpha t_1 + \beta t_2)(e_k) = \alpha \sum_l A_{lk}^{(1)} e_l + \beta \sum_l A_{lk}^{(2)} e_l = \sum (\alpha A^{(1)} + \beta A^{(2)})_{lk} e_l,$$

so that, in view of Eq. (4.21),

$$(\alpha t_1 + \beta t_2) \mapsto \alpha A^{(1)} + \beta A^{(2)}.$$

This shows that our map f is a morphism. Further, by the construction (4.21) and the uniqueness of expansion of vectors, the map f defined by $t \mapsto A$ is injective. Theorem 4.3b(2) then assures us that it is also surjective,† hence, as claimed, f is an isomorphism.

---

†Actually, given a basis and the matrix A, we define the corresponding linear transformation t_A by setting $t_A(e_k) = \sum_l A_{lk} e_l$ for $k = 1, \ldots, n$ and then extend t_A linearly to all of $\mathscr{L}$. This illustrates the surjectiveness by exhibiting how to associate with any given matrix A a (unique) linear transformation.

We summarize our results by the following isomorphism theorem:

THEOREM 4.3b(7).   *If $\mathscr{L}$ is a finite n-dimensional linear space, then the linear space $\mathscr{C}(\mathscr{L})$ of all linear operators on $\mathscr{L}$ is isomorphic to the linear space of all n by n matrices.*†

In passing we note that, since M^{n^2} is n^2-dimensional, our result illustrates the dimension Theorem 4.3b(6). As a matter of fact, Theorem 4.3b(7) follows, without any calculation, directly from the facts that (a) $\mathscr{C}(\mathscr{L}) \equiv \mathscr{C}(\mathscr{L}, \mathscr{L})$ must be n^2-dimensional on account of Theorem 4.3b(6), (b) we know that M^{n^2} is n^2-dimensional, (c) Theorem 4.3b(3) assures us that these two equal-dimensional spaces are isomorphic. The major merit of our discussion preceding the formulation of Theorem 4.3b(7) is that it *explicitly* constructs the claimed isomorphism, so that we know how to associate a specific matrix to any given linear operator and vice versa.

The importance of Theorem 4.3b(7) is obvious to every physicist: it justifies the acquired habit of thinking about linear operators in terms of matrices. However, the student should be very conscious of the fact that the theorem holds only for finite dimensional spaces. As we shall see in detail in Chapter 12, for infinite dimensional spaces the representation of a linear operator in terms of "infinite matrices" is possible only in rare and rather special cases.

As a matter of fact, apart from aiding our visualization of operators, Theorem 4.3b(7) is not particularly useful even in the finite dimensional case, because, clearly, the correspondence $t \mapsto A$ is *not unique*, it depends on the arbitrary choice of a basis in $\mathscr{L}$.

After having discussed the special case of $\mathscr{C}(\mathscr{L}_1, \mathscr{L}_2)$ when $\mathscr{L}_1 = \mathscr{L}_2$ (i.e. when $\mathscr{C}$ is the class of linear operators on $\mathscr{L}$) we now briefly comment on another, equally important special case. Let $\mathscr{L}$ be a linear space and let K be its field of scalars (i.e. let K be either the field of real numbers or that of the complex numbers). Since K is a (one-dimensional) linear space over itself (cf. Example β on p. 122), we surely can take $\mathscr{L}_1 = \mathscr{L}$, $\mathscr{L}_2 = K$, and consider the linear space $\mathscr{C}(\mathscr{L}, K)$ of all linear transformations of $\mathscr{L}$ into its own field of scalars. A linear transformation $t: \mathscr{L} \to K$ is usually called a *linear functional* so that $\mathscr{C}(\mathscr{L}, K)$ is the *linear space of all linear functionals* on $\mathscr{L}$. This space is also frequently referred to as the *algebraic dual* of $\mathscr{L}$, and is often denoted by the special symbol $\mathscr{L}^*$.

---

†It should be obvious that if $\mathscr{L}$ is real (complex), then M^{n^2} is the real (complex) matrix space.

We will not discuss here the linear space of linear functionals, primarily because $\mathscr{L}^*$ attains special importance when $\mathscr{L}$ has also a topological structure. Thus, the discussion will be postponed until Chapter 11. However, we give a couple of examples for linear functionals.

Example ε.   Let $\mathscr{L}$ be the linear space M^{n^2} of all n by n matrices. If the matrix A has matrix elements A_{kl}, we define its *trace* by

$$\text{Tr } A = \sum_{k=1}^{n} A_{kk},$$

which is obviously a scalar. The map

$$F: M^{n^2} \to \mathbf{R} \quad \text{given by} \quad A \mapsto \text{Tr } A$$

is easily verified to be a linear function, hence it is, in our present terminology, a linear functional on M^{n^2}, i.e. a member of $M^{n^{2^*}}$.

Example ζ.   Let X be an arbitrary set, and let $\mathscr{L}$ be the linear space of *all* functions f from X to $\mathbf{R}$ (cf. Example ε on p. 123). Let x_0 be an arbitrary but fixed point of X and consider the map $F: \mathscr{L} \to \mathbf{R}$ given by $f \mapsto f(x_0)$ for all functions $f \in \mathscr{L}$. Clearly, under F, the function $\alpha f + \beta g \mapsto \alpha f(x_0) + \beta g(x_0)$, so F is a linear map from $\mathscr{L}$ to $\mathbf{R}$, and hence, a linear functional. Thus $F \in \mathscr{L}^*$.

We add some comments on the algebraic dual space $\mathscr{L}^*$ of *finite dimensional* linear spaces $\mathscr{L}$. First, from Theorem 4.3b(6) we obtain that

$$\dim \mathscr{L}^* = \dim \mathscr{L},$$

since $\dim K = 1$. Therefore, using Theorem 4.3b(3) the reader will immediately deduce the following

THEOREM 4.3b(8).   *If $\mathscr{L}$ is a finite n-dimensional real (complex) linear space, then the algebraic dual space $\mathscr{L}^*$ is isomorphic to the linear space $\mathbf{R}^n (\mathbf{C}^n)$.*

Recalling again that any n-dimensional space $\mathscr{L}$ is isomorphic to $\mathbf{R}^n (\mathbf{C}^n)$, we can reexpress this theorem by the interesting statement that *every finite dimensional linear space $\mathscr{L}$ is isomorphic to its own dual space $\mathscr{L}^*$.*

The reader is urged to explicitly derive Theorem 4.3b(8) by choosing a basis in $\mathscr{L}$ and exhibiting the isomorphism $F \mapsto x$ (with $F \in \mathscr{L}^*$, $x \in \mathscr{L}$) in a way similar to the explicit derivation of the isomorphism expressed in Theorem 4.3b(7).

By now, it may hardly be necessary to once again alert the reader that Theorem 4.3b(8) is, in general, not true for infinite dimensional spaces: far from being isomorphic to each other, $\mathscr{L}$ and $\mathscr{L}^*$ may differ quite considerably and it is not an easy task to identify the dual $\mathscr{L}^*$ of a given $\mathscr{L}$. Some of these problems will be taken up in Chapter 11.

We conclude this subsection on morphisms of linear spaces by surveying, as we did for other algebraic systems, the associated quotient structures.

Since a linear manifold $\mathscr{M}$ of a linear space $\mathscr{L}$ is an additive subgroup, we can define the coset $\mathscr{M} + x$ generated by an element $x \in \mathscr{L}$ as the set

$$\mathscr{M} + x = \{m + x \,|\, m \in \mathscr{M}\}.$$

The distinct cosets form a partition of $\mathscr{L}$, and $\cup_x (\mathscr{M} + x) = \mathscr{L}$. The class of all distinct cosets is the quotient set $\mathscr{L}/\mathscr{M}$. We now endow this set with an algebraic structure by defining the sum and scalar multiple of cosets:

$$(\mathscr{M} + x) \boxplus (\mathscr{M} + y) = \mathscr{M} + (x + y), \tag{4.22a}$$

$$\alpha \cdot (\mathscr{M} + x) = \mathscr{M} + \alpha x. \tag{4.22b}$$

It is very easy to verify that these definitions are in fact independent of the choice of coset representatives, because $\mathscr{M}$ is a *manifold*. Thus, we have indeed well-defined composition laws $\boxplus$ and $\cdot$ on $\mathscr{L}/\mathscr{M}$. A trivial check shows that these laws obey the axioms of a linear space. The set $\mathscr{L}/\mathscr{M}$ endowed with this linear space structure is called the *quotient space of $\mathscr{L}$ by $\mathscr{M}$*. Clearly, the null vector of the quotient space is the coset $\mathscr{M} + o = \mathscr{M}$, and the negative of $\mathscr{M} + x$ is the coset $\mathscr{M} - x$.

It is very easy to visualize finite dimensional coset spaces. Take, for example, $\mathscr{L} = \mathbf{R}^2$. A manifold $\mathscr{M}$ is a straight line through the origin. The construction of Fig. 4.8 clearly shows that the coset $\mathscr{M} + x$ is a straight line parallel to $\mathscr{M}$ and passing through the point x. The set of all such parallel lines is the quotient space $\mathscr{L}/\mathscr{M}$. The figure also illustrates the construction of coset sums and scalar products of cosets.

We already know that the kernel of any morphism is a linear manifold. As expected, the converse is also true. More precisely: let $\mathscr{M}$ be an arbitrary manifold of $\mathscr{L}$, then *the canonical map $c: \mathscr{L} \to \mathscr{L}/\mathscr{M}$, given by $x \mapsto \mathscr{M} + x$, is a homomorphism, whose kernel is precisely $\mathscr{M}$*. It then follows that, *if $f: \mathscr{L} \to \mathscr{L}'$ is an arbitrary morphism, then $\mathscr{L}/\mathrm{Ker}\,(f)$ is isomorphic to* $\mathrm{Im}\,(f)$. In particular, *if $f: \mathscr{L} \to \mathscr{L}'$ is a homomorphism, then $\mathscr{L}/\mathrm{Ker}\,(f)$ is isomorphic to $\mathscr{L}'$*, where the isomorphism is given by the map $\mathscr{M} + x \mapsto f(x)$.

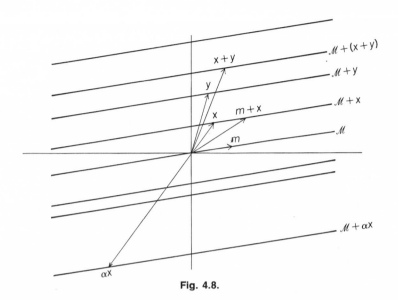

Fig. 4.8.

To illustrate the last isomorphism theorem, consider the projection $p : \mathbf{R}^2 \to \mathbf{R}$ given by $(\xi_1, \xi_2) \mapsto \xi_1$. We know already that this is a homomorphism. The kernel is the set $\{\ldots (0, \xi_2) \ldots\}$, i.e. the axis $\mathcal{M}_2$ (the vertical line passing through the origin). The quotient space $\mathbf{R}^2/\mathcal{M}_2$ is the set of all "vertical lines." The isomorphism $\mathbf{R}^2/\mathcal{M}_2 \approx \mathbf{R}$ is given by associating with each vertical line the point of the $\mathbf{R}$-axis through which it passes.

It may be expected that the quotient space is, in a sense, smaller than the original space. This is borne out by

THEOREM 4.3b(9). *If $\mathcal{L}$ is a finite dimensional linear space, then* dim $\mathcal{L}/\mathcal{M}$ = dim $\mathcal{L}$ − dim $\mathcal{M}$.

Proof. Consider the canonical homomorphism $c : \mathcal{L} \to \mathcal{L}/\mathcal{M}$. Here Im $(c) = \mathcal{L}/\mathcal{M}$ and Ker $(c) = \mathcal{M}$ (because the null coset $\mathcal{M} + o$ is the image of $\mathcal{M}$ under $x \mapsto \mathcal{M} + x$). Hence, Sylvester's law gives dim $\mathcal{L}$ = dim $\mathcal{L}/\mathcal{M}$ + dim $\mathcal{M}$, q.e.d.

At this point we conclude our introductory study of linear transformations and also, more generally, the survey of vector spaces. Much more will be said about these structures when also topological properties of the underlying sets are taken into account. Thus, Chapters 9, 10, 11, 12, and 13 will not only make use of the already clarified purely algebraic properties of linear spaces and their transformations, but will also add some further algebraic material.

PROBLEMS

4.3b-1. Determine which of the maps $\mathbf{R}^3 \to \mathbf{R}^3$ are morphisms, and in particular, which are isomorphisms:

(a) $(\xi_1, \xi_2, \xi_3) \mapsto (\xi_2, -\xi_1, \xi_3)$,

(b) $(\xi_1, \xi_2, \xi_3) \mapsto (\xi_1, 0, \xi_3)$,

(c) $(\xi_1, \xi_2, \xi_3) \mapsto (\xi_1^2, \xi_2, \xi_3)$,

(d) $(\xi_1, \xi_2, \xi_3) \mapsto (1, 1, 1)$,

(e) $(\xi_1, \xi_2, \xi_3) \mapsto (0, 0, 0)$.

(*Remark*: You may make some general statement that will enable you to decide, by simple inspection, what type of assignment between triplets is a morphism. It may be useful to recall the correspondence of linear transformations and matrices. This may give you also a criterion for isomorphism.)

4.3b-2. Let P the linear space of polynomials, cf. Problem 4.3-3. Denote by p_n an arbitrary polynomial of degree n, i.e. let $p_n = a_0 + a_1 x + \cdots + a_n x^n$. Determine which of the maps listed below is an endomorphism of P, i.e. a linear map of P into itself:

(a) $p_n \mapsto a_0 + a_1 x^2 + a_2 x^4 + \cdots + a_n x^{2n}$,

(b) $p_n \mapsto a_0 x + \dfrac{a_1}{2} x^2 + \cdots + \dfrac{a_n}{n+1} x^{n+1}$,

(c) $p_n \mapsto a_1 + 2a_2 x + \cdots + na_n x^{n-1}$.

(*Remark*: The second and third maps are called the integration map and the differentiation map, respectively.)

4.3b-3. Let $f \in \mathscr{F}[-1, 1]$ (cf. Problem 4.3-7) and consider the mapping $I(f) = \int_{-1}^{+1} f(x)\,dx$. Show that I is a morphism from $\mathscr{F}$ to the linear space $\mathbf{R}$ of all real numbers. Is it an isomorphism?

4.3b-4. Let $f: \mathbf{R}^4 \to \mathbf{R}^3$ be given by

$$(\xi_1, \xi_2, \xi_3, \xi_4) \mapsto (\xi_2 - \xi_3, \xi_3 - \xi_4, \xi_4 - \xi_1).$$

Show that this is a linear function. Illustrate Sylvester's law.

4.3b-5. Let f be an endomorphism of a linear space $\mathscr{L}$. Let A be the set of vectors that are left fixed by f, i.e.

$$A = \{x \in \mathscr{L} \mid f(x) = x\}.$$

Show that A is a linear manifold of $\mathscr{L}$.

4.3b-6. Show that the injective elements of $\mathscr{C}(\mathscr{L}_1, \mathscr{L}_2)$ are exactly those linear transformations t which map linearly independent sets of $\mathscr{L}_1$ into linearly independent sets of $\mathscr{L}_2$.

4.3b-7. Let $\mathscr{C}(\mathscr{L})$ be the linear space of all linear operators on a finite dimensional linear space. Determine which subset of the M^{n^2} matrix space is isomorphic to the class of all *auto*morphisms of $\mathscr{L}$.

4.3b-8. Show that any "change of basis"† in a linear space $\mathscr{L}$ is an automorphism of $\mathscr{L}$. Conclude from this that, for a finite dimensional space, any change of basis can be represented by writing $e'_k = Se_k$, $(k = 1, \ldots, n)$, where S is a nonsingular n by n matrix. (*Remark on notation*: If $x = (\alpha_1, \ldots, \alpha_n)$ is a vector of $\mathscr{L}$ and M is an n by n matrix, Mx means the vector whose lth component is given by $\Sigma_{j=1}^{n} M_{lj}\alpha_j$.)

4.3b-9. Let $\mathscr{L}$ be a finite n-dimensional linear space. Let F_i $(i = 1, 2, \ldots, n)$ be linear functionals with the property that $F_i(e_k) = \delta_{ik}$ for all $k = 1, 2, \ldots, n$. Show that the F_i are uniquely defined. Show that the F_i $(i = 1, 2, \ldots, n)$ form a basis for the dual space $\mathscr{L}^*$.

4.3b-10. Show that the subset $\mathscr{M} = \{(\xi, \eta, \xi + \eta) | \xi, \eta \in \mathbf{R}\}$ is a linear manifold of $\mathbf{R}^3$. Characterize the quotient space $\mathbf{R}^3/\mathscr{M}$. What is the dimension of $\mathbf{R}^3/\mathscr{M}$?

4.3b-11. Show *directly* (without recourse to the canonical homomorphism as was done in the text) that

$$\dim \mathscr{L}/\mathscr{M} = \dim \mathscr{L} - \dim \mathscr{M}.$$

(*Hint*: Construct a basis for $\mathscr{L}/\mathscr{M}$.)

4.3b-12. Suppose $\mathscr{L} = \mathscr{M} \oplus \mathscr{N}$ and consider the map $t: \mathscr{N} \to \mathscr{L}/\mathscr{M}$ defined by $t(n) = \mathscr{M} + n$ for all $n \in \mathscr{N}$. Prove that t is an isomorphism. (*Remark: This theorem is true also in the infinite dimensional case.* If one were content to consider only the finite dimensional case, the proof would be rather trivial from dimensional considerations. But, as it is, you should be able to prove that t is onto and one-to-one.)

4.3b-13. The essence of the preceding problem is that, if $\mathscr{L} = \mathscr{M} \oplus \mathscr{N}$, then $\mathscr{N}$ is isomorphic to the quotient space $\mathscr{L}/\mathscr{M}$. Using this result, prove the following useful theorem: If $\mathscr{L} = \mathscr{M} \oplus \mathscr{N}$ and $\mathscr{M}$ is given, then $\mathscr{N}$ is uniquely determined, up to an isomorphism. (*Hint*: Suppose that $\mathscr{L} = \mathscr{M} \oplus \mathscr{N}$ and that also $\mathscr{L} = \mathscr{M} \oplus \mathscr{K}$.)

4.4 LINEAR ALGEBRAS

So far we have studied algebraic systems with one and with two internal composition laws (groups and rings, respectively) and systems with one internal plus one external law (vector spaces). In this section we shall study more complicated systems that have two internal composition laws *and* one external composition law. As a matter of fact, we shall consider systems which *combine the features of a ring and of a linear space*. Such a system is called a *linear algebra* or just an *algebra*, for short.

---

†That is, any linear map which maps one basis onto another.

By definition, *a linear algebra is an algebraic system* ($\mathscr{A}$, K, $\cdot$), *where* $\mathscr{A}$ *is a ring, K is a commutative field, the external composition law* $\cdot$ *(scalar multiplication) is an action of K on* $\mathscr{A}$, *the internal composition law of addition in* $\mathscr{A}$ *is related to scalar multiplication by mixed distributive laws,*† *and the internal composition law of multiplication in* $\mathscr{A}$ *is related to scalar multiplication by the mixed associativity law* $\alpha(xy) = (\alpha x)y = x(\alpha y)$.

To make this condensed definition clear, we write down all defining axioms of an algebra‡ $\mathscr{A}$. In doing so, we take the liberty to use the symbol + to denote both the sum of ring elements $x, y, \ldots$ as well as the sum of elements $\alpha, \beta, \ldots$ of the field K. Similarly, we use simple juxtaposition of symbols to indicate a product xy in the ring, a product $\alpha\beta$ of two scalars, and a scalar product αx of a scalar and of an element of $\mathscr{A}$. These conventions should cause no confusion. Thus, the axioms are as follows:

(ia) $x + (y + z) = (x + y) + z$ for all $x, y, z \in \mathscr{A}$.

(ib) There exists an element $o \in \mathscr{A}$ such that $o + x = x + o = x$ for all $x \in \mathscr{A}$.

(ic) For every $x \in \mathscr{A}$ there exists an element $-x \in \mathscr{A}$ such that $x + (-x) = o$.

(id) $x + y = y + x$ for all $x, y \in \mathscr{A}$.

(iia) $x(yz) = (xy)z$ for all $x, y, z \in \mathscr{A}$.

(iib) $x(y + z) = xy + xz$ and $(x + y)z = xz + yz$ for all $x, y, z \in \mathscr{A}$.

(iiia) $\alpha(\beta x) = (\alpha\beta)x$ for all $\alpha, \beta \in K$ and all $x \in \mathscr{A}$.

(iiib) $1x = x$ for all $x \in \mathscr{A}$ (where 1 is the unit of K).

(iva) $\alpha(x + y) = \alpha x + \alpha y$ for all $\alpha \in K$ and all $x, y \in \mathscr{A}$.

(ivb) $(\alpha + \beta)x = \alpha x + \beta x$ for all $\alpha, \beta \in K$ and all $x \in \mathscr{A}$.

(ivc) $\alpha(xy) = (\alpha x)y = x(\alpha y)$ for all $\alpha \in K$ and all $x, y \in \mathscr{A}$.

We remark that o and $-x$ are unique. All simple relations that were noted for linear spaces (p. 121) naturally hold now too. In addition, we also have, from axiom (ivc),

$$(\alpha\beta)(xy) = (\alpha x)(\beta y).$$

In all our future work *we shall be concerned only with cases when K is either the field of real numbers or the field of complex numbers.* Correspondingly, we shall speak of *real* and *complex algebras*, respectively. The

---

†The same mixed distributive laws apply as for linear spaces.

‡Without causing confusion, we may use the symbol $\mathscr{A}$ to denote the underlying set, the ring structure, the linear space structure (i.e. the additive group $\mathscr{A}$ combined with the field K), and the linear algebra structure as a whole.

zero element and unit element of K will be denoted by 0 and 1, respectively.

If, considered as a ring, $\mathscr{A}$ is commutative (i.e. if $xy = yx$ for all $x, y \in \mathscr{A}$), then we call $\mathscr{A}$ a *commutative algebra*. If, as a ring, $\mathscr{A}$ has a unit element ϵ, we call $\mathscr{A}$ an *algebra with unit*.† Then ϵ is unique and $\epsilon x = x\epsilon = x$ for all $x \in \mathscr{A}$. Finally, if, as a ring, $\mathscr{A}$ is a field, i.e. if it has a unit and every $x \neq o$ has an inverse x^{-1}, then we say that $\mathscr{A}$ is a *division algebra*. The inverse x^{-1} is unique.

Since $\mathscr{A}$ has a linear space structure, we may consider sets of linearly independent elements of $\mathscr{A}$. We say that the algebra $\mathscr{A}$ has *order n* (sometimes called *dimension n*) if, as a linear space, it has dimension n. In a similar sense we speak of infinite dimensional algebras, i.e. of algebras with infinite order.

We give now some examples.

Example α.   The one-element set $\{o\}$ with $o + o = o$, $oo = o$, $\alpha o = o$ is the trivial *zero algebra*.

Example β.   The real numbers **R**, with ordinary addition and multiplication as ring sum and ring product and with $K = \mathbf{R}$ is a real, commutative division algebra of order one if scalar multiplication αx is also defined as ordinary product of real numbers. Similarly, the set **C** of complex numbers with $x + y$ and xy being ordinary sum and product and with $K = \mathbf{C}$, and αx again ordinary product, is a complex, commutative division algebra of order one. If one takes here $K = \mathbf{R}$, then one gets a real, commutative division algebra of order two (1 and i form a basis).

Example γ.   Let $\mathscr{A}$ consist of all ordered pairs of real numbers and let $K = \mathbf{R}$. Define

$$(x, y) + (u, v) = (x + u, y + v),$$
$$(x, y)(u, v) = (xu, yv),$$
$$\alpha(x, y) = (\alpha x, \alpha y).$$

A routine calculation shows that the algebra axioms are satisfied and we have an algebra of order two. The null element is $(0, 0)$. The algebra is commutative, and has a unit element $(1, 1)$. But it is not a division algebra, since elements of the form $(0, y)$ (or elements $(x, 0)$) cannot have an inverse, because $0a = 1$ cannot be satisfied with any a.

---

†Even though we use Greek letters to denote the scalars, we wish to retain ϵ for denoting the unit of the ring $\mathscr{A}$, so as to conform with our notation in Section 4.2. Hopefully, no confusion will arise.

Example δ.   The preceding example can be generalized as follows. Let $\mathscr{A}$ and $\mathscr{B}$ be two algebras over the same field of scalars K. Consider the Cartesian product set $\mathscr{A} \times \mathscr{B}$ (consisting of pairs (a, b)) and define on this set the laws

$$(a, b) + (a', b') = (a + a', b + b'),$$
$$(a, b)(a', b') = (aa', bb'),$$
$$\alpha(a, b) = (\alpha a, \alpha b), \qquad \alpha \in K.$$

Clearly, the first and third laws make the set $\mathscr{A} \times \mathscr{B}$ into the direct sum of the corresponding linear spaces, and the second law makes it the direct product of the two ring structures (cf. p. 106). All one has to check to establish that we have an algebra is the mixed associative law. The algebra so constructed is usually called the *direct sum* of $\mathscr{A}$ and $\mathscr{B}$ (and is denoted by $\mathscr{A} \oplus \mathscr{B}$). If both $\mathscr{A}$ and $\mathscr{B}$ are commutative, so is $\mathscr{A} \oplus \mathscr{B}$, and if both have a unit, then $\epsilon = (\epsilon_{\mathscr{A}}, \epsilon_{\mathscr{B}})$ is a unit of $\mathscr{A} \oplus \mathscr{B}$. But $\mathscr{A} \oplus \mathscr{B}$ will not be a division algebra, even if $\mathscr{A}$ and $\mathscr{B}$ are such.

Example ε.   Let Q be the ring of quaternions discussed in Section 4.2. If we think of the elements $q \in Q$ as quadruplets of real numbers (cf. p. 109), we realize that Q *may be considered as a linear space over the real numbers as scalars*.† Indeed, the special elements ϵ, j, k, l form a basis and any element q may be written as $q = \alpha\epsilon + \beta j + \gamma l + \delta k$, where $\alpha, \beta, \gamma, \delta$ are arbitrary real scalars. Thus, we have a linear space of dimension four. Since we also have defined the product of quaternions,‡ and since we trivially see that $\mu(q_1 q_2) = (\mu q_1)q_2 = q_1(\mu q_2)$ for any real μ and any pair q_1, q_2 of quaternions,§ it is clear that *the quaternions can be looked upon as a real algebra of order four*. This algebra is not commutative. But it has a unit (namely $\epsilon = (1, 0, 0, 0)$) and it is a division algebra, since Q, as a ring, is a field. We may "complexify" the quaternion algebra by taking, instead of the reals, the field of complex numbers as scalars.

---

†Componentwise addition and componentwise scalar product are the composition laws. The first was already stipulated as the sum in the quaternion ring.

‡Note that writing a quaternion as a linear combination of the basis quaternions, we can represent a product as follows:

$$q_1 q_2 \equiv (\alpha\epsilon + \beta j + \gamma l + \delta k)(\alpha'\epsilon + \beta' j + \gamma' l + \delta' k)$$
$$= (\alpha\alpha' - \beta\beta' - \gamma\gamma' - \delta\delta')\epsilon + (\beta\alpha' + \alpha\beta' + \delta\gamma' - \gamma\delta')j$$
$$+ (\gamma\alpha' - \delta\beta' + \alpha\gamma' + \beta\delta')l + (\delta\alpha' - \beta\gamma' + \gamma\beta' + \alpha\delta')k.$$

This is obvious from looking at the original definition of product in terms of quadruplets.

§Note that writing a quaternion as a linear combination of the basis quaternions, we obviously have $\mu q = \mu\alpha\epsilon + \mu\beta j + \mu\gamma l + \mu\delta k$.

This *complex quaternion algebra* has also order four, is noncommutative, possesses a unit, but it is no longer a division algebra (cf. remarks on p. 110).

Example ζ. Let X be some arbitrary infinite set and let $\mathscr{A}$ be the collection of all functions from X to the real (complex) numbers. Let K be the field of real (complex) numbers. We already know (cf. Example ϵ on p. 123) that with pointwise sum

$$(f + g)(x) = f(x) + g(x)$$

and pointwise scalar product

$$(\alpha f)(x) = \alpha(f(x))$$

we have a linear space of functions. If we also define "pointwise product" of two functions† by

$$(fg)(x) = f(x)g(x),$$

an easy check shows that we have constructed a real (complex) algebra of infinite order. It is a commutative algebra and has a unit ($\epsilon(x) = 1$ for all x).

Example η. We recall that the set of all real (complex) n by n matrices with $K = \mathbf{R}(\mathbf{C})$ as scalars is a linear space with respect to standard matrix sum and scalar multiplication of matrices. If we define in this set ordinary matrix multiplication as a ring product (this was also discussed previously in Section 4.2), then, as an easy check verifies, we have an algebra of order n^2. It has a unit, but it is not commutative and is not a division algebra. The set of n by n matrices of the form

$$\begin{pmatrix} 0 & 0 & \dots & 0 \\ \cdot & & & \cdot \\ \cdot & & & \cdot \\ \cdot & & & \cdot \\ \alpha & \beta & \dots & \nu \end{pmatrix}$$

is also an algebra, of order n. It is not commutative, and has no unit.

Example ϑ. In many of our later studies, the *algebra of linear operators* plays a central role. Let $\mathscr{L}$ be an arbitrary linear space and let, as in Subsection 4.3b, $\mathscr{C}(\mathscr{L})$ denote the class of all linear transformations t from $\mathscr{L}$ to $\mathscr{L}$, i.e. let $\mathscr{C}(\mathscr{L})$ be the class of all linear space endomorphisms

---

†The pointwise product of functions must not be confused with the composite of functions. In fact, $f \circ g$ does not make any sense here if $X \neq \mathbf{R}$ (or $\mathbf{C}$).

of $\mathscr{L}$. We recall that under pointwise sum

$$(t_1 + t_2)(x) = t_1(x) + t_2(x)$$

and pointwise scalar product

$$(\alpha t)(x) = \alpha t(x)$$

the set $\mathscr{C}(\mathscr{L})$ is a linear space, usually called the space of linear operators on $\mathscr{L}$. Now we define in $\mathscr{C}(\mathscr{L})$ a product, by declaring $t_1 t_2$ to be the usual composition† maps $t_1 \circ t_2$. That is, define

$$(t_1 t_2)(x) = t_1(t_2(x)).$$

If t_1 and t_2 are linear maps, so is $t_1 t_2$, because

$$(t_1 t_2)(\alpha x + \beta y) = t_1[\alpha t_2(x) + \beta t_2(y)]$$
$$= \alpha t_1[t_2(x)] + \beta t_1[t_2(y)] = \alpha (t_1 t_2)(x) + \beta (t_1 t_2)(y).$$

Hence, $\mathscr{C}(\mathscr{L})$ is a closed algebraic system also under product of transformations. With very little labor the reader will verify that the three composition laws on $\mathscr{C}(\mathscr{L})$ satisfy all algebra axioms. (For example, $[(t_1 + t_2)t_3](x) = (t_1 + t_2)[t_3(x)] = t_1[t_3(x)] + t_2[t_3(x)] = t_1 t_3(x) + t_2 t_3(x) = (t_1 t_3 + t_2 t_3)(x)$, and so on.) Thus, we have the remarkable result that *the set $\mathscr{C}(\mathscr{L})$ of all linear operators on $\mathscr{L}$ is an algebra with respect to the composition laws as defined above.* This algebra is *not commutative,*‡ but it has a unit (the map $\epsilon = 1_{\mathscr{L}}$). It is *not* a division algebra, since $\mathscr{C}(\mathscr{L})$ contains maps that are not bijective, so that these cannot have multiplicative inverses. We shall have more to say about the algebra $\mathscr{C}(\mathscr{L})$ later on.

After these examples, we, as usual, take up the topic of subsystems. A sub*system* $\mathscr{V}$ of an algebra $\mathscr{A}$ is called a *subalgebra* if it satisfies the algebra axioms, i.e. if it is an algebra in its own right. Recalling the criteria of a subset to be a subring and to be a linear manifold, it is immediately seen that a *subset $\mathscr{V}$ of an algebra $\mathscr{A}$ is a subalgebra iff $x, y \in \mathscr{V}$ and $\alpha \in K$ imply $x + y \in \mathscr{V}$, $\alpha x \in \mathscr{V}$ and $xy \in \mathscr{V}$.* Thus, closure under all composition laws guarantees that the subset is a subalgebra, similarly as we found for linear spaces. For example, the order two real algebra of complex numbers is a subalgebra of the real quaternion algebra. The one-dimensional complex algebra of complex numbers is a subalgebra of the complex quaternion algebra. Any maximal subset of linear operators on $\mathscr{L}$ which

---

†Since the range of any t is a subset of $\mathscr{L}$, function composition makes sense. No such operation could be defined in general for $\mathscr{C}(\mathscr{L}_1, \mathscr{L}_2)$ if $\mathscr{L}_2 \neq \mathscr{L}_1$.

‡Unless dim $\mathscr{L} = 1$.

mutually commute (i.e. have the property that $(t_1t_2)(x) = (t_2t_1)(x)$ for all $x \in \mathscr{L}$) is a subalgebra of $\mathscr{C}(\mathscr{L})$.

We now discuss some tools which make the handling of algebras often rather easy. The starting point of the following discussion is the observation that, every algebra being a linear space, there will exist a basis $\{\dots e_\alpha \dots\}$ in terms of which any element x of $\mathscr{A}$ can be uniquely expanded.† Now, let

$$x = \sum_i \alpha_i e_i \quad \text{and} \quad y = \sum_k \beta_k e_k.$$

Then, in view of the axioms for algebras, we get

$$xy = \sum_{i,k} \alpha_i \beta_k (e_i e_k). \tag{4.23}$$

Since $e_i e_k \in \mathscr{A}$, we can expand uniquely and write

$$e_i e_k = \sum_l c_{ikl} e_l, \tag{4.24}$$

where c_{ikl} are well-defined scalars. If the space is n-dimensional, then of course, the summation goes from $l = 1$ to $l = n$, and we have n^2 such equations (one for each ordered pair ik of indices). Equation (4.24) can be considered as a multiplication table for the basis vectors. The coefficients c_{ikl} are called the *structure constants* of the algebra. *They determine uniquely the product of arbitrary algebra elements.* Indeed, substituting Eq. (4.24) into Eq. (4.23) we have

$$xy = \sum_{i,k,l} \alpha_i \beta_k c_{ikl} e_l, \tag{4.25}$$

and we recall that, since $\mathscr{A}$ has a linear space structure, the representation of an element $xy \in \mathscr{A}$ in terms of a linear combination of the basis members is unique. Thus, *the multiplicative structure of an algebra is fully characterized by its structure constants.*

There are, clearly, n^3 structure constants for an algebra of order n. However, they are *not* independent. The associativity of products gives the constraints $e_i(e_k e_l) = (e_i e_k)e_l$ for any three basis elements, which, in view of Eq. (4.24), give the relations

$$\sum_{l=1}^{n} c_{ikl} c_{ljm} = \sum_{l=1}^{n} c_{ilm} c_{kjl}. \tag{4.26}$$

We can invert the above described procedure in the following sense.

†We admit infinite dimensional algebras in this context with the same proviso as was used in the discussion of infinite dimensional linear spaces.

Suppose we are given a linear space $\mathscr{L}$. We choose a basis and then define, in any way we wish, products of the e_k, writing

$$e_i e_k = \sum_l c_{ikl} e_l$$

with the only *restriction* that the coefficients c_{ikl} obey the relations (4.26). Then, for any pair $x, y \in \mathscr{L}$ we define the product by linearly extending the product composition on the basis to all of $\mathscr{L}$, and by adopting the mixed distributive and associative laws, i.e. by writing for any x, y with given components α_k and β_k (relative to the chosen basis), $xy = \sum_{i,k,l} \alpha_i \beta_k c_{ikl} e_l$. It is fairly obvious that in this way we converted $\mathscr{L}$ into an algebra $\mathscr{A}$.

As a matter of fact, implicitly we already used this process when we discussed the real quaternion algebra. Q, as a linear space, has the basis ϵ, j, k, l. In Section 4.2 we defined associative products between these elements, cf. Eqs. (4.15a, b). Writing then $q = \alpha\epsilon + \beta j + \gamma l + \delta k$, we are led to the product of any two quaternions as was given in the footnote on p. 152.

Another example of this kind is the well-known *Dirac algebra*. Let $\mathscr{D}_{16}$ be a 16-dimensional complex linear space (i.e. $\mathscr{D}_{16}$, as a linear space, is isomorphic to $\mathbf{C}^{16}$). Instead of denoting a basis by $\{e_1, \ldots, e_{16}\}$, we use special symbols and label the 16 basis elements as follows:

$$\epsilon,$$

$$\gamma_1, \gamma_2, \gamma_3, \gamma_4,$$

$$\gamma_{12}, \gamma_{13}, \gamma_{14}, \gamma_{23}, \gamma_{24}, \gamma_{34},$$

$$\gamma_{123}, \gamma_{124}, \gamma_{134}, \gamma_{234},$$

$$\gamma_{1234}.$$

It is also convenient to use a generic symbol γ_A ($A = 1, 2, \ldots, 16$) for an arbitrary element in this basis array. Then, an arbitrary element of the space $\mathscr{D}_{16}$ can be written as

$$x = \sum_{A=1}^{16} \alpha_A \gamma_A,$$

where the α_A are arbitrary complex numbers. Now define, to start with, the products

$$\epsilon\gamma_A = \gamma_A, \qquad (A = 1, \ldots, 16).$$

$$\gamma_\mu \gamma_\nu = -\gamma_\nu \gamma_\mu = \gamma_{\mu\nu}, \qquad (\mu, \nu = 1, 2, 3, 4 \quad \text{and} \quad \mu < \nu),$$

$$\gamma_\mu^2 \equiv \gamma_\mu \gamma_\mu = \epsilon, \qquad (\mu = 1, 2, 3, 4).$$

In order to complete the multiplication table between the basis vectors, we simply satisfy at each step associativity. Thus, we define, taking note of the already established products,

$$\gamma_{\mu\rho\sigma} = \gamma_\mu \gamma_{\rho\sigma} \equiv \gamma_\mu (\gamma_\rho \gamma_\sigma) = (\gamma_\mu \gamma_\rho)\gamma_\sigma \equiv \gamma_{\mu\rho}\gamma_\sigma,$$

$$(\mu, \rho, \sigma = 1, \ldots, 4 \quad \text{and} \quad \mu < \rho < \sigma),$$

and so on. A simple way to summarize this is to say that $\gamma_{\mu\nu} = \gamma_\mu \gamma_\nu$, $\gamma_{\mu\nu\rho} = \gamma_\mu \gamma_\nu \gamma_\rho$, $\gamma_{\mu\nu\rho\sigma} = \gamma_\mu \gamma_\nu \gamma_\rho \gamma_\sigma$, whereby one should take note of $\gamma_\mu \gamma_\nu = -\gamma_\nu \gamma_\mu$ $(\mu \neq \nu)$ and of $\gamma_\mu^2 = \epsilon$. In this way, we have defined in $\mathcal{D}_{16}$ a multiplication; if

$$x = \sum_{A=1}^{16} \alpha_A \gamma_A \quad \text{and} \quad y = \sum_{B=1}^{16} \beta_B \gamma_B,$$

then

$$xy = \sum_{A,B,D}^{16} \alpha_A \beta_B c_{ABD} \gamma_D,$$

where c_{ABD} is the structure constant defined by

$$\gamma_A \gamma_B = \sum_{D=1}^{16} c_{ABD} \gamma_D.$$

By construction, the structure constants† obey the restrictions (4.26), since we define the γ-products in an associative way. Hence, we have *constructed an algebra* $\mathcal{D}_{16}$, commonly called the Dirac algebra. This algebra has order 16, it possesses a unit (namely ϵ), but it is not commutative (for example, $\gamma_1 \gamma_2 = -\gamma_2 \gamma_1$). There are many divisors of zero in $\mathcal{D}_{16}$, for example, $\gamma_1 + \epsilon$ is one, since $(\gamma_1 + \epsilon)(\gamma_1 - \epsilon) = \epsilon + \epsilon \gamma_1 - \gamma_1 \epsilon - \epsilon = 0$ (so that, of course, $\gamma_1 - \epsilon$ is also a divisor of zero). Consequently, $\mathcal{D}_{16}$ is not a division algebra. But many elements do have inverses. In particular, $\gamma_\mu^{-1} = \gamma_\mu$ $(\mu = 1, 2, 3, 4)$ (because $\gamma_\mu^2 = \epsilon$), and also any other basis element γ_A has an inverse which is either γ_A itself or $-\gamma_A$. (For example, the inverse of γ_{123} is $-\gamma_{123}$, since $\gamma_{123}(-\gamma_{123}) = -\gamma_1 \gamma_2 \gamma_3 \gamma_1 \gamma_2 \gamma_3 = \epsilon$, as easily checked.) Some other elements are also invertible; e.g.,

$$(\alpha\gamma_1 + \beta\epsilon)^{-1} = \frac{1}{\alpha^2 - \beta^2}(\alpha\gamma_1 - \beta\epsilon).$$

We observe that, in consequence of our method of defining the products of basis elements, the Dirac algebra is completely specified by the

---

†Incidentally, it is easily seen that all structure constants are either $+1$ or -1.

defining equations

$$\gamma_\mu \gamma_\nu + \gamma_\nu \gamma_\mu = 2\delta_{\mu\nu}\epsilon, \qquad (\mu, \nu = 1, 2, 3, 4)$$

(where $\delta_{\mu\nu}$ is the Kronecker symbol). These equations contain the crucial informations $\gamma_\mu^2 = \epsilon$ and $\gamma_\mu \gamma_\nu = -\gamma_\nu \gamma_\mu$, and the remaining eleven basis elements can be considered as being *constructed* by the rule $\gamma_{\mu\nu...} = \gamma_\mu \gamma_\nu ...$. The four elements γ_μ ($\mu = 1, 2, 3, 4$) are called the *generators* of $\mathcal{D}_{16}$.

The Dirac algebra is a special case of a well-known class of algebras, called *Clifford algebras*. The Clifford algebra $\mathcal{D}_{2^r}$ is an algebra of order 2^r, which has r generators $\gamma_1, \gamma_2, \ldots, \gamma_r$ (in the above sense), obeying the defining equations

$$\gamma_i \gamma_k + \gamma_k \gamma_i = 2\delta_{ik}\epsilon, \qquad (i, k = 1, 2, \ldots, r),$$

where ϵ is the unit element. The basis elements (other than $\epsilon, \gamma_1, \ldots, \gamma_r$) are the associatively defined products $\gamma_{kl...m} = \gamma_k \gamma_l \ldots \gamma_m$. The Dirac algebra is the special case with $r = 4$. The simplest Clifford algebra is $\mathcal{D}_4$ (i.e. $r = 2$). The generators are γ_1 and γ_2, the basis is $\epsilon, \gamma_1, \gamma_2, \gamma_{12}$, where $\gamma_{12} = \gamma_1 \gamma_2$. Commonly, one uses the symbol γ_0 for ϵ and γ_3 for $\gamma_1 \gamma_2$. The algebra with elements $x = \sum_{k=0}^{3} \alpha_k \gamma_k$ is customarily called the *Pauli algebra* and is very familiar in physical applications. The multiplication table reads

	γ_0	γ_1	γ_2	γ_3
γ_0	γ_0	γ_1	γ_2	γ_3
γ_1	γ_1	γ_0	γ_3	γ_2
γ_2	γ_2	$-\gamma_3$	γ_0	$-\gamma_1$
γ_3	γ_3	$-\gamma_2$	γ_1	γ_0

The Clifford algebra with $r = 3$ (i.e. $\mathcal{D}_8$) also occurs in some applications.

PROBLEMS

4.4-1. Let $\mathcal{A}$ be the set of all n by n strictly upper triangular matrices, i.e. matrices A with n rows and n columns for which $A_{kl} = 0$ if $k \geq l$. Show that $\mathcal{A}$ is an algebra of order $\frac{1}{2}(n^2 - n)$ with respect to the usual matrix composition laws.

4.4-2. Let $\mathcal{A}$ be an algebra and define the *center* $\mathcal{C}$ of $\mathcal{A}$ as the subset $\mathcal{C} = \{x | xy = yx \text{ for all } y \in \mathcal{A}\}$. Show that $\mathcal{C}$ is a subalgebra of $\mathcal{A}$. Show also the following: If $\mathcal{A}$ has a unit and $a \in \mathcal{A}$ is an invertible element which belongs to $\mathcal{C}$, then also $a^{-1} \in \mathcal{C}$.

4.4-3. Let $\mathcal{A}$ be an algebra with unit. Let x be an invertible element of $\mathcal{A}$ and

$\lambda \neq 0$ a scalar. Show that λx is invertible and in fact $(\lambda x)^{-1} = \lambda^{-1} x^{-1}$. Show also that if λx is invertible then so is x.

4.4-4. Let $\mathscr{C}(\mathscr{L})$ be the algebra of linear operators on some arbitrary linear space. Show that the subset of all invertible elements of $\mathscr{C}(\mathscr{L})$ is a group with respect to the multiplication defined in $\mathscr{C}(\mathscr{L})$. (*Remark*: This group is called the *general linear group* $GL(\mathscr{L})$ on the linear space $\mathscr{L}$.)

4.4-5. Let $\mathscr{A}$ be a real linear space over the reals, of dimension three. Denote the elements of a basis by ϵ, u, v and define among these a product such that ϵ acts as unit and $uu = u$, $uv = v$, $vu = o$, $vv = o$, where o is the zero vector. Show that in this way $\mathscr{A}$ becomes a real algebra of order 3. Is it commutative? Is it a division algebra?

4.4-6. Consider the linear space P of all polynomials introduced in Problem 4.3-3 and the basis given in Problem 4.3a-4. Define the structure constants $c_{ikl} = \delta_{l,i+k}$ (where δ is the Kronecker delta), check that these satisfy the necessary restrictions, and construct in this way *the algebra of polynomials*. Show that, even though this algebra has no divisors of zero, it is not a division algebra, because with the exception of polynomials of order zero, no other polynomial has an inverse.

4.4-7. Let $\mathscr{D}_{2^n}$ be a Clifford algebra. Show that its center (cf. Problem 4.4-2) is an algebra of order one or of order two, depending on whether n is even or odd. (*Hint*: Show that if n is even, the only basis element which commutes with all others is ϵ, and if n is odd, then both ϵ and $\gamma_{12...n}$ commute with all basis elements.)

4.4-8. Construct all algebras of order two.

4.4a. Morphisms of Algebras; Quotient Algebras

A map $f: \mathscr{A} \to \mathscr{A}'$ between algebras over the same field of scalars is called a morphism if

$$f(x + y) = f(x) + f(y), \qquad f(\alpha x) = \alpha f(x), \qquad f(xy) = f(x)f(y).$$

Thus every algebra morphism is a linear transformation from $\mathscr{A}$ to $\mathscr{A}'$, but *in addition* it has the property of preserving products.

The usual statements about the image and kernel of a morphism hold and surely need not be repeated.

It will also suffice to give one, nontrivial example. Let $\mathscr{A}$ be the Pauli algebra, i.e. the Clifford algebra $\mathscr{D}_4$ (with the basis γ_0, γ_1, γ_2, γ_3, where γ_0 is the unit and $\gamma_3 = \gamma_1 \gamma_2$). Let $\mathscr{A}'$ be the complex quaternion algebra (with the basis ϵ, j, k, l). These two algebras are isomorphic. Indeed, since they are both linear spaces of the same dimension (namely $n = 4$), they are surely isomorphic as linear spaces. Furthermore, the assignments

$$\gamma_0 \mapsto \epsilon \qquad \gamma_1 \mapsto ij \qquad \gamma_2 \mapsto -ik \qquad \gamma_3 \mapsto l$$

(where i stands for the usual imaginary number $\sqrt{-1}$) show, when one compares the γ_k ($k = 0, 1, 2, 3$) multiplication table with the quaternion multiplication table (4.15a, b) (cf. p. 108) that products also are conserved under the morphism, q.e.d. Incidentally, noting that the quaternion algebra is isomorphic to the algebra spanned by the Pauli matrices,† we see that the Clifford algebra $\mathcal{D}_4$ deserves the name "Pauli algebra."

By way of a further example of algebra morphisms, we give a simple theorem that completes our knowledge concerning the relation between the space $\mathcal{C}(\mathcal{L})$ of all linear operators on a finite dimensional space $\mathcal{L}$ on the one hand, and the linear space of all n by n matrices, on the other. We recall Theorem 4.3b(7) which assured us that these spaces are isomorphic. But by now we also know from Example ϑ on p. 154 that $\mathcal{C}(\mathcal{L})$ is an algebra and from Example η on p. 153 we know that M^{n^2} is also an algebra. We now claim that the linear space isomorphism between these spaces is also an algebra isomorphism:

THEOREM 4.4a(1).   *If $\mathcal{L}$ is a finite n-dimensional linear space, then the algebra $\mathcal{C}(\mathcal{L})$ of all linear operators on $\mathcal{L}$ is isomorphic to the algebra of all n by n matrices.*

Proof.   We recall that the linear space morphism between $\mathcal{C}(\mathcal{L})$ and M^{n^2} was defined by the assignment $t \mapsto A$ constructed with Eq. (4.21), i.e. by identifying the matrix elements of A from

$$t(e_k) = \sum_{l=1}^{n} A_{lk} e_l.$$

Now suppose

$$t_1 \mapsto A^{(1)} \quad \text{and} \quad t_2 \mapsto A^{(2)}.$$

Then

$$t_1 t_2 (e_k) = t_1 \left(\sum_{l=1}^{n} A_{lk}^{(2)} e_l \right) = \sum_{l=1}^{n} A_{lk}^{(2)} t_1(e_l)$$

$$= \sum_{l=1}^{n} \sum_{j=1}^{n} A_{lk}^{(2)} A_{jl}^{(1)} e_j \equiv \sum_{l,j=1}^{n} A_{jl}^{(1)} A_{lk}^{(2)} e_j$$

---

†This follows trivially from the original definition of quaternions, in terms of 2 by 2 matrices, as was given on p. 108.

$$= \sum_{j=1}^{n} (A^{(1)}A^{(2)})_{jk}e_j.$$

This shows that indeed

$$t_1 t_2 \to A^{(1)}A^{(2)},$$

which completes the proof of our theorem.

Our next topic is a generalization of Cayley's theorem to algebras. Like the former did for groups, our present study will greatly help the visualization of algebras.

Let $\mathscr{A}$ be an arbitrary algebra, and let $x \in \mathscr{A}$ be an arbitrary, fixed element. Let us define the map

$$t_x : \mathscr{A} \to \mathscr{A} \quad \text{given by} \quad t_x(y) = xy \quad \text{for all} \quad y \in \mathscr{A}.$$

We call this map "*left multiplication by x.*" Consider now the class $\mathscr{T}$ of all left multiplications on $\mathscr{A}$. We first show that $\mathscr{T}$ is a set of *linear transformations* on $\mathscr{A}$, when the latter is considered as a linear space. Indeed, for any t_x we find

$$t_x(\alpha y + \beta z) = x(\alpha y + \beta z) = \alpha xy + \beta xz = \alpha t_x(y) + \beta t_x(z),$$

which proves the assertion. Thus, $\mathscr{T}$ is a subset of the class $\mathscr{C}(\mathscr{A})$ of all linear operators on $\mathscr{A}$. Next we show that, actually, $\mathscr{T}$ is a *subalgebra* of the algebra $\mathscr{C}(\mathscr{A})$. All we have to show is that we have closure for all three composition laws. But indeed, if t_x and t_y belong to $\mathscr{T}$, then

$$(t_x + t_y)(z) = xz + yz = (x + y)z = t_{x+y}(z)$$

so $t_x + t_y = t_{x+y}$, and thus belongs to $\mathscr{T}$. Similarly, $\alpha t_x(z) = \alpha xz = t_{\alpha x}(z)$, so that together with t_x, the map $\alpha t_x = t_{\alpha x}$ also belongs to $\mathscr{T}$. Finally, $t_x t_y(z) = t_x(yz) = xyz = t_{xy}(z)$, hence $t_x, t_y \in \mathscr{T}$ implies that $t_x t_y = t_{xy}$ is in $\mathscr{T}$. Thus, indeed, $\mathscr{T}$ *is an algebra consisting of linear transformations on the linear space* $\mathscr{A}$; actually it is a subalgebra of $\mathscr{C}(\mathscr{A})$, as claimed.

Our last step in the program is to relate the original algebra $\mathscr{A}$ and the algebra $\mathscr{T}$ of linear operators. It is easy to see that $\mathscr{A}$ *and* $\mathscr{T}$ *are always homomorphic.* Indeed, consider the map

$$f : \mathscr{A} \to \mathscr{T} \quad \text{given by} \quad x \mapsto t_x \quad \text{for all} \quad x \in \mathscr{A}.$$

This is a *morphism* of algebras, since

$$x + y \mapsto t_{x+y} = t_x + t_y,$$
$$\alpha x \mapsto t_{\alpha x} = \alpha t_x,$$
$$xy \mapsto t_{xy} = t_x t_y,$$

where we used our previously established results on compositions of left multiplications. Furthermore, the morphism is surely surjective, since $\mathcal{T}$ consists of nothing else but elements t_x. This establishes our claim that $\mathcal{A}$ and $\mathcal{T}$ are homomorphic.

It would be good to go further and see whether we could have an *isomorphism*. Suppose $\mathcal{A}$ is an algebra which has at least one regular element, say r. Then $x \neq y$ will imply that $t_x \neq t_y$, because if we had $t_x = t_y$, then $t_x(r) = t_y(r)$, i.e. $xr = yr$, which by cancellation of r, would give $x = y$. We shall call an algebra which has at least one regular element, a "nonpathologic" algebra. In particular, *if $\mathcal{A}$ has a unit ϵ, this ϵ is always regular, so $\mathcal{A}$ is nonpathologic.* Now, from the above observation it follows that for nonpathologic algebras the map $x \mapsto t_x$ is one-to-one, hence, in this case, the homomorphism between $\mathcal{A}$ and $\mathcal{T}$ is actually an isomorphism, as desired.

We formalize our results:

THEOREM 4.4a(2). *Every nonpathologic algebra $\mathcal{A}$ is isomorphic to an algebra $\mathcal{T}$ of linear operators. This $\mathcal{T}$ is the class of left multiplications on $\mathcal{A}$ and it is a subalgebra of $\mathscr{C}(\mathcal{A})$. The isomorphism between $\mathcal{A}$ and $\mathcal{T}$ is given by the assignment $x \mapsto t_x$ for all $x \in \mathcal{A}$.*

The algebra $\mathcal{T}$ of linear operators on $\mathcal{A}$ is usually called the *left regular representation of $\mathcal{A}$*. It is obvious that, in a similar manner, one can construct a right regular representation.

At this point we digress a bit and comment briefly on *representations of algebras*. Let $\mathcal{A}$ be an arbitrary algebra and let $\mathcal{O}$ be an algebra consisting of linear operators on some linear space $\mathscr{L}$. If there exists a homomorphism from $\mathcal{A}$ to $\mathcal{O}$, then we say that we have a *linear representation* of $\mathcal{A}$. The space $\mathscr{L}$ is said to be the *representation space* (or carrier space of the representation). If $\mathscr{L}$ is a finite n-dimensional space, then the representation furnished by the operators of $\mathcal{O}$ on $\mathscr{L}$ is said to be n-dimensional. If actually $\mathcal{A}$ and $\mathcal{O}$ are isomorphic, the representation is said to be faithful.

Having this terminology in mind, we see that the regular representation

$\mathcal{T}$ is a linear representation of $\mathcal{A}$ *on the linear space $\mathcal{A}$ itself.* In particular, if $\mathcal{A}$ is nonpathologic, then the regular representation is faithful.

Some additional comments can be made if $\mathcal{A}$ is *finite dimensional and nonpathologic.* Since then $\mathcal{T}$ is isomorphic to $\mathcal{A}$, as a linear space $\mathcal{A}$ and $\mathcal{T}$ must have the same dimension. Hence, in that case, *the order of the algebra $\mathcal{A}$ and the order of the operator algebra $\mathcal{T}$ is the same.* More important is the fact that, since $\mathcal{T}$ acts on the linear space $\mathcal{A}$, the *regular representation of an algebra of order n is n-dimensional.* Finally, noting that (a) $\mathcal{T}$ is a subalgebra of $\mathcal{C}(\mathcal{A})$, (b) recalling that $\mathcal{C}(\mathcal{A})$ is isomorphic to the full matrix algebra M^{n^2}, we see that *every nonpathologic algebra $\mathcal{A}$ of finite order n is isomorphic to a subalgebra Σ^n of the matrix algebra of n by n matrices.*† This last observation is the analog of the simpler fact that every finite n-dimensional linear space is isomorphic to $\mathbf{R}^n$ (or $\mathbf{C}^n$).

We now set out to *explicitly construct the matrices that represent the elements of $\mathcal{A}$ in the (left) regular representation.*

Let $x \in \mathcal{A}$ and let t_x be the element of $\mathcal{T}$ that corresponds to x in the isomorphism $\mathcal{A} \approx \mathcal{T}$. Let $\{e_1, \ldots, e_n\}$ be a basis of $\mathcal{A}$. Then, by the definition of t_x, we have

$$t_x(e_k) = xe_k = \sum_i \alpha_i e_i e_k = \sum_{i,l} \alpha_i c_{ikl} e_l,$$

where c_{ikl} is a structure constant. Let us introduce the notation

$$S_{lk}(x) \equiv \sum_{i=1}^{n} \alpha_i c_{ikl}. \tag{4.27}$$

Then we have

$$t_x(e_k) = \sum_{l=1}^{n} S_{lk}(x) e_l. \tag{4.28}$$

This means that the linear operator $t_x \in \mathcal{T}$ is mapped on the matrix $S(x)$,

$$t_x \mapsto S(x), \tag{4.29}$$

under the isomorphism $\mathcal{T} \approx \Sigma^n$ (where Σ^n is a subalgebra of the M^{n^2} matrix algebra). Here the entry in the lth row and kth column of $S(x)$ is determined via Eq. (4.27) by the structure constants of the group and the components of x. Let us now take, in particular, x to be one of the basis elements, say $x = e_r$. The components of e_r relative to the basis $\{e_1, \ldots, e_n\}$ are, of course $\alpha_i = \delta_{ir}$ (since $e_r = \Sigma_{i=1}^{n} \delta_{ir} e_i$). Therefore, Eq. (4.27) tells us that the matrix $S(e_r)$ associated to e_r has the matrix

---

†It goes without saying that the matrix algebra in question is that of real (complex) matrices depending on whether $\mathcal{A}$ is a real (complex) algebra.

elements

$$S_{lk}(e_r) = \sum_{i=1}^{n} \delta_{ir} c_{ikl} = c_{rkl}. \tag{4.30}$$

This suggests to define a *matrix* c_r for each $r = 1, 2, \ldots, n$, whose element in the lth row and kth column is precisely the number c_{rkl}. Thus, we can write

$$S(e_r) = c_r \qquad (r = 1, 2, \ldots, n),$$

with

$$(c_r)_{lk} = c_{rkl}. \tag{4.31}$$

With this notation, Eq. (4.27) gives

$$S(x) = \sum_{i=1}^{n} \alpha_i c_i. \tag{4.32}$$

This matrix $S(x)$ is precisely the image of x under the isomorphism $\mathcal{A} \approx \mathcal{T}$. All this means the following. Let $x \in \mathcal{A}$. Then in the left regular representation $\mathcal{T}$ of $\mathcal{A}$, the matrix corresponding to x is given by the n by n matrix $S(x)$ of Eq. (4.32). This $S(x)$ is a linear combination of n by n matrices c_i, with the components α_i of x acting as coefficients. In turn, each matrix c_i can be calculated from the structure constants via Eq. (4.31).

In summary, we can say that the regular representation is "furnished" by the structure constants. We also see that the basis of $\mathcal{T}$ is the set $\{c_1, \ldots, c_n\}$ of n by n matrices, each of these being determined by Eq. (4.31) from the structure constants.

We give a simple example for the calculation of the regular representation. Let $\mathcal{A}$ be the algebra consisting of all 2 by 2 matrices of the form

$$A = \begin{pmatrix} a & b \\ 0 & c \end{pmatrix}.$$

(It is easy to check that this is an algebra of order 3.) Let us choose a basis, for example

$$e_1 = \begin{pmatrix} 1 & 0 \\ 0 & 0 \end{pmatrix}, \qquad e_2 = \begin{pmatrix} 0 & 1 \\ 0 & 0 \end{pmatrix}, \qquad e_3 = \begin{pmatrix} 0 & 0 \\ 0 & 1 \end{pmatrix}.$$

Calculating all products, $e_1 e_2 = e_2$, $e_1 e_3 = 0$, etc. we easily find the structure constants relative to this basis:

$$c_{111} = c_{122} = c_{232} = c_{333} = 1,$$

all others vanishing. According to the recipe (4.31), we then obtain

$$
c_1 = \begin{pmatrix} 1 & 0 & 0 \\ 0 & 1 & 0 \\ 0 & 0 & 0 \end{pmatrix}, \quad
c_2 = \begin{pmatrix} 0 & 0 & 0 \\ 0 & 0 & 1 \\ 0 & 0 & 0 \end{pmatrix}, \quad
c_3 = \begin{pmatrix} 0 & 0 & 0 \\ 0 & 0 & 0 \\ 0 & 0 & 1 \end{pmatrix}.
$$

These are the representatives of e_1, e_2, e_3 in $\mathcal{T}$. Consequently, according to Eq. (4.32), the representative of the arbitrary algebra element A in the regular representation is the 3 by 3 matrix

$$
S(A) = \begin{pmatrix} a & 0 & 0 \\ 0 & a & b \\ 0 & 0 & c \end{pmatrix}.
$$

The reader will easily check out, if he wishes, that this is a faithful representation of $\mathcal{A}$.

There is one more remark. Obviously, the explicit form of the regular representation depends on the choice of basis in $\mathcal{A}$. However, it is easy to see that different choices of basis lead to equivalent representations, i.e. to representations $\mathcal{T}^{(1)}, \mathcal{T}^{(2)}, \ldots$ which are isomorphic to each other and which, in fact, can be obtained from each other by a change of basis in the matrix algebra Σ^n.

It would lead outside the scope of this book to go into further details† and with regrets, we pass on now to a new topic. To conclude the discussion of algebra morphisms, we very briefly discuss the quotient structures associated with algebras.

Let $\mathcal{A}$ be an algebra and let $\mathcal{Y}$ be a subalgebra of $\mathcal{A}$ which is also an ideal of $\mathcal{A}$ (considered as a ring). Such a subalgebra $\mathcal{Y}$ is called an *algebra ideal* of $\mathcal{A}$. The definition may be paraphrased by saying that an algebra ideal is both an ideal with respect to the ring structure of $\mathcal{A}$ as well as a linear manifold with respect to the linear space structure of $\mathcal{A}$. Thus, the *subset* $\mathcal{Y}$ of $\mathcal{A}$ is an algebra ideal if

$$ j_1, j_2 \in \mathcal{Y} \quad \text{implies} \quad j_1 + j_2 \in \mathcal{Y}, $$

$$ j \in \mathcal{Y} \quad \text{and} \quad \alpha \in K \quad \text{imply} \quad \alpha j \in \mathcal{Y}, $$

$$ j \in \mathcal{Y} \quad \text{implies} \quad xj \in \mathcal{Y} \text{ and } jx \in \mathcal{Y} \quad \text{for all} \quad x \in \mathcal{A}. $$

Note that a *ring* ideal of $\mathcal{A}$ need not necessarily be an *algebra* ideal.

---

†For example, the question of reducibility is very interesting. As inspection of the above example shows, the regular representation is, in general, a direct sum of representations with lower dimension.

Every algebra has the trivial algebra ideals {o} and $\mathscr{A}$. An algebra which has no other algebra ideals is said to be *simple*.

Let now $\mathscr{Y}$ be an algebra ideal of $\mathscr{A}$ and construct the additive cosets

$$\mathscr{Y} + x = \{j + x \,|\, j \in \mathscr{Y}\}.$$

The coset space $\mathscr{A}/\mathscr{Y}$ can be endowed with the structure of a linear algebra if we define

$$(\mathscr{Y} + x) \boxplus (\mathscr{Y} + y) = \mathscr{Y} + (x + y),$$
$$(\mathscr{Y} + x) \odot (\mathscr{Y} + y) = \mathscr{Y} + xy,$$
$$\alpha \cdot (\mathscr{Y} + x) = \mathscr{Y} + \alpha x.$$

(The uniqueness of these composition laws follows from the fact that $\mathscr{Y}$ is an algebra ideal.) The space $\mathscr{A}/\mathscr{Y}$ with these composition laws is called the *quotient algebra*. The zero element is $\mathscr{Y} + o = \mathscr{Y}$, the negative of $\mathscr{Y} + x$ is $\mathscr{Y} - x$. If $\mathscr{A}$ has a unit ϵ, then $\mathscr{A}/\mathscr{Y}$ has also a unit, namely $\mathscr{Y} + \epsilon$.

As expected, the kernel of an algebra morphism $f: \mathscr{A} \to \mathscr{A}'$ is an algebra ideal of $\mathscr{A}$. Conversely, if $\mathscr{Y}$ is an algebra ideal, then the canonical map $c: \mathscr{A} \to \mathscr{A}/\mathscr{Y}$ (given by $x \mapsto \mathscr{Y} + x$) is a homomorphism with Ker $(c) = \mathscr{Y}$. Furthermore, if $f: \mathscr{A} \to \mathscr{A}'$ is a morphism with kernel $\mathscr{Y}$, then $\mathscr{A}/\mathscr{Y} \to$ Im (f) (given by $\mathscr{Y} + x \mapsto f(x)$) is an isomorphism.

The reader may well suspect that the theory of algebras is extremely rich and diverse, and that we did hardly more than present the basic concepts.†

PROBLEMS

4.4a-1. Show that the set of matrices of the form

$$M = \begin{pmatrix} \alpha_1 & \alpha_2 & \alpha_3 \\ 0 & \alpha_1 + \alpha_2 & \alpha_3 \\ 0 & 0 & \alpha_1 \end{pmatrix}$$

is an algebra which is isomorphic to the algebra defined in Problem 4.4-5.

4.4a-2. Show that the algebra of complex numbers over the real scalars is isomorphic to the algebra of all real matrices of the form

$$M = \begin{pmatrix} a & b \\ -b & a \end{pmatrix}.$$

---

†A well-readable and sufficiently detailed account on linear algebras, geared strongly to the physicist's interest, may be found in Falk[9].

4.4a-3. Let $\mathscr{L}$ be a finite n-dimensional linear space. Show that the corresponding general linear group $GL(\mathscr{L})$ (cf. Problem 4.4-4) is isomorphic to the group of all nonsingular n by n matrices.

4.4a-4. Construct the left-regular representation of the real quaternion algebra.

4.4a-5. Find the subalgebra of all 2×2 matrices for which the set of matrices of the form

$$N = \begin{pmatrix} 0 & \alpha \\ 0 & 0 \end{pmatrix}$$

is an algebra ideal.

4.4a-6. Find all algebra ideals of the algebra $\mathscr{A}$ defined in Problem 4.4-5.

4.4a-7. Let $\mathscr{A}$ be an algebra *with identity* and show that if $\mathscr{Y}$ is an ideal with respect to the ring structure of $\mathscr{A}$, then it is automatically a linear manifold of $\mathscr{A}$, so that in this case it is really an algebra ideal. (*Hint*: Note that if ϵ is the unit element of $\mathscr{A}$, then $\alpha\epsilon \in \mathscr{A}$ for any scalar α.)

4.4a-8. Let $\mathscr{Y}$ be an algebra ideal of $\mathscr{A}$. Show that the subsets of $\mathscr{A}$ defined by

$\mathscr{Y}^2 = $ linear manifold spanned by the set $\{j_1 j_2 | j_1 \in \mathscr{Y}, j_2 \in \mathscr{Y}\}$,

.
.
.

$\mathscr{Y}^n = $ linear manifold spanned by the set $\{j_1 j_2 \ldots j_n | j_k \in \mathscr{Y}, k = 1, 2, \ldots, n\}$,

.
.
.

are all algebra ideals of $\mathscr{A}$.

4.4a-9. An algebra $\mathscr{A}$ is said to be *nilpotent* if there exists some n such that $\mathscr{A}^n = $ linear manifold spanned by the set $\{x_1 x_2 \ldots x_n | x_k \in \mathscr{A}, k = 1, 2, \ldots, n\} = \{o\}$. Similarly, an ideal $\mathscr{Y}$ is nilpotent if there exists some m such that $\mathscr{Y}^m = \{o\}$. Prove the following theorem: If $\mathscr{Y}$ is a nilpotent algebra ideal of $\mathscr{A}$ and if the quotient algebra $\mathscr{A}/\mathscr{Y}$ is nilpotent, then $\mathscr{A}$ itself is nilpotent. (*Hint*: Show that the nilpotence of $\mathscr{A}/\mathscr{Y}$ means that there exists some integer s such that $\mathscr{A}^s \subset \mathscr{Y}$.)

4.4a-10. Let $\mathscr{A}$ be the algebra of all real triangular 2×2 matrices,

$$A = \begin{pmatrix} a_{11} & a_{12} \\ 0 & a_{22} \end{pmatrix}.$$

Show that the map $A \mapsto a_{11}$ is a homomorphism $f: \mathscr{A} \to \mathbf{R}$. Find the kernel, check that Ker (f) is an ideal and illustrate that $\mathscr{A}/\mathrm{Ker}\,(f)$ is isomorphic to $\mathbf{R}$.

4.5 NONASSOCIATIVE ALGEBRAS

All algebraic systems so far studied by us had the property that *all* composition laws of the system were associative. In this section we shall briefly discuss a class of rather important systems where *one* of the internal composition laws is not required to be associative.

DEFINITION 4.5(1).   *A nonassociative algebra is an algebraic system which obeys all axioms of a linear algebra* † *except that* $x(yz) = (xy)z$ *need not hold.*

It should be obvious that such systems surely exist. Indeed, let $\mathcal{V}$ be a linear space with a basis $\{\ldots e_k \ldots\}$ and let us define products $e_i e_k = \Sigma_l\, c_{ikl} e_l \in \mathcal{V}$ with completely *arbitrary* structure constants. We then get a system which will obey all axioms (i)–(iv) of an algebra (cf. p. 150) with the possible exception of axiom (iia).

Completely general nonassociative algebras are of little interest. If, however, one imposes some additional identities to be satisfied by the elements of a nonassociative algebra,‡ then several types of useful systems, with a rich theory and many applications, ensue.

It should be clear that all concepts, properties, and theorems relating to associative algebras which are independent of the associative law, have their complete counterpart in any nonassociative algebra. Thus, without any further explanation we may speak of order, subalgebras, direct sums, ideals, quotient algebras, morphisms of nonassociative algebras.

4.5a. Lie Algebras

The most familiar class of nonassociative algebras is known by the name of Lie algebras. The product of two elements x, y of a Lie algebra L we denote by the symbol $[x, y]$ and we postulate that the Lie product satisfies the following two relations:

$$[x, x] = o \quad \text{for all} \quad x \in L \tag{4.33}$$

$$[x, [y, z]] + [y, [z, x]] + [z, [x, y]] = o \quad \text{for all} \quad x, y, z \in L. \tag{4.34}$$

Any nonassociative algebra where the product satisfies these two rules is called a Lie algebra. If the field of scalars is that of the real (complex) numbers, we talk of a real (complex) Lie algebra. Of course, all usual axioms (except associativity) are satisfied by Lie products. For example, the mixed distributivity reads $[x, (y + z)] = [x, y] + [x, z]$, and the mixed associativity assumes the form $\alpha[x, y] = [\alpha x, y] = [x, \alpha y]$, where α is an arbitrary scalar.

†Cf. the axioms on p. 150.
‡In other words, we talk about algebras where the associativity postulate of multiplication is *replaced* by some other postulate(s).

Equation (4.33) is *equivalent* to the following:

$$[x, y] = -[y, x] \quad \text{for all} \quad x, y \in L. \tag{4.33a}$$

This can be seen as follows. Let $z = x + y$ in Eq. (4.33). Then $[x + y, x + y] \equiv [x, x] + [x, y] + [y, x] + [y, y] = 0$, from which (using again Eq. (4.33)), the claimed relation (4.33a) follows. Conversely, taking in Eq. (4.33a) $x = y$, we obviously get Eq. (4.33).

The property (4.33a) of a Lie product is called *antisymmetry*. The other fundamental relation (4.34) is known as the *Jacobi identity*. Thus, by definition, *a Lie algebra is a nonassociative algebra with an antisymmetric product that obeys the Jacobi identity*. In passing we note that, using the antisymmetry, Eq. (4.34) may be written as

$$[x, [y, z]] = [[x, y], z] - [y, [z, x]].$$

The presence of the second term on the r.h.s. shows that, in general, a Lie product *cannot* be associative.

A very simple example of a Lie algebra is the following. Let $\mathbf{R}^3$ be the usual three-dimensional linear vector space. Consider the familiar "vector product" of two vectors, i.e. if $x = (\xi_1, \xi_2, \xi_3)$ and $y = (\eta_1, \eta_2, \eta_3)$, set

$$x \times y = (\xi_2\eta_3 - \xi_3\eta_2, \xi_3\eta_1 - \xi_1\eta_3, \xi_1\eta_2 - \xi_2\eta_1).$$

As is well known, $x \times x = 0$ (i.e. $x \times y = -y \times x$) and $(x \times y) \times z = x \times (y \times z) + y \times (z \times x)$. Thus, the vector product obeys the postulates of a Lie product.

We now proceed to characterize a Lie algebra in terms of its *structure constants*. If we write any $x \in L$ as $x = \Sigma_k \alpha_k e_k$, where $\{\ldots e_k \ldots\}$ is some basis, and define the constants c_{ikl} by writing

$$[e_i, e_k] \equiv \sum_l c_{ikl} e_l,$$

then a simple calculation based on Eqs. (4.33) and (4.34) shows that the structure constants obey the following relations:

$$c_{ikl} = -c_{kil}, \tag{4.35a}$$

$$\sum_r (c_{ijr}c_{rks} + c_{jkr}c_{ris} + c_{kir}c_{rjs}) = 0. \tag{4.35b}$$

These relations replace Eq. (4.26) which was valid for any associative algebra.

Conversely, if $\mathscr{L}$ is a linear space, we can turn it into a Lie algebra by defining in any way we wish products of the e_k, setting

$$[e_i, e_k] = \sum_l c_{ikl} e_l,$$

with the only *restriction* that the coefficients c_{ikl} obey the relations
(4.35a, b). Indeed, if $x, y \in \mathscr{L}$, we then find that the Lie product $[x, y]$,
defined by writing

$$[x, y] = \sum_{i,k,l} \alpha_i \beta_k [e_i, e_k] \equiv \sum_{i,k,l} \alpha_i \beta_k c_{ikl} e_l,$$

obeys the antisymmetry and the Jacobi identity postulates. Putting it in
another way: if the basis vectors e_k obey the antisymmetry postulate and
the Jacobi identity, then the linear space spanned by them is a Lie algebra.
It is customary to call the basis vectors of a Lie algebra L the *generators*
of L. Clearly, the Lie brackets of the generators of a Lie algebra
completely determine the Lie algebra.

In passing we note that it is perfectly possible to further restrict the
structure constants by the relations

$$c_{ikl} = c_{kil}.$$

Together with Eq. (4.35a), these relations yield $c_{ikl} = 0$ for *all* structure
constants. Then $[x, y] = o$ for any pair $x, y \in L$. A Lie algebra where any
pair of elements has a vanishing Lie product is called an *Abelian* or
commutative Lie algebra.†

We now show that *from any given associative algebra $\mathscr{A}$ one can
construct a Lie algebra*. Indeed, since in $\mathscr{A}$ the ordinary associative products $e_i e_k$ between two basis elements are already defined, we can set

$$[e_i, e_k] \equiv e_i e_k - e_k e_i$$

with a meaningful r.h.s. An elementary calculation shows that with this
definition we obtain

$$[e_i, e_k] = -[e_k, e_i],$$

and also

$$[e_i, [e_k, e_l]] + [e_k, [e_l, e_i]] + [e_l, [e_i, e_k]] = o.$$

Thus, the basis vectors satisfy the antisymmetry and Jacobi axioms, so
that by what we said above, we have constructed a Lie algebra. For

---

†It should be noted that if L is an Abelian Lie algebra, then, trivially, $[x, [y, z]] = [[x, y], z]$
for all $x, y, z \in L$. Thus, an Abelian Lie algebra happens to be associative. For this reason, it
may be preferable to use the term "not necessarily associative algebra" rather than just
"nonassociative algebra."

arbitrary elements $x, y \in \mathscr{A}$ we then have, with an easy calculation,

$$[x, y] = xy - yx. \tag{4.36}$$

Actually, we may use this basis-independent relation as the *definition* of a Lie algebra constructed from the given associative algebra $\mathscr{A}$. The r.h.s. of Eq. (4.36) is usually called the *commutator* of the elements x, y of the associative algebra $\mathscr{A}$. The Lie algebra with the Lie product being defined as the commutator, is called *the commutator algebra of the associative algebra* $\mathscr{A}$ and will be denoted by $\mathscr{A}^-$.

   Remarks: (a) As a *linear space*, $\mathscr{A}$ and $\mathscr{A}^-$ are identical.

   (b) If $\mathscr{A}$ is a commutative algebra, then $\mathscr{A}^-$ is an Abelian Lie algebra.

   (c) Since associative products yz are defined, it makes sense to consider in $\mathscr{A}^-$ Lie products such as $[x, yz]$ and associative products in $\mathscr{A}$ like $x[y, z]$. One actually finds that

$$[x, yz] = [x, y]z + y[x, z].$$

Since every associative algebra $\mathscr{A}$ gives rise to a Lie algebra $\mathscr{A}^-$, the reader may now wonder whether there is some kind of converse relation. The answer is given by the fundamental result of Poincaré–Birkhoff–Witt:

THEOREM 4.5a(1).   *Every Lie algebra is isomorphic to a* SUBALGEBRA *of some commutator Lie algebra* $\mathscr{A}^-$ *of an associative algebra* $\mathscr{A}$.

Next we take up the topic of Lie algebras whose elements are linear operators. We recall that if $\mathscr{L}$ is any given linear space then the set $\mathscr{C}(\mathscr{L})$ consisting of all linear operators on $\mathscr{L}$ (i.e. the set of all endomorphisms of $\mathscr{L}$) is an associative algebra with respect to the usual compositions of maps. Naturally, we can now form $\mathscr{C}^-(\mathscr{L})$, i.e. the commutator algebra of $\mathscr{C}(\mathscr{L})$. Any subalgebra of $\mathscr{C}^-(\mathscr{L})$ is called an *operator Lie algebra* or a Lie algebra of linear transformations on $\mathscr{L}$. The elements of an operator Lie algebra are linear operators t acting on $\mathscr{L}$, and the Lie product is defined by

$$[t_1, t_2](x) = (t_1 t_2)(x) - (t_2 t_1)(x) \quad \text{for all} \quad x \in \mathscr{L}.$$

The following theorem exhibits the importance of operator Lie algebras:

THEOREM 4.5a(2).   *Every Lie algebra is isomorphic to a Lie algebra of linear operators.*

The proof of this deep structural theorem follows from the above quoted Poincaré–Birkhoff–Witt theorem and from our previous result, Theorem 4.4a(2).

If, in particular, we deal with a Lie algebra of finite order n, we have the special case:

THEOREM 4.5a(3). *Every finite n-dimensional Lie algebra is isomorphic to a subalgebra of a Lie algebra consisting of n by n matrices, with the Lie product being the usual commutator of matrices.*

We illustrate some of the last statements on an example.

Example α. Consider the set of all 2×2 real matrices. We recall that this is an associative algebra, namely $\mathscr{C}(\mathbf{R}^2)$, the algebra of all linear operators on $\mathbf{R}^2$. Let us take for a convenient basis the matrices

$$\sigma_0 = \begin{pmatrix} 1 & 0 \\ 0 & 1 \end{pmatrix}, \qquad \sigma_1 = \begin{pmatrix} 0 & 1 \\ 1 & 0 \end{pmatrix}, \qquad \sigma_2 = \begin{pmatrix} 0 & 1 \\ -1 & 0 \end{pmatrix}, \qquad \sigma_3 = \begin{pmatrix} 1 & 0 \\ 0 & -1 \end{pmatrix}.$$

Define the Lie products as the commutators,

$$[\sigma_0, \sigma_\mu] = 0, \qquad [\sigma_1, \sigma_2] = -2\sigma_3, \qquad [\sigma_2, \sigma_3] = -2\sigma_1, \qquad [\sigma_3, \sigma_1] = 2\sigma_2.$$

(All other commutators are of course also determined: $[\sigma_\nu, \sigma_\mu] = -[\sigma_\mu, \sigma_\nu]$ and $[\sigma_\mu, \sigma_\mu] = 0$.) In this way we are led to the Lie algebra $\mathscr{C}^-(\mathbf{R}^2)$. The elements are of the form $x = \Sigma_{\mu=0}^{} \alpha_\mu \sigma_\mu$, and arbitrary Lie products are easily evaluated. In fact we note that the structure constants are

$$c_{123} = c_{231} = -2, \qquad c_{312} = 2.$$

All other structure constants are trivially obtained from these by the antisymmetry in the first pair of indices (i.e. for example, $c_{213} = +2$, etc.) or are zero. Now consider the subset $\mathscr{D}(\mathbf{R}^2)$ consisting of matrices of the form

$$\hat{x} = \sum_{k=1}^{3} \alpha_k \sigma_k = \begin{pmatrix} \alpha_3 & \alpha_1 + \alpha_2 \\ \alpha_1 - \alpha_2 & -\alpha_3 \end{pmatrix}.$$

This subset is a sub-Lie algebra, because σ_k $(k = 1, 2, 3)$ span a linear manifold which is closed under *Lie* products.† $\mathscr{D}(\mathbf{R}^2)$ is, like $\mathscr{C}^-(\mathbf{R}^2)$, a Lie algebra of linear operators. We observe that $\mathscr{D}(\mathbf{R}^2)$ *cannot* be obtained as

---

†Note that the subset $\mathscr{D}(\mathbf{R}^2)$ is *not* closed under ordinary associative products, since, for example, $\sigma_1^2 = \sigma_0$.

the commutator algebra $\mathscr{A}^-$ of some associative algebra $\mathscr{A}$. Indeed, the linear space consisting of matrices $\hat{x}$ with the form spelled out above is *not* an associative algebra (because, as pointed out in the footnote, it does not close under ordinary product-taking of matrices). Thus, we have here a good illustration of Theorem 4.5a(1) (because $\mathscr{D}(\mathbf{R}^2)$ is a subalgebra of the commutator Lie algebra $\mathscr{C}^-(\mathbf{R}^2)$ obtained from $\mathscr{C}(\mathbf{R}^2)$).

In the following we shall study an analog of our result for associative algebras, concerning the regular representation. We recall that any associative algebra $\mathscr{A}$ was found to be homomorphic to the algebra $\mathscr{T}$ of (left) multiplications on $\mathscr{A}$. In the present case of Lie algebras, we start with the following observation. Let L be an arbitrary Lie algebra and let $x \in L$ be an arbitrary, fixed element. Define the so-called *adjoint map by* x (to be denoted as ad_x):

$$ad_x : L \to L \quad \text{given by} \quad ad_x(y) = [x, y] \quad \text{for all} \quad y \in L.$$

Consider now the class $\operatorname{Ad} L$ of *all* adjoint maps on L. First of all, this class is a set of *linear operators* on L. Indeed,

$$ad_x(\alpha y + \beta z) = [x, \alpha y + \beta z] = \alpha[x, y] + \beta[x, z]$$
$$= \alpha ad_x(y) + \beta ad_x(z).$$

Next we show that $\operatorname{Ad} L$ has actually the structure of a Lie algebra, if we define, naturally enough,

$$(ad_x + ad_y)(z) = ad_x(z) + ad_y(z), \tag{4.37a}$$

$$(\alpha ad_x)(z) = \alpha ad_x(z), \tag{4.37b}$$

$$[ad_x, ad_y](z) = (ad_x ad_y - ad_y ad_x)(z). \tag{4.37c}$$

To verify this statement, we first observe that closure under sum and scalar product is fairly obvious. Actually, for any $z \in L$, $ad_x(z) + ad_y(z) = [x, z] + [y, z] = [x + y, z] = ad_{x+y}(z)$, so that

$$ad_x + ad_y = ad_{x+y}. \tag{4.38a}$$

Similarly, we see that

$$\alpha ad_x = ad_{\alpha x}. \tag{4.38b}$$

Finally, to see that $\operatorname{Ad} L$ is closed also under the third composition law (4.37c) we observe that, for any $z \in L$,

$$(ad_x ad_y - ad_y ad_x)(z) = ad_x[y, z] - ad_y[x, z] = [x, [y, z]] - [y, [x, z]]$$
$$= -[z, [x, y]] = [[x, y], z],$$

where in the last but one step we used the Jacobi identity, and in the last step the antisymmetry. Now, the r.h.s. is nothing but $ad_{[x,y]}(z)$, so we have

$$(ad_x ad_y - ad_y ad_x) = ad_{[x,y]}. \tag{4.38c}$$

Thus, in summary, with respect to the composition laws (4.37a, b, c) the set Ad L is closed. Finally, we verify that Ad L is a Lie algebra. Clearly, in view (4.38c),

$$[ad_x, ad_x] = ad_{[x,x]} = ad_o$$

which is the zero map on L (since $ad_o(z) = [o, z] = o$). Furthermore,

$$[ad_x, [ad_y, ad_z]] = [ad_x, ad_{[y,z]}] = ad_{[x,[y,z]]},$$

and therefore the satisfaction of the Jacobi identity follows from its fulfillment in L itself.†

Thus, in summary, we see that Ad L *is a Lie algebra of linear operators acting on the linear space of L itself.* Actually, Ad L is a *subalgebra* of $\mathscr{C}^-(L)$.

Our last endeavor in this direction is to show that L *and* Ad L *are homomorphic.* Indeed, consider the map

$$f: L \rightarrow \text{Ad } L \quad \text{given by} \quad x \mapsto ad_x \quad \text{for all} \quad x \in L.$$

This is a *morphism* of Lie algebras, since

$$x + y \mapsto ad_{x+y} = ad_x + ad_y,$$
$$\alpha x \mapsto ad_{\alpha x} = \alpha\, ad_x,$$
$$[x, y] \mapsto ad_{[x,y]} = [ad_x, ad_y],$$

where we used Eqs. (4.38a, b, c). Furthermore, the morphism is surjective, since Ad L consists of nothing else but elements ad_x. This proves the statement about the homomorphism between L and Ad L.

A further moment's thought shows that the kernel of the morphism $f: L \rightarrow \text{Ad } L$ consists precisely of those elements $x \in L$ which have vanishing Lie brackets $[x, y] = o$ for all $y \in L$. That is, Ker $(f) = $ center of L. Hence, if the center of L consists of the single element $o \in L$, then and only then the homomorphism between L and Ad L is actually an isomorphism.

---

†The sophisticated reader should note that the set Ad L is *not* closed under the associative product of its elements ad_x, hence Ad L is *not* an associative algebra, so that the Lie algebra Ad L is *not* a commutator algebra of some known associative algebra. That is why we had to prove explicitly the fulfillment of the Lie axioms for the composition law $[ad_x, ad_y]$.

The homomorphism of L to Ad L is called the *adjoint representation* of the Lie algebra L. It is faithful iff Cent $L = \{o\}$. Clearly Ad L is analogous to the regular representation $\mathscr{T}$ of associative algebras.

We conclude this discussion with a closer consideration of the *adjoint representation of Lie algebras of finite order*. To start with, we observe that, since the map $f: L \to$ Ad L is onto, the dimension of Ad L as a linear space is at most that of L. From this it follows that the adjoint representation will consist of a set of n by n matrices if L had order n. We now wish to construct this Ad L (i.e. the matrices which are isomorphic to Ad L) from the structure constants of L. We first note that, with a basis $\{e_1, \ldots, e_n\}$ in L, we have

$$ad_x(e_k) = [x, e_k] = \sum_i \alpha_i [e_i, e_k] = \sum_{i,l} \alpha_i c_{ikl} e_l.$$

If we use the notation

$$\Omega_{lk}(x) = \sum_{i=1}^{n} \alpha_i c_{ikl}, \tag{4.39}$$

then

$$ad_x(e_k) = \sum_{l=1}^{n} \Omega_{lk}(x) e_l. \tag{4.40}$$

Thus, the operator $ad_x \in$ Ad L corresponds to the n by n matrix $\Omega(x)$, defined through Eq. (4.39). If, in particular, we take $x = e_r$, we have, from Eq. (4.39),

$$\Omega_{lk}(e_r) = c_{rkl}. \tag{4.41}$$

If we now define a *matrix c_r* for each $r = 1, 2, \ldots, n$, whose element in the lth row and kth column is given by

$$(c_r)_{lk} = c_{rkl}, \tag{4.42}$$

then Eq. (4.39) gives

$$\Omega(x) = \sum_{i=1}^{n} \alpha_i c_i, \tag{4.43}$$

and this matrix is precisely the image of $x \in L$ in the adjoint matrix representation Ad L.

The construction of the adjoint matrix representation is, as we see, very much the same as was the case for the construction of the regular representation of an associative algebra. We give an example.

Example β.   Let L be the Lie algebra $\mathscr{D}(\mathbf{R}^2)$ discussed in Example α above. With the structure constants there calculated we find, from Eq.

(4.42), the basis matrices in Ad $\mathscr{D}(\mathbf{R}^2)$:

$$c_1 = \begin{pmatrix} 0 & 0 & 0 \\ 0 & 0 & -2 \\ 0 & -2 & 0 \end{pmatrix}, \qquad c_2 = \begin{pmatrix} 0 & 0 & -2 \\ 0 & 0 & 0 \\ 2 & 0 & 0 \end{pmatrix}, \qquad c_3 = \begin{pmatrix} 0 & 2 & 0 \\ 2 & 0 & 0 \\ 0 & 0 & 0 \end{pmatrix}.$$

One easily checks that, for example, $[ad_{\sigma_1}, ad_{\sigma_2}] \approx [c_1, c_2] = -2c_3 \approx -2ad_{\sigma_3}$ as it should be. The image of a general element $\hat{x} \in \mathscr{D}(\mathbf{R}^2)$ in Ad $\mathscr{D}(\mathbf{R}^2)$ is represented by the matrix

$$\Omega(\hat{x}) = \begin{pmatrix} 0 & 2\alpha_3 & -2\alpha_2 \\ 2\alpha_3 & 0 & -2\alpha_1 \\ 2\alpha_2 & -2\alpha_1 & 0 \end{pmatrix}.$$

The last topic on Lie algebras that we wish to cover is the brief discussion of *ideals*. Following the general definition of an algebra ideal, we will call a subalgebra D of a Lie algebra an ideal if, with respect to the Lie product, we have a "ring-ideal" behavior, i.e. if $d \in D$ implies $[d, x] \in D$ and $[x, d] \in D$. Since the first follows from the second, (because of antisymmetry) we may say that a sub*set* D of a Lie algebra L is an ideal of L if

$$d_1, d_2 \in D \quad \text{imply} \quad d_1 + d_2 \in D,$$
$$d \in D \text{ and } \alpha \in K \quad \text{imply} \quad \alpha d \in D,$$
$$d \in D \quad \text{implies} \quad [x, d] \in D \quad \text{for all} \quad x \in L.$$

Ideals of Lie algebras are very frequently called *invariant subalgebras*. For brevity, we shall not use this term in the following.

Every Lie algebra has the trivial ideals $\{o\}$ and L. If L (with order >1) has no other (proper) ideals, then it is said to be simple. If L has no (nonzero) Abelian ideal, then L is said to be *semisimple*.

We give a few examples of ideals.

Example γ. The Lie algebra $\mathscr{C}^-(\mathbf{R}^2)$ of Example α has two proper ideals. The first is simply its center,† $C = \{\alpha\sigma_0 | \alpha \in \mathbf{R}\}$. The second is the subalgebra $\mathscr{D}(\mathbf{R}^2)$. Indeed, if

$$d = \sum_{k=1}^{3} \alpha_k \sigma_k \in \mathscr{D}(\mathbf{R}^2) \quad \text{and} \quad x = \sum_{\mu=0}^{3} \beta_\mu \sigma_\mu \in \mathscr{C}^-(\mathbf{R}^2),$$

then

$$[x, d] = \sum_{l=1}^{3} \gamma_l \sigma_l \in \mathscr{D}(\mathbf{R}^2).$$

†The center of every Lie algebra is an ideal, cf. Problem 4.5a-8.

Thus, $\mathscr{C}^-(\mathbf{R}^2)$ is not simple, nor is it semisimple (the ideal C *is* Abelian). On the other hand, $\mathscr{D}(\mathbf{R}^2)$ itself is simple, because it has no proper ideals at all.

Example δ. Consider the associative algebra T_3 of all 3×3 strictly triangular real matrices, i.e. the set of all matrices of form

$$x = \begin{pmatrix} 0 & \alpha_1 & \alpha_3 \\ 0 & 0 & \alpha_2 \\ 0 & 0 & 0 \end{pmatrix}.$$

A convenient basis is given by

$$e_1 = \begin{pmatrix} 0 & 1 & 0 \\ 0 & 0 & 0 \\ 0 & 0 & 0 \end{pmatrix}, \qquad e_2 = \begin{pmatrix} 0 & 0 & 0 \\ 0 & 0 & 1 \\ 0 & 0 & 0 \end{pmatrix}, \qquad e_3 = \begin{pmatrix} 0 & 0 & 1 \\ 0 & 0 & 0 \\ 0 & 0 & 0 \end{pmatrix}.$$

The commutator Lie algebra T_3^- is then fully characterized by the Lie brackets

$$[e_1, e_2] = e_3, \qquad [e_1, e_3] = \mathrm{o}, \qquad [e_2, e_3] = \mathrm{o}.$$

The structure constants are: $c_{123} = -c_{213} = 1$, all others zero. By simple inspection of the Lie brackets, we see that there are the following ideals:

$$D_1 = \{\alpha e_3\}, \qquad \text{(this is the center of } T_3^-)$$
$$D_2 = \{\alpha e_2 + \beta e_3\},$$
$$D_3 = \{\alpha e_1 + \beta e_3\}.$$

(The reader is urged to check out these statements.) Thus, our T_3^- Lie algebra is not simple, neither is it semisimple, since D_1, D_2, and D_3 are easily seen to be Abelian ideals.

Example ε. Let L be the direct sum of the Lie algebra $\mathscr{D}(\mathbf{R}^2)$ with itself, $L = \mathscr{D}(\mathbf{R}^2) \oplus \mathscr{D}(\mathbf{R}^2)$. This means, of course, that L has order 6, and the basis can be written as $\sigma_1, \sigma_2, \sigma_3, \tau_1, \tau_2, \tau_3$ with the Lie products

$$[\sigma_1, \sigma_2] = -2\sigma_3, \qquad [\sigma_2, \sigma_3] = -2\sigma_1, \qquad [\sigma_3, \sigma_1] = 2\sigma_2,$$
$$[\tau_1, \tau_2] = -2\tau_3, \qquad [\tau_2, \tau_3] = -2\tau_1, \qquad [\tau_3, \tau_1] = 2\tau_2,$$
$$[\sigma_k, \tau_l] = \mathrm{o}, \qquad (k, l = 1, 2, 3).$$

One sees immediately that there are two proper ideals, namely

$$D_1 = \left\{\sum_{k=1}^{3} \alpha_k \sigma_k\right\} \quad \text{and} \quad D_2 = \left\{\sum_{k=1}^{3} \beta_k \tau_k\right\}.$$

However, neither of these is Abelian, hence L, even though not simple, is nevertheless semisimple.

In the following we briefly study ideals of special importance. For this (and some other) purposes, the following notation is useful. Let L_1 and L_2 be linear manifolds of a Lie algebra L. Then define†

$[L_1, L_2]$ = linear manifold spanned by the set $\{[x_1, x_2] | x_1 \in L_1, x_2 \in L_2\}$.

With this notation, for example, the criterion that a subalgebra D of L be an ideal, assumes the form

$$[L, D] \subset D.$$

Let now L be a Lie algebra and consider the set

$$L^{(1)} = [L, L], \tag{4.44a}$$

i.e. the manifold spanned by all Lie products $[x, y]$. From the definition it trivially follows that $L^{(1)}$ is a subalgebra, and since $[L^{(1)}, L]$ is clearly a subset of $L^{(1)}$, we see that $L^{(1)}$ defined by Eq. (4.44a) is an ideal of L. It is called the *first derived ideal* of L. A little consideration shows that

$$L^{(2)} = [L^{(1)}, L^{(1)}] \tag{4.44b}$$

is also an ideal of L which actually is contained in $L^{(1)}$. Proceeding in this manner, i.e. defining

$$L^{(k)} = [L^{(k-1)}, L^{(k-1)}] \tag{4.45}$$

we obtain the *series of derived ideals*,

$$L \supset L^{(1)} \supset L^{(2)} \supset \cdots \supset L^{(k)} \supset \cdots.$$

Suppose that there exists a k such that $L^{(k)} = \{o\}$. Then the Lie algebra L is said to be *solvable*. *Every Abelian Lie algebra is solvable*, since then $L^{(1)} = [L, L] = \{o\}$. On the other hand, *a solvable Lie algebra with $k > 1$ cannot be simple, nor even semisimple.*‡ Indeed, a solvable Lie algebra surely has an ideal $L^{(k-1)}$, and since $[L^{(k-1)}, L^{(k-1)}] = \{o\}$, this ideal is Abelian.

---

†A similar notation is sometimes used also for associative algebras. If $\mathcal{A}_1, \mathcal{A}_2$ are linear manifolds of the algebra $\mathcal{A}$, then

$$\mathcal{A}_1 \mathcal{A}_2 = \text{linear manifold spanned by the set}$$
$$\{x_1 y_2 | x_1 \in \mathcal{A}_1, y_2 \in \mathcal{A}_2\}.$$

Actually, we made use of this notation in Problems 4.4a-8 and 4.4a-9.

‡We note at this point that Abelian Lie algebras are not semisimple.

We illustrate these concepts on our previous examples. The Lie algebra $L = \mathcal{D}(\mathbf{R}^2)$ is not solvable, since $[L, L] = L$, so that $L^{(1)}, L^{(2)}, \ldots$ are all equal to L and hence $\neq \{o\}$. The same holds for $L = \mathcal{D}(\mathbf{R}^2) \oplus \mathcal{D}(\mathbf{R}^2)$. On the other hand, the triangular Lie algebra $L = T_3^-$ is solvable because $L^{(1)} = [L, L] = \{\alpha e_3\}$, so that $L^{(2)} = [L^{(1)}, L^{(1)}] = \{o\}$.

Another class of interesting ideals can be constructed in a similar manner. Define

$$L^2 = [L, L]$$

(this happens to coincide, of course, with $L^{(1)}$ above, and is an ideal); then set

$$L^3 = [L^2, L], \ldots, L^k = [L^{k-1}, L], \ldots . \qquad (4.46)$$

(Here L^3 consists of the linear manifold spanned by elements of the form $[[x, y], z]$, and so on.) A moment's reflection shows that each L^k is again an ideal, and we have the *lower central series* of ideals

$$L \supset L^2 \supset L^3 \supset \ldots L^k \supset \ldots .$$

Suppose there exists a k such that $L^k = \{o\}$. Then the Lie algebra L is said to be *nilpotent*. One easily shows that *every nilpotent Lie algebra is solvable* (but the converse is not true). It then follows that a nilpotent Lie algebra cannot be simple (if $k > 2$) nor semisimple. An example of a nilpotent Lie algebra is again $L = T_3^-$, because $L^2 = [L, L] = \{\alpha e_3\}$, and so $L^3 = [L^2, L] = \{o\}$.

A family-tree of Lie algebras may be now given as follows

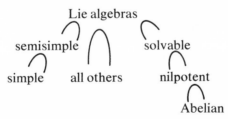

We conclude this section with a brief mention of *quotient structures* for Lie algebras. Let D be an ideal of L and let us form the additive cosets

$$D + x = \{d + x \mid d \in D\}.$$

On the corresponding quotient set L/D we define three composition laws,

$$(D + x) \boxplus (D + y) = D + (x + y),$$
$$\alpha \cdot (D + x) = D + \alpha x,$$
$$[\![(D + x), (D + y)]\!] = D + [x, y].$$

Since D is an ideal, these are valid definitions, independent of representatives, and it is easy to check that L/D is a Lie algebra, called the *quotient Lie algebra* modulo D. We leave it to the reader to formulate and illustrate all usual structural theorems that relate quotient structures and kernels of morphisms.

Concerning some additional basic topics about Lie algebras we call the reader's attention to Problems 4.5a–11 through 4.5a–13.

The usual apologies for not having gone too deep into the subject† must be here amended with one more remark of regret. The major importance of Lie algebras resides in their intimate connection with Lie groups. We cannot even mention this topic within the framework of our survey.‡

PROBLEMS

4.5a-1. Let $\mathscr{A}$ be an associative algebra with structure constants c_{ikl}. Show that the structure constants of the commutator Lie algebra $\mathscr{A}^-$ are given by $\gamma_{ikl} = c_{ikl} - c_{kil}$. Conversely, show that by this formula, the structure constants of any associative algebra $\mathscr{A}$ give rise to those of a Lie algebra.

4.5a-2. Construct all Lie algebras of order two and order three. Discuss their properties. (*Remark*: Attempt the case of order 3 only if you have lots of time.)

4.5a-3. Let T_3^- be the commutator Lie algebra of all 3×3 *strictly* triangular matrices

$$A = \begin{pmatrix} 0 & \alpha & \beta \\ 0 & 0 & \gamma \\ 0 & 0 & 0 \end{pmatrix}$$

and compute the adjoint representation. Use your result to show that T_3^- is *homo*morphic to an Abelian Lie algebra of order two.

4.5a-4. Let $\mathscr{A}$ and $\mathscr{B}$ be associative algebras. Denote the corresponding Lie algebras by $\mathscr{A}^-$ and $\mathscr{B}^-$. Suppose $f: \mathscr{A} \to \mathscr{B}$ is a homomorphism. Show that f is also a homomorphism from $\mathscr{A}^-$ to $\mathscr{B}^-$.

4.5a-5. Let $\mathscr{A}$ be an arbitrary (either associative or not associative) algebra. Denote the product by $x * y$. Let $d: \mathscr{A} \to \mathscr{A}$ be some map with the property that

$$d(x * y) = d(x) * y + x * d(y).$$

Such a map on $\mathscr{A}$ is called a *derivation*.§ Consider now the class Der $\mathscr{A}$ of all

---

†A clear and thorough coverage of Lie algebras is given in Jacobson[18].

‡A simple introduction to these questions may be found in Cohn[4].

§An example is the following. Let $\mathscr{A}$ be the associative function algebra of infinitely differentiable real valued functions on $\mathbf{R}$ and let $d: \mathscr{A} \to \mathscr{A}$ be given by $d(f) = df/dx$.

derivations on $\mathscr{A}$ and show that this set is a linear manifold of the space $\mathscr{C}(\mathscr{A})$ of all linear operators on $\mathscr{A}$. (Define the usual pointwise sum and scalar product for two derivations.) Introduce now the additional composition law

$$[d_1, d_2](x) = (d_1 d_2 - d_2 d_1)(x)$$

(where, of course, $d_1 d_2$ means composition of maps) and show that Der $\mathscr{A}$ is actually a Lie algebra. It is called the *derivation algebra on* $\mathscr{A}$.

4.5a-6. Let L be a Lie algebra and show that Ad L is a sub-Lie algebra of Der L. (That is, show that every map ad_x on L is a derivation.) Show that, furthermore, Ad L is an ideal of Der L.

4.5a-7. Let $\mathscr{A}$ be an associative algebra and let $\mathscr{A}^-$ be the corresponding commutator Lie algebra. Let d be a derivation on $\mathscr{A}$ and show that d is also a derivation on $\mathscr{A}^-$. The converse statement is not true. Construct an example where a derivation on $\mathscr{A}^-$ is not a derivation on $\mathscr{A}$. (*Hint*: Consider the simplest, nontrivial $\mathscr{A}^-$, which is an Abelian Lie algebra of order two.)

4.5a-8. Show that, for any Lie algebra L, the center $C = \{c \in L \,|\, [c, x] = o$ for all $x \in L\}$ is an Abelian ideal of L. (*Remark*: Of course, it may happen that $C = \{o\}$.)

4.5a-9. Show that the set of triangular n by n matrices is a Lie algebra. (M is triangular if $M_{kl} = 0$ for all $l < k$). Determine the derived series and the lower central series. Is this Lie algebra solvable? Is it nilpotent?

4.5a-10. Consider a linear space of dimension three and define for the basis vectors the independent Lie brackets by setting

$$[e_1, e_2] = - e_3, \qquad [e_1, e_3] = e_2, \qquad [e_2, e_3] = o.$$

Show that this is a Lie algebra. Compute the adjoint representation. Show that the Lie algebra is solvable but not nilpotent.

4.5a-11. Let L be a Lie algebra and let L_1 and L_2 be two invariant subalgebras (i.e. ideals) such that, as a linear space, L is the direct sum of L_1 and L_2. We then say that *the Lie algebra* L *is the direct sum of the Lie algebras* L_1 *and* L_2, and write $L = L_1 \oplus L_2$. Show that if $L = L_1 \oplus L_2$, then $[L_1, L_1] \subset L_1$, $[L_2, L_2] \subset L_2$, and $[L_1, L_2] = \{o\}$. Express these results in terms of the generators of L_1 and L_2. Assuming that L has finite order, express the condition for L to be a direct sum in terms of the structure constants. Give an example of a Lie algebra that is the direct sum of Lie algebras.

4.5a-12. Let L be a Lie algebra, let L_1 be an arbitrary subalgebra and let L_2 be an invariant subalgebra (ideal), and suppose that L, as a linear space, is the direct sum of L_1 and L_2. We then say that *the Lie algebra* L *is the semidirect sum of the Lie algebras* L_1 *and* L_2, and write $L = L_1 \uplus L_2$. Show that if $L = L_1 \uplus L_2$, then $[L_1, L_1] \subset L_1$, $[L_2, L_2] \subset L_2$, and $[L_1, L_2] \subset L_2$. Express these results in terms of the generators of L_1 and L_2. Assuming that L has finite order, express the condition for L to be a semidirect sum in terms of the structure constants.

4.5a-13. Let $L = K \oplus T$ where T is an invariant subalgebra. Show that

$$\frac{K \oplus T}{T} \approx K.$$

(*Hint*: Prove first that L is homomorphic to K.)

4.5a-14. Let E be a Lie algebra of order 6, with generators denoted by J_k ($k = 1, 2, 3$) and P_k ($k = 1, 2, 3$), obeying the Lie bracket relations

$$[J_1, J_2] = -J_3, \qquad [J_2, J_3] = -J_1, \qquad [J_3, J_1] = -J_2,$$
$$[J_1, P_2] = -P_3, \qquad [J_2, P_3] = -P_1, \qquad [J_3, P_1] = -P_2,$$

and all other (independent) brackets zero. Illustrate the theorem of the preceding problem for the Lie algebra E.

4.5b. Some Other Nonassociative Algebras

In this subsection we merely wish to mention two nonassociative algebras which, along with Lie algebras, are often met in physics.

(α) *Jordan algebras.* These are nonassociative algebras J where the Jordan product $x \cdot y$ of elements obeys the following two axioms:†

$$x \cdot y = y \cdot x \quad \text{for all} \quad x, y \in J,$$
$$x^2 \cdot (y \cdot x) = (x^2 \cdot y) \cdot x \quad \text{for all} \quad x, y \in J.$$

If $\mathcal{A}$ is an associative algebra and if we define

$$x \cdot y = \frac{1}{2}(xy + yx),$$

we can check by a trivial calculation that we obtained a Jordan algebra, to be denoted by $\mathcal{A}^+$. In contrast to Lie algebras, however, we do not have an analog of the Poincaré–Birkhoff–Witt theorem: there exist Jordan algebras which are *not* subalgebras of any $\mathcal{A}^+$ algebra.

(β) *Poisson algebras.* Let $\mathcal{P}$ be an associative algebra. Let us introduce an *additional*, nonassociative internal composition law, called "Poisson bracket formation" and denoted by $[x, y]_p$. We set the following axioms (in addition to the already existing ones on $x + y$, αx, xy):

(i) $[x, x]_p = o$ for all $x \in \mathcal{P}$,
(ii) $[x, (y + z)]_p = [x, y]_p + [x, z]_p$ and $[(x + y), z]_p = [x, z]_p + [y, z]_p$ for all $x, y, z \in \mathcal{P}$,
(iii) $\alpha[x, y]_p = [\alpha x, y]_p = [x, \alpha y]_p$ for all $x \in \mathcal{P}$ and all $\alpha \in K$,
(iv) $[x, yz]_p = [x, y]_p z + y[x, z]_p$ for all $x, y, z \in \mathcal{P}$.

†We use, for brevity, the notation x^2 to indicate $x \cdot x$.

Axioms (ii), (iii), and (iv) are mixed distributive and associative laws. We observe that axiom (i) is equivalent to

$$[x, y]_p = -[y, x]_p \quad \text{for all} \quad x, y \in \mathscr{P}.$$

As a matter of fact, we already met an example of a Poisson algebra. Indeed, if $\mathscr{A}$ is an associative algebra and if we define $[x, y]_p = xy - yx$, then all axioms above are fulfilled, as a direct calculation shows. On the other hand, we recall that the commutator $xy - yx$ defines also a Lie algebra $\mathscr{A}^-$. Hence, every $\mathscr{A}^-$ Lie algebra *is* a Poisson algebra, and actually, besides the required axioms, we *also* have the Jacobi identity for the Poisson bracket (which coincides with the Lie bracket).

Another familiar example of Poisson algebras comes from classical mechanics. Let $\mathscr{P}$ be the set of all infinitely differentiable real valued functions from $\mathbf{R} \times \mathbf{R} \to \mathbf{R}$. (That is, $\mathscr{P}$ consists of real functions $f(x, y)$ of two variables, and all partial derivatives like $\partial f / \partial x$, $\partial f / \partial y$, $\partial^2 f / \partial x^2$, $\partial^2 f / \partial y^2$, $\partial^2 f / \partial x \partial y$, etc. exist.) Make this set into an ordinary associative algebra by pointwise sum, scalar product, and product:

$$(f + g)(x, y) = f(x, y) + g(x, y),$$
$$(\alpha f)(x, y) = \alpha f(x, y),$$
$$(fg)(x, y) = f(x, y) g(x, y).$$

Define now the Poisson bracket

$$[f, g]_p = \frac{\partial f}{\partial x} \frac{\partial g}{\partial y} - \frac{\partial f}{\partial y} \frac{\partial g}{\partial x}.$$

A direct check shows that all axioms of a Poisson algebra are fulfilled. *In addition*, it so happens that the Jacobi identity again holds true. Hence, the Poisson algebra of classical mechanics is also a Lie algebra (but of course it is not a commutator algebra).

PROBLEMS

4.5b-1. Let J be a Jordan algebra and let D be an ideal. Show that the quotient space J/D is also a Jordan algebra.

4.5b-2. Let $\mathscr{A}$ be an associative algebra and let $f: \mathscr{A} \to \mathscr{A}$ be a linear bijective map, with the property that $f(xy) = f(y)f(x)$. Show that f is an automorphism of the Jordan algebra $\mathscr{A}^+$. Let $\mathscr{B} = \{x \in \mathscr{A} \mid f(x) = x\}$ be the set of elements that are left fixed under f and show that $\mathscr{B}$ is a Jordan subalgebra of $\mathscr{A}^+$. In particular, let $\mathscr{A}$ be the algebra of all n by n matrices and take for f the map which sends each matrix A to its transpose $\tilde{A}$. Show that f satisfies the criteria of the above definition and identify the Jordan subalgebra $\mathscr{B}$ of $\mathscr{A}^+$.

4.5b-3. Let $\mathscr{P}$ be an arbitrary Poisson algebra. Define for a given $f \in \mathscr{P}$, the map $d_f: \mathscr{P} \to \mathscr{P}$ given by $d_f(g) = [f, g]_p$ for all $g \in \mathscr{P}$. Show that this map is a derivation† with respect to the associative product structure and, if $\mathscr{P}$ happens to also obey the Jacobi identities, then d_f is a derivation even with respect to the Poisson bracket product. Consider the Poisson algebra of classical mechanics and construct explicitly the linear operator d_f.

---

†Regarding the definition of derivations, cf. Problem 4.5a-5.

IIB: Topological Structures

5

Topological Spaces

When we started our exploration of algebraic systems, we noted that a child's first experience with mathematical structures is the observation that numbers can be combined in several ways to get other numbers and that these composition rules are intimately interrelated. Now, in his further development, the child will become aware of the fact that the "real line" possesses also a completely different kind of structure. He will realize (starting probably from the notion of distance) that points have neighborhoods, that these neighborhoods are related to intervals, and that sequences of points sometimes converge to a limit. Later on, he will learn about functions on the real line, become aware of their properties, in particular of continuity and perhaps with its relation to convergence. Topology is the generalization of these elementary concepts of geometry and analysis, and we shall explore these structures in the following. Once again, we shall rely not on the historical development of the subject, but rather on the deductive, "axiomatic" method.

Let X be an arbitrary (universal) set. Let $\tau = \{\ldots V_\alpha \ldots\}$ be any collection of subsets† of X which obeys the following axioms:

(i) $\emptyset$ belongs to τ,
(ii) X belongs to τ,
(iii) Unions $\cup_\beta V_\beta$ of arbitrary subcollections of τ belong to τ,
(iv) Intersections $\cap_{k=1}^n V_k$ of finite subcollections of τ belong to τ.

Any collection $\tau = \{\ldots V_\alpha \ldots\}$, $V_\alpha \subset X$, which obeys these axioms is

† That is, let τ be a subset of the power set $\mathscr{P}(X)$.

called *a topology* τ *on* X (or a topology *for X*). The essential point is that *a topology on X is a collection of subsets of X which is closed under the formation of arbitrary unions and of finite intersections.*† A set X equipped with a topology τ on X is called a *topological space* and is denoted by the ordered pair (X, τ). Thus, a topological space is a system, or structure, very much like an algebraic system $(X, \Box)$. The elements V_α of the topology τ are called, by definition, the *open sets of X*. Whenever it is clear from the context what the open sets are (i.e. what the topology τ is), we shall take the liberty to talk simply of the "topological space X." Obviously, on a given set X one may define several (maybe even infinitely many) different topologies (i.e. collections of open sets) and *each* definition τ gives rise, in general, to a *different* topological space.

5.1 EXAMPLES; METRIC SPACES

Before we explore the general characteristics of topological spaces, it will be useful to have a selection of examples on hand.

Example α.   Let X be an arbitrary set. Define $\tau = \mathscr{P}(X)$, i.e. declare every subset of X to be open. The axioms (i)–(iv) are trivially fulfilled, hence we have a topological space. This space is called the *discrete space* (X, τ) and its topology is the *discrete topology.*

Example β.   Another extreme is the following. Let X be arbitrary and define $\tau = \{\emptyset, X\}$, i.e. call only the empty set and X itself open. This is called the *indiscrete topology* for X and (X, τ) is an *indiscrete space.*

Example γ.   Let $X = \{a, b, c\}$. Define the topology (i.e. the class of open sets of X) as

$$\tau = \{\emptyset, \{a\}, \{a, b\}, \{a, c\}, X\}.$$

A direct check shows that the axioms are fulfilled so that (X, τ) is a topological space.

Example δ.   Let X be an infinite set and define $V \in \tau$ if either $V = \emptyset$ or if V^c is a finite set. It is an easy exercise to show that τ is a topology. It is called the *finite complement topology.*

---

†It can be seen that axioms (i) and (ii) can be omitted since, strictly speaking, they follow from axioms (iii) and (iv). Indeed, the union of an "empty subcollection" of sets is $\emptyset$; and the intersection of an empty subcollection of sets is X (since the requirement "$x \in X$ belongs to all sets that are in the subcollection" is vacuously fulfilled for all x). Nevertheless, we shall ignore this subtlety and always explicitly verify the truth or otherwise of axioms (i) and (ii).

Historically the first, and probably the most important, examples of topological spaces were a type of space called *metric spaces*. In the rest of this section we shall discuss this notion, from a completely general point of view.

Let X be an arbitrary (universal) set. Let

$$d: X \times X \to \mathbf{R}$$

be a given function from $X \times X$ to the reals which obeys the following axioms:†

(i) $d(x, y) \geq 0$ for all $x, y \in X$,
(ii) $d(x, y) = 0$ iff $x = y$,
(iii) $d(x, y) = d(y, x)$ for all $x, y \in X$,
(iv) $d(x, y) \leq d(x, z) + d(z, y)$ for all $x, y, z \in X$.

Any function d which obeys these axioms is called a *metric on X*. The numerical value of d for a given pair (x, y) of points of X is referred to as the *distance* between x and y. Thus, the essence of the definition of a metric is that the distance of any two points is a *nonnegative* real number; the distance between two points vanishes iff the two points coincide; the distance is a *symmetric* function of the points; and finally the distance obeys the *triangular relation* (iv) for any three points. Since these are precisely the obvious characteristics of the familiar "distance in the Euclidean plane," the terminology is a natural generalization of a simple intuitive concept.

A set X equipped with a metric on X is called a *metric space* and is denoted by the ordered pair (X, d). If, from the context, it is clear what metric d we are talking of, we often take the liberty to speak simply of "the metric space X." Obviously, on a given set X one may define several (maybe even infinitely many) different metrics, and *each d* gives rise in general to a *different* metric space (X, d).

In passing we note that from the axioms of a metric one easily deduces the following relation:

$$d(x, y) \geq |d(x, z) - d(z, y)| \text{for all}   x, y, z \in X.$$

Even though this is not a separate axiom, it is worthwhile memorizing, since in many calculations it plays an important role.

We now give examples of metric spaces.

---

†Since the elements of $X \times X$ are the ordered pairs (x, y) with $x \in X, y \in X$, the notation $d(x, y)$ for the value of d at (x, y) is standard.

Example ε. Let X be an arbitrary set and define

$$d(x, y) = \begin{cases} 0 & \text{if } x = y \\ 1 & \text{if } x \neq y. \end{cases}$$

The axioms are obviously fulfilled, and the metric d is the *discrete metric* for the *discrete metric space* (X, d).

Example ζ. Let $X = \mathbf{R}$ be the real line and define the distance $d(x, y) = |x - y|$. An elementary calculation shows that this is a metric, called the *usual metric* for $\mathbf{R}$, and $(\mathbf{R}, d)$ is the simplest metric space we are familiar with.

Example η. Let $X = \mathbf{R}^n$, with points

$$x = (\xi_1, \ldots, \xi_n), \qquad y = (\eta_1, \ldots, \eta_n); \qquad \xi_k, \eta_k \quad \text{real.}$$

Define

$$d(x, y) = \left[\sum_{k=1}^{n} (\xi_k - \eta_k)^2 \right]^{1/2}.$$

All requisite properties for a metric are clearly satisfied, only the triangular relation is not obvious. The proof of the latter follows, however, from the Minkowski inequality, cf. Appendix I. Taking there $p = 2$ and writing

$$(\xi_k - \eta_k) = (\xi_k - \gamma_k + \gamma_k - \eta_k),$$

we have

$$\left[\sum_{k=1}^{n} (\xi_k - \eta_k)^2 \right]^{1/2} \leq \left[\sum_{k=1}^{n} (\xi_k - \gamma_k)^2 \right]^{1/2} + \left[\sum_{k=1}^{n} (\gamma_k - \eta_k)^2 \right]^{1/2},$$

which shows that $d(x, y) \leq d(x, u) + d(u, y)$ for any $u = (\gamma_1, \ldots, \gamma_n)$. Thus, we have a metric, which will be referred to as the *Pythagorean* or *Euclidean metric* for $\mathbf{R}^n$. The corresponding metric space $(\mathbf{R}^n, d)$ will be denoted in the following by E^n and, for short, will be referred to as the *Euclidean n-space*.† Two special, and easily visualized cases are, of course, E^2 and E^3, i.e. the Euclidean 2- and 3-space, respectively. Furthermore, $E^1 \equiv E$ is simply the space described in Example $ζ$.

Example ϑ. The previous example has an often occurring generalization. Let again $X = \mathbf{R}^n$ and set

---

†Often the term "*n-dimensional Euclidean space*" is also used. However, in the context of pure metric space theory, we ignore the familiar algebraic (linear space) structure of $\mathbf{R}^n$ and therefore the word "dimension" is without significance.

$$d(x, y) = \left[\sum_{k=1}^{n} |\xi_k - \eta_k|^p\right]^{1/p}, \quad \text{with} \quad p \geq 1 \quad \text{fixed real number.}$$

The proof that this is a metric follows from the general form of the Minkowski inequality. The corresponding metric space $(\mathbf{R}^n, d)$ will be denoted for short by E_p^n. Of course, the Euclidean n-space E^n is a special case, $E^n = E_2^n$. Observe also that, for $X = \mathbf{R}$, all E_p^1 spaces coincide, and become identical to the usual metric space of the real line, discussed in Example ζ. We then simply use the notation E.

Example κ. The spaces E_p^n whose underlying sets $\mathbf{R}^n$ are n-tuples of real numbers generalize, without any problem, to the sets $\mathbf{C}^n$, whose elements are n-tuples $z = (\zeta_1, \dots, \zeta_n)$ of complex numbers. Again, we define, for any fixed $p \geq 1$,

$$d(z, w) = \left[\sum_{k=1}^{n} |\zeta_k - \omega_k|^p\right]^{1/p}.$$

The corresponding metric spaces will be denoted by C_p^n. In particular, for $p = 2$ and so with†

$$d(z, w) = \left[\sum_{k=1}^{n} |\zeta_k - \omega_k|^2\right]^{1/2}$$

we have the complex Euclidean n-spaces, more frequently called *unitary n-spaces* and denoted simply by C^n (instead of C_2^n).

Example λ. Let X be the set of all (infinite) real sequences $x = (\xi_1, \dots, \xi_k, \dots)$ *subject to the condition* that‡

$$\sum_{k=1}^{\infty} \xi_k^2 < \infty.$$

Define

$$d(x, y) = \left[\sum_{k=1}^{\infty} (\xi_k - \eta_k)^2\right]^{1/2}.$$

First we must show that the r.h.s. really exists, i.e. that we do have a well-defined function. By the Minkowski inequality

$$\left[\sum_{k=1}^{\infty} (\xi_k - \eta_k)^2\right]^{1/2} \leq \left(\sum_{k=1}^{\infty} \xi_k^2\right)^{1/2} + \left(\sum_{k=1}^{\infty} \eta_k^2\right)^{1/2},$$

†Note that for the *real* $E_2^n = E^n$ space (as well as for all E_p^n spaces with $p = $ even), we can write $|\xi_k - \eta_k|^p = (\xi_k - \eta_k)^p$ as we did in Example η.

‡The conventional notation simply means that the infinite series $\sum_{k=1}^{\infty} \xi_k^2$ converges.

and since, by the assumption on the elements x, y of our underlying set, both terms on the r.h.s. are finite numbers, the infinite series on the l.h.s. converges, so that $d(x, y)$ is well defined. The proof that $d(x, y)$ is a metric, follows the same pattern as for finite sequences, i.e. the triangular inequality is checked again by the use of the Minkowski inequality. Thus, we have a metric space of "square summable" infinite real sequences. This space will be denoted by E^∞ and is sometimes called "the *infinite dimensional Euclidean space*." The procedure generalizes without any problem to "p-summable" real sequences, i.e. to sequences $x = (\xi_1, \ldots, \xi_k, \ldots)$ which obey the restriction

$$\sum_{k=1}^{\infty} |\xi_k|^p < \infty.$$

We then define

$$d(x, y) = \left[\sum_{k=1}^{\infty} |\xi_k - \eta_k|^p\right]^{1/p}$$

and denote the space by E_p^∞. Needless to say, we can now further generalize to p-summable complex sequences and correspondingly get the spaces C_p^∞. In particular, the space consisting of complex sequences $z = (\zeta_1, \ldots, \zeta_k, \ldots)$ subject to

$$\sum_{k=1}^{\infty} |\zeta_k|^2 < \infty$$

and with

$$d(z, w) = \left[\sum_{k=1}^{\infty} |\zeta_k - \omega_k|^2\right]^{1/2}$$

is called the "*infinite dimensional unitary space*" and is denoted by C^∞ (rather than by C_2^∞).

Example μ. Lest the reader think that we cannot make interesting metric spaces from the set $\mathbf{R}^n$ (or $\mathbf{C}^n$) other than the E_p^n (or C_p^n) spaces, we give yet another example. Using the same underlying set of real (or complex) n-tuples, we define

$$d(x, y) = \max_{1 \le k \le n} \{|\xi_k - \eta_k|\}.$$

We leave it to the reader to verify that this is a metric.† As a matter of

---

†It can be shown that

$$\max\{|\xi_k - \eta_k|\} = \lim_{p \to \infty} \left[\sum_{k=1}^{n} |\xi_k - \eta_k|^p\right]^{1/p}.$$

Therefore, these spaces are often denoted by E_∞^n (or C_∞^n), respectively.

fact, if we have a space of infinite real (or complex) sequences $x = (\xi_1, \ldots, \xi_k, \ldots)$ subject to the boundedness condition

$$|\xi_k| \le N, \qquad N > 0, \quad k = 1, 2, \ldots$$

then

$$d(x, y) = \sup_k \{|\xi_k - \eta_k|\}$$

also defines a metric.

Example v. In later applications, we shall have frequently to do with metric spaces whose elements are functions. We discuss here the simplest case of such function spaces. Let $X = [a, b]$ denote a "closed and bounded interval" of the real numbers, i.e. the set of reals such that $a \le x \le b$, with a, b finite. Consider the class of all real (or complex) valued continuous† functions on $[a, b]$. Define "the distance of two functions f and g" by

$$d(f, g) = \sup_{x \in [a, b]} |f - g| \equiv \sup_{a \le x \le b} \{|f(x) - g(x)|\}.$$

Again, it is only the triangular inequality that is not obvious. But, if h is any function in our set, we have

$$|f(x) - g(x)| \le |f(x) - h(x)| + |h(x) - g(x)|,$$

from which the desired inequality follows easily. The metric space so obtained will be denoted by $C[a, b]$. Figure 5.1 illustrates the distance $d(f, g)$.

After all these examples, the reader may now wonder what have metric spaces to do with topological spaces? In the following we shall demonstrate that if we have a metric d on some set X, then this metric induces a topology τ on X, so that every metric space may be looked upon as a topological space. Our discussion leading to this result will be a bit tortuous.

First of all, given a metric space (X, d), we define an *open ball*‡ centered at x_0 and having radius $r > 0$ to be the subset

$$S_r(x_0) = \{y \in X | d(y, x_0) < r\}. \tag{5.1}$$

We emphasize that, by definition, r is *strictly* positive so that an open ball has always a *finite* radius.

---

†We anticipate here that the reader is familiar with the elementary concept of continuous real (or complex) functions on the real line.

‡The term open *sphere* is also frequently used, but it may give rise to confusion. An open ball is "what is inside a balloon." The term "sphere" should be reserved to denote precisely the skin of a balloon.

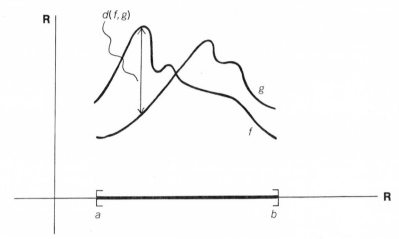

Fig. 5.1.

The notion of open balls is rather simple and Fig. 5.2 shows the "shape" of open balls for three different metrics† in $\mathbf{R}^2$ (namely for $E_2^2 = E^2$, E_1^2, and E_∞^2). Similarly, Fig. 5.3 illustrates an open ball in the space $C[a, b]$. Note that only those functions g belong to the open ball with radius r about the center f_0 which do not touch the "parallel" curves even in one single point (since we must have $d(g, f_0)$ strictly smaller than r). Finally, we record that an open ball about x_0 with radius r on the real line with the usual metric is precisely the set of real numbers x satisfying the inequality $x_0 - r < x < x_0 + r$.

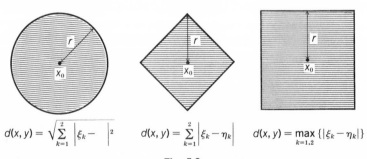

$$d(x, y) = \sqrt{\sum_{k=1}^{2} \left|\xi_k - \right|^2} \qquad d(x, y) = \sum_{k=1}^{2} \left|\xi_k - \eta_k\right| \qquad d(x, y) = \max_{k=1,2} \{|\xi_k - \eta_k|\}$$

Fig. 5.2.

†The reader should be aware that the concept of an open ball depends both on the underlying set X and on the metric d.

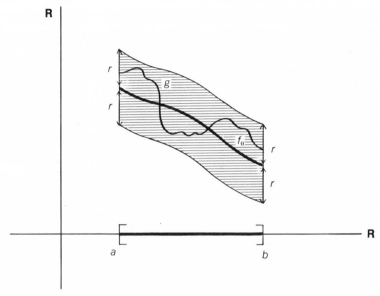

Fig. 5.3.

Now let V be a subset of X. We agree to say that "V is open" iff *for every point $x \in V$ there exists some $r > 0$ such that the open ball $S_r(x)$ is completely contained in V.* Loosely speaking, we call a subset V of a metric space open iff each of its points can be surrounded by *some* open ball, which lies in V. For example, on the real line with the usual metric, every familiar "open interval"† (a, b), i.e. every set of real numbers x such that $a < x < b$, is an open set according to our present definition, since given any x in (a, b) we can find a (sufficiently small) $r > 0$ such that the set $S_r(x) = (x - r, x + r)$ is completely inside the set (a, b). On the other hand, the semiclosed interval $[a, b)$ is not open, because the "endpoint" a has the property that every open ball centered about it has points that do not belong to $[a, b)$. For the closed interval $[a, b]$, both endpoints are "wrong." Furthermore, it will be worthwhile to remember that the singleton set $\{x\}$, consisting of the single real number x, is not open in E. Same applies to any finite subset $\{x_1, x_2, \ldots, x_k\}$. The reader should concoct other examples, using the spaces discussed above, and

---

†In conformity with elementary practice, we shall denote open, lower and upper semiclosed, closed intervals of the real line by (a, b), $[a, b)$, $(a, b]$, and $[a, b]$, respectively, meaning the sets $a < x < b$, $a \le x < b$, $a < x \le b$, and $a \le x \le b$, respectively.

keeping in mind that the definition of openness in X is relative to the agreed-upon metric d.

From our present definition of openness it immediately follows that $\emptyset$ is an open subset of any metric space (the criterion is vacuously fulfilled) and X itself is open (for any point $x \in X$ there exists an open ball, necessarily contained in X). To show that the term "open set" is advisedly chosen (i.e. that the open sets of (X, d) as defined above are members of a topology) we need the following

LEMMA 5.1(1).   *A set $V \subset (X, d)$ is open iff it is the union of open balls.*

We leave the simple proof as an exercise,† but wish to make three remarks. First, the lemma does *not* claim that an open set of a metric space is a finite union of open balls. We may need infinitely many, maybe even uncountably many open balls to build up an open set. Second, the lemma assures us that an open ball *is* an open set (in fact, the simplest type of an open set). This justifies the consistency of nomenclature. Third, it should be rather obvious that unions of open balls are, in general not open balls, so that the "shape" of an open set may be quite wild.

Now we are ready to show that arbitrary unions and finite intersections of our open sets of (X, d) are again open. Actually, to show closure under unions, we simply realize that the union of unions of open balls is again a union of open balls, so the lemma gives the desired result. To show closure under finite intersections, let $V = \cap_{k=1}^{n} V_k$ with each V_k open. Let $x \in V$, hence x is in *each* V_k. Each V_k being open, there exists for each k an $r_k > 0$ such that $S_{r_k}(x) \subset V_k$. Let r be the smallest number in the set $\{r_1, \ldots, r_n\}$. Then $S_r(x) \subset S_{r_k}(x)$ for each k, so that $S_r(x) \subset V_k$ for each k, and hence $S_r(x) \subset V$. Thus, $V = \cap_{k=1}^{n} V_k$ is open, because for every $x \in V$ we could find an open ball fully contained in V.‡

We summarize our insight in the following manner. *If (X, d) is an arbitrary metric space, then, defining open balls and declaring all arbitrary unions of open balls to be the class of open sets, we obtain a topology τ on X.* Thus, metric spaces are topological spaces where the "*natural topol-*

---

†If you are unwilling to spend time on the proof or run into difficulty, see, for example, Simmons[36], p. 61.

‡The reader will observe that the lemma was needed only to prove the closure under unions. The closure under finite intersections followed directly from the definition of open sets. He also should observe that the proof of closure breaks down for infinite intersections. (Why?) For example, an infinite class of balls in $\mathbf{R}^2$, centered about a fixed point x_0 and having radii $1, 1/2, 1/3, \ldots, 1/k, \ldots$ has, as its intersection, the set $\{x_0\}$ which is not open.

ogy" is the one which is derived from the metric via the concept of open balls. Of course, the same underlying set may be given some other topology, but when we talk about a metric space as a topological space, we assume that the natural topology is implied.

At this point, the reader is perhaps led to ask as to whether the notion of a topological space is really broader than that of a metric space? In other words, he may wonder whether it is possible or not to devise for an arbitrary *given* topological space (X, τ) a suitable metric d which, via the construction of open balls, leads to the originally prescribed topology τ. The answer is negative: *there exist topological spaces which are not "metrizable."* We shall not, in this book, discuss all sufficient and necessary criteria on a topological space which are conditions for metrizability, but content ourselves with a few examples.

Example ξ. Any discrete topological space is metrizable. The discrete metric (Example **ε**) reproduces the discrete topology, as the reader will easily verify.

Example π. Let **R** be the real line and define the topology by setting $\tau = \{$all unions of subsets $(a, b) \equiv \{a < x < b\}\}$. It is easy to see that this is a topology, and in fact it is recovered by the usual metric $d(x, y) = |x - y|$, because for this, the open balls are precisely the sets $S_r(x_0) = \{y | x_0 - r < y < x_0 + r\}$. Thus, the topology originally defined is exactly the natural topology of **R** considered as a metric space E.

Example ρ. An indiscrete topological space is not metrizable. To see this, let $x \neq y$ so that by the axioms of a metric, $d(x, y) \equiv \rho > 0$. Therefore, the open ball $S_\rho(x)$ (which is an open metric set) would contain x but not y. But in the indiscrete space, no open set can contain one point without also containing all other points (because the only open sets are $\emptyset$ and X itself). Hence, no metric can reproduce the indiscrete topology.

We conclude these considerations with the interesting remark that, if a topological space is metrizable at all, then there are actually infinitely many metrics on X which yield the same prescribed topology τ.

PROBLEMS

5.1-1. Describe all possible topologies on a set consisting of three points.

5.1-2. Let X be a totally ordered poset. Define the family of "open intervals" as follows: if $x \neq y$, let $]x, y[$ be the set of points z such that $x < z < y$; if $x = y$, then $]x, y[= \emptyset$. Show that the class of sets consisting of all possible unions of open intervals is a topology for X.

5.1-3. Let τ_1 and τ_2 be two topologies on X. Show that $\tau \equiv \tau_1 \cap \tau_2$ is also a topology on X. Generalize your result as follows: if (τ_λ) is an indexed family of topologies on X, then $\cap_\lambda \tau_\lambda$ is also a topology on X. In contrast, show that $\cup_\lambda \tau_\lambda$ need not be a topology.

5.1-4. Let (X, d) be a metric space. Define the map $d': X \times X \rightarrow \mathbf{R}$ as follows:

$$d'(x, y) = \begin{cases} d(x, y) & \text{if } d(x, y) \leq 1, \\ 1 & \text{if } d(x, y) \geq 1. \end{cases}$$

Show that d' is also a metric on X.

5.1-5. Let (X, d) be a metric space. Show that $d': X \times X \rightarrow \mathbf{R}$ given by

$$d'(x, y) = \frac{d(x, y)}{1 + d(x, y)}$$

is also a metric on X.

5.1-6. Define the *diameter* of an arbitrary metric space by

$$d(X) = \sup \{d(x, y) | x \in X, \quad y \in X\}.$$

A metric space is said to be bounded if the set $\{\ldots d(x, y) \ldots\}$ is a bounded subset of the reals, i.e. if $d(X)$ is finite. Show that the metrics d' of Problems 5.1-4 and 5.1-5 make X a bounded metric space, even if (X, d) (with the original metric d) was not a bounded metric space.

5.1-7. Let (X_1, d_1) and (X_2, d_2) be two metric spaces. Consider the set $X_1 \times X_2$ and show that the following three definitions are metrics on the product set:
(a) $d[(x_1, x_2), (x'_1, x'_2)] = \max_{k=1,2} \{d_k(x_k, x'_k)\}$,
(b) $d[(x_1, x_2), (x'_1, x'_2)] = d_1(x_1, x'_1) + d_2(x_2, x'_2)$,
(c) $d[(x_1, x_2), (x'_1, x'_2)] = [d_1^2(x_1, x'_1) + d_2^2(x_2, x'_2)]^{1/2}$.

5.1-8. Let E be the real line with its usual metric. Give an example of a countable class of open sets V_k such that $\cap_{k=1}^\infty V_k$ is not open.

5.1-9. Show that a discrete topological space is metrizable and that the discrete metric does the trick. Can you find another, equally good metric?

5.1-10. Let $X = \{a, b, c\}$ and let $\tau = \{\emptyset, \{a\}, \{a, b\}, \{a, c\}, X\}$. Show that τ is a topology on X but (X, τ) is not metrizable.

5.1-11. Show that a space with the finite complement topology (Example δ) is not metrizable.

5.2 GENERAL STRUCTURE OF TOPOLOGICAL SPACES

In this section we discuss a number of overall characteristics pertaining to topological spaces.

Knowing already that on a given set X there can exist many distinct topologies, we may wish to compare them. To this end we introduce a partial order† into the class of all topologies on X. We shall say that a topology τ_1 on X is *smaller* than a topology τ_2 on X if $\tau_1 \subset \tau_2$. We often indicate this writing $\tau_1 < \tau_2$. If $\tau_1 < \tau_2$, we often say also that τ_1 is a *weaker* and τ_2 is a *stronger* topology‡ on X.

The weakest topology on any X is the indiscrete topology $\tau = \{\emptyset, X\}$. The strongest topology is the discrete one, $\tau = \mathcal{P}(X)$. It should be noted that we have only a *partial* ordering of topologies. For example, on the real line **R** the finite complement topology τ_1 (Example δ in Section 5.1) is weaker than the usual topology τ_2 (unions of all open intervals), but if we define

$$\tau_3 = \{V \subset \mathbf{R} |\ V = \emptyset \quad \text{or} \quad V = \text{an open set in the usual topology which contains the number 0}\},$$

then we have a topology such that $\tau_3 < \tau_2$ but which is not contained in, nor contains the finite complement topology τ_1. On the other hand, it can be shown that the class of all topologies on any X is a lattice, with a first and last member.

Our next inquiry concerns the construction of new topological spaces from given ones. Let (X, τ) be a topological space and A be a subset of X. The collection of subsets of A consisting of the intersections of A with all open sets V of (X, τ) is easily seen to obey the criteria for open sets relative to A. Thus,

$$\tau_A = \{A \cap V | V \in \tau\}$$

is a topology for A. We call τ_A the *relative topology* on A (induced by the topology τ of X) and the topological system (A, τ_A) is called a *topological subspace* of (X, τ). Thus, if we talk of a subspace A of a topological space X, we mean a subset equipped not with an arbitrary, but precisely with the relative topology. For example, if (X, τ) is an indiscrete topological space and $A \subset X$ is a subspace of (X, τ), then it is also an indiscrete space with $\tau_A = \{\emptyset, A\}$.

Let, in particular, (X, d) be a metric space and $A \subset X$. Consider the restriction of $d: X \times X \to \mathbf{R}$ to the function $d_A: A \times A \to \mathbf{R}$. It is clear that (A, d_A) is a metric space, called a metric subspace. A further study reveals that the natural topology on (A, d_A) is precisely the relative

---

†We recall that each topology τ is a subset of the power set $\mathcal{P}(X)$ and that $\mathcal{P}(X)$ has a natural order relation, namely the one defined by set inclusion.

‡The pair of terms "coarser-finer" is also used occasionally but not always consistently and should be avoided.

topology induced by the natural topology of (X, d). Hence, any metric subspace of a metric space is a topological subspace in the above sense.

We now turn our interest in the other direction. Let (X, τ_X) and (Y, τ_Y) be given topological spaces. We can construct in a very natural way a "big" new space. Let $Z = X \times Y$ be the Cartesian product of the underlying sets. Let G and H be arbitrary open sets of (X, τ_X) and (Y, τ_Y), respectively. Define on Z the class τ of open sets to consist of all possible unions made up from all possible sets $G \times H$ with $G \in \tau_X$ and $H \in \tau_Y$. It is not difficult to check that this class τ (often denoted by $\tau_X \times \tau_Y$) has indeed all requisite properties of a topology. Thus, $(Z, \tau) \equiv (X \times Y, \tau_X \times \tau_Y)$ is a topological space, called the *product space* with the *product topology*.

To give an example, let $X = \mathbf{R}$, $Y = \mathbf{R}$, and $Z = \mathbf{R} \times \mathbf{R} = \mathbf{R}^2$. Defining the topology on $\mathbf{R}$ to consist of all possible unions made up of all possible open intervals (cf. Example π in Section 5.1), it will suffice to consider all subsets of $\mathbf{R}^2$ which are of the form $(a, b) \times (c, d)$ (where (a, b) stands for the subset $a < x < b$ of $\mathbf{R}$). Such subsets of $\mathbf{R}^2$ are called *open rectangles* and are illustrated in Fig. 5.4α. Note that some open rectangles (which

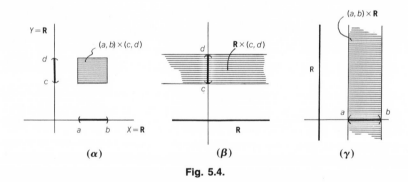

(α) (β) (γ)

Fig. 5.4.

have the form $\mathbf{R} \times (c, d)$ or $(a, b) \times \mathbf{R}$) are "unbounded" and deserve a special name: they are called *open strips* (see Figs. 5.4β and γ). Now, by the general definition of a product space, the class of open sets of $\mathbf{R}^2$ will be the class of all possible unions made up from all possible open rectangles. A simple consideration† shows that this product topology coincides with the usual topology of $\mathbf{R}^2$, namely the one derived from the

†See, for example, McCarty[28], p. 98.

metric

$$d(x, y) = \left[\sum_{k=1}^{2} (\xi_k - \eta_k)^2 \right]^{1/2}.$$

There is a broad generalization of this example. Let (X, d_X) and (Y, d_Y) be metric spaces. If $z_1 = (x_1, y_1)$ and $z_2 = (x_2, y_2)$ denote two points of $X \times Y$, we define the *product metric* d: $(X \times Y) \times (X \times Y) \rightarrow \mathbf{R}$ by setting

$$d(z_1, z_2) = [d_X^2(x_1, x_2) + d_Y^2(y_1, y_2)]^{1/2}.$$

It is not difficult to show that this is a metric for $Z = X \times Y$, thus (Z, d) is a metric space. Then, it can be proved that the natural topology of (Z, d) is precisely the product topology arising from the natural topologies of (X, d_X) and (Y, d_Y).

The product topology can be generalized to products of arbitrary (not necessarily countable) collections of spaces, but this is a rather subtle business.†

After having studied the construction of new topological spaces from old ones, we now concentrate on a fixed given space (X, τ) and try to find a way to visualize the topology in a more convenient manner than by thinking all the time of the entire class of open sets. Let $\mathscr{B}$ be a subclass of open sets of (X, τ) such that any open set V of X can be represented as a union (not necessarily countable) of sets taken from $\mathscr{B}$. Such a class $\mathscr{B}$ is called a *base for the topology* τ. Thus,

$$\mathscr{B} \subset \tau \quad \text{is a base iff} \quad \tau = \{ \cup_\beta B_\beta | B_\beta \in \mathscr{B} \}.$$

Members of a base $\mathscr{B}$ are called *basic open sets*.

We are already familiar with an example of a base: in any metric space the class of all open balls is obviously a base. Thus, for example, on the real line the class of all open intervals is a base for the natural topology.‡ But one can easily find "smaller" bases. For example, let $\mathscr{B}$ consist of all open intervals (ρ, σ) which have *rational* endpoints. This is a base, because if (a, b) is an *arbitrary* open interval, we can represent it as $\cup_{k=1}^{\infty} (\rho_k, \sigma_k)$, where§ $\rho_k \rightarrow a$, $\sigma_k \rightarrow b$, and then a completely arbitrary open set is, as we know, a further union of some open intervals (a, b).

Another interesting example of a base is given by the product topology

†See, for example, Simmons[36], p. 117.

‡This is the reason why the topology for **R** defined in Example π of Section 5.1 leads precisely to the natural topology.

§We anticipate here that the reader is familiar with the convergence of sequences of numbers and knows that any real number is the limit of a sequence of rational numbers.

of $\mathbf{R}^2$: the class of all open rectangles is a base. As a matter of fact, this has an important generalization. By its very definition given above, *the product topology $\tau_X \times \tau_Y$ on $X \times Y$ can be characterized in the simplest manner by saying that it is a topology whose base is the class $\mathscr{B} = \{G \times H \mid G \in \tau_X, H \in \tau_Y\}$.*

As these examples illustrate, bases are useful if their members are very simple or very few in number. We now go a step further and introduce the concept of a subbase. A *subbase for a topology* is a subclass $\mathscr{S}$ of open sets with the property that each member of some base $\mathscr{B}$ is a finite intersection of sets taken from $\mathscr{S}$.

A simple example of a subbase is given again by considering the product topology of $\mathbf{R}^2$: It is evident that the class of all open strips is a subbase.

We now actually turn round the argument and show how one can construct from an *arbitrary* collection $\mathscr{A}$ of given subsets of X a topology for X which contains, among its open sets, all members of $\mathscr{A}$. A little consideration shows that if $\mathscr{A} = \{\ldots A_\alpha \ldots\}$ and if we declare its elements A_α together with $\emptyset$ and X itself† to be a subbase, form all possible finite intersections, declare the collection so obtained to be a base, and form all possible unions, then in this manner we get a class τ which serves as a topology for X and which, obviously, contains the given sets A_α. The merit of this construction is that it leads to the "topologization" of a set X in such a way that some desired subsets are guaranteed to be open and also such that (X, τ) is not bigger than absolutely necessary, i.e. we get the weakest topology which contains the given sets.

We illustrate this procedure with the following example. Let $X = \{a, b, c, d\}$. We wish to introduce on X the weakest topology which contains the subsets $\{a, b\}$ and $\{a, c\}$. To this end we first form all (finite) intersections as prescribed, getting the collection $\{\emptyset, X, \{a, b\}, \{a, c\}, \{a\}\}$. This is taken to be a base, so that we are led to the topology $\tau = \{\emptyset, X, \{a\}, \{a, b\}, \{a, c\}, \{a, b, c\}\}$.

PROBLEMS

5.2-1. Show that three topologies on $X_1 \times X_2$ derived from the metrics (a), (b), and (c) of Problem 5.1-7 are of equal fineness, i.e. $\tau_1 = \tau_2 = \tau_3$. (Note that, nevertheless, the open balls are different in the three metric spaces.)

5.2-2. Let (X, τ) be a topological space and let (A, τ_A) and (B, τ_B) be subspaces. Suppose $A \subset B$. Show that τ_A is the same topology as the one which would be the

†Strictly speaking, one need not explicitly include $\emptyset$ and X, cf. footnote on p. 188.

relative topology on A induced by the topology τ_B of B (which, in turn, is of course the relative topology on B induced by the topology τ on X). In other words, show that every subspace A of every subspace B of X is a subspace of X. For an illustration, consider the discrete relative topologies on **Z**, **R**, **C**.

5.2-3. Show that the natural topology on $X_1 \times X_2$ defined by any of the metrics (a), (b), and (c) of Problem 5.1-7 is the same as the product topology on $X_1 \times X_2$. (*Note*: See also Problem 5.2-1.)

5.2-4. Let $X_1 = \{a_1, b_1\}$, $X_2 = \{a_2, b_2\}$, each space having the discrete topology. Construct the topological product space and determine its topology.

5.2-5. Consider the Euclidean metric space E^n and show that the collection of open balls with rational centers and rational radii form a base for the natural topology of E^n. (*Remark*: A point in **R**^{n} is called rational if all its components $\xi_1, \ldots, \xi_n$ are rational numbers.)

5.2-6. Suppose (X, τ) possesses a countable base. Show that then every possible base must contain a countable subset which is itself a base.

5.2-7. Let (X, τ_X) and (Y, τ_Y) be spaces with bases $\mathscr{B}_X$ and $\mathscr{B}_Y$. Show that

$$\mathscr{B} = \{B_X \times B_Y | B_X \in \mathscr{B}_X \quad \text{and} \quad B_Y \in \mathscr{B}_Y\}$$

is a base for the product topology of $X \times Y$. (*Remark*: This new base is "smaller" than the base $\{G \times H | G \in \tau_X, H \in \tau_Y\}$ originally used for defining the product topology.)

5.2-8. Consider the real line **R** with its usual topology and show that the collection of all open intervals of length 1 is a subbase.

5.3 NEIGHBORHOODS; SPECIAL POINTS; CLOSED SETS

In this section we shall endeavor to obtain a good view of the internal characteristics of a given topological space.

A very simple but crucial concept for visualizing the structure of a space is that of a *neighborhood*. A neighborhood $N(x)$ of a point $x \in (X, \tau)$ is any open set† that contains the point x. By its definition, a neighborhood is never empty, and every point x has some neighborhood (in the worst case, its only neighborhood is X itself). Observe also that if $N(x)$ is a neighborhood of x and $y \in N(x)$, then $N(x)$ is also a neighborhood of y.‡

The simplest example of a neighborhood is found in metric spaces. Any open ball centered about x is a neighborhood of x. Thus, the

---

†Many authors use the term in a more general sense, and call any set $M(x)$ a neighborhood of x if there exists an open subset $V \subset M(x)$ such that $x \in V$. However, in our usage, any neighborhood is *open*.

‡Thus, our notation $N(x)$ is somewhat misleading.

neighborhood concept is the straightforward generalization of the very pictorial notion of a ball but completely liberated from the concept of "distance." We also note that if $N(x)$ is an *arbitrary* neighborhood of x in a metric space, then $N(x)$ will contain a subneighborhood which is simply an open ball about x.

Neighborhoods can be used to characterize open sets. If V is an open set of (X, τ), then every $x \in V$ has some neighborhood $N(x)$ which is entirely contained in V (in the worst case, $N(x) = V$). Conversely, if A is some subset such that for every $x \in A$ there exists a neighborhood $N(x)$ which is entirely contained in A, then A is open (this follows from noting that A may be written as the union of all neighborhoods of all its points). These observations are a generalization of the fact that, in a metric space, a set is open iff it is the union of open balls.

One of the most important applications of the neighborhood concept is that it enables one to classify certain interesting points of fixed subsets.

Let A be a subset of some topological space (X, τ). A point $x \in A$ is called an *isolated point* of the set A if it has a neighborhood which contains no other point of A than x itself.†

Next we consider the "antithesis" of an isolated point. Let again A be a subset of some topological space (X, τ). A point $x \in X$ is called an *accumulation point* ‡ of the set A if *every* neighborhood of x contains at least one point $a \in A$, where it is understood that $a \neq x$. If x is an accumulation point of A, it may or may not belong to A (the accumulation point is defined as a point of X (not necessarily in A)). The definition of an accumulation point for metric spaces becomes particularly simple: if $A \subset (X, d)$, then $x \in X$ is an accumulation point of A iff for every $\epsilon > 0$ there exists at least one $a \in A$ $(a \neq x)$ such that $d(a, x) < \epsilon$. Since ϵ is arbitrary, it then follows that, *in a metric space*, an accumulation point x of A contains actually infinitely many distinct points of A in each of its neighborhoods. This reveals that the points of A different from x sort of "pile up" at x.

We give two preliminary examples for these new concepts.

Examples α.   In the indiscrete topological space let A be an arbitrary set (containing at least two points). It has no isolated points because if $x \in A$, its only neighborhood is X which contains actually all points of A.

---

†One has to be careful here. The neighborhoods of x are defined as certain subsets of X (not A) and an isolated point x of A may very well have many points $y \in X$ but $y \notin A$ in even *all* of its neighborhoods.

‡The somewhat misleading term "limit point" is also frequently used.

This same argument shows that every point of A is an accumulation point, moreover *any* point $x \in X$ is an accumulation point of A.

Example β.   Let X be the real line with the usual metric. Let A be the subset of reals consisting of the semiclosed interval $0 < x \leqslant 1$, the numbers 2 and 3, the rational numbers that lie between 4 and 5 inclusive, and the number $\sqrt{10}$. Here, the isolated points of A are the numbers 2, 3, and $\sqrt{10}$. The accumulation points of A are: all reals $0 \leqslant x \leqslant 1$ and all real numbers $4 \leqslant x \leqslant 5$. We note that out of these accumulation points the numbers $0 < x \leqslant 1$ and all rationals between 4 and 5 belong to A itself, whereas the number 0 and all irrationals between 4 and 5 do not belong to A.

We turn now our interest to sets which have the property that they contain all their accumulation points. These promise to be interesting sets because they are "closed up." It is not difficult to see how one can fabricate such a set. Suppose V is an open set and consider its complement V^c. This set must contain all its accumulation points. For let x be an accumulation point of V^c, and suppose $x \notin V^c$, i.e. $x \in V$. Then every $N(x)$ contains at least one point of V^c, hence $x \in V$ has no neighborhood entirely in V, so that V could not be open,† contrary to assumption. Hence $x \notin V$, so that we must have $x \in V^c$: every accumulation point of V^c belongs to V^c if V is open. The converse is also easily established: if a set A contains all its accumulation points, then its complement A^c is open. Thus, *the class of sets defined by the requirement that they contain all their accumulation points is precisely the class of sets which consists of the complements of all open sets.* This leads us to the following

DEFINITION 5.3(1).   *A subset C of (X, τ) is said to be a closed set iff its complement is open, i.e. iff $C^c \in \tau$.*

From this definition we obtain

THEOREM 5.3(1).   (a) $\emptyset$ *is a closed set,* (b) X *is a closed set,* (c) *Finite unions* $\cup_{k=1}^{n} C_k$ *of closed sets are closed,* (d) *Arbitrary intersections* $\cap_\alpha C_\alpha$ *of closed sets are closed.*

Proof.   Parts (a) and (b) follow from the fact that $\emptyset = X^c$, $X = \emptyset^c$ and that X and $\emptyset$ are open. Parts (c) and (d) follow from the deMorgan laws

$$\overset{n}{\underset{k=1}{\cup}} C_k = \left(\overset{n}{\underset{k=1}{\cap}} C_k^{\,c} \right)^c, \qquad \underset{\alpha}{\cap} C_\alpha = \left(\underset{\alpha}{\cup} C_\alpha^{\,c} \right)^c$$

†Cf. the discussion following the definition of neighborhoods.

and from the observation that finite intersections and arbitrary unions of open sets are open.

The theorem shows that the properties of closed sets parallel closely those of open sets. Actually, a topological space could be defined by axiomatically prescribing the class of closed sets on X (as a class of subsets closed under finite unions and under arbitrary intersections). Open sets would then arise as a derived concept, namely as the complements of closed sets.

Parts (a) and (b) of the above theorem merits special attention. Since $\emptyset$ and X are open by definition of τ and also closed by Theorem 5.3(1), we see that openness and closedness are not mutually exclusive attributes. Apart from $\emptyset$ and X there are, in general, many other sets which are simultaneously open and closed. See, for example, Problem 5.3-6. (These questions will be studied in detail in Chapter 6, Section 1.) It is also obvious that there are many sets which are neither open nor closed (e.g., a semiclosed interval $(a, b]$ of the real numbers).

The discussion preceding the formal definition of closed sets taught us that closed sets contain all their accumulation points. One may ask: what else do they contain? The answer is rather surprising: *apart from the set of its accumulation points, a closed set contains nothing else but its isolated points, if there are any.* We omit the proof and appeal to visualization by diagrams. Note also that the set of accumulation points is disjoint from the set of isolated points.

We give a few simple examples.

Example γ. Referring back to Example α, we see that in an indiscrete topological space no closed set can exist (except $\emptyset$ and X itself), since any $A \subset X$ has accumulation points that do not belong to it. The same result would follow also from noting that in this space the only open sets are $\emptyset$ and X.

Example δ. Let (X, d) be a metric space and define a *closed ball* with radius $r \geqslant 0$ about the center x_0 as the set

$$S_r[x_0] = \{y \,|\, d(y, x_0) \leqslant r\}.$$

Even though intuition shows us right away that this should be a closed set, we give a detailed proof. Using the definition of closed sets, we must show that the complement of $S_r[x_0]$, i.e. the set

$$K = \{z \,|\, d(z, x_0) > r\}$$

is open. Let, for some $z \in K$, $d(z, x_0) \equiv a$, and consider the open ball $S_{a-r}(z) = \{v \mid d(v, z) < a - r\}$. Any point of this open ball belongs to K. For indeed, $d(v, x_0) \geqslant |d(v, z) - d(z, x_0)| = |a - r - \epsilon - a| = r + \epsilon$ (where ϵ is an arbitrary positive number), so that $d(v, x_0) > r$, i.e. $v \in K$ as claimed. But, since we showed that for any $z \in K$ there exists an open ball that is entirely contained in K, it follows that K is open. Thus, $K^c = S_r[x_0]$ is closed, q.e.d.

Example ϵ. Let (X, τ) be an arbitrary space and let $C \subset X$ have no accumulation points. Then C is closed, because the criterion that it contains all its accumulation points is vacuously fulfilled. A particularly important consequence of this amusing observation is that, *in a metric space*, every finite subset is closed. (Actually, every such set consists of isolated points only.) In particular, every singleton set $\{x\}$ of a metric space is closed.

Example ζ. Let E^m be the Euclidean m-space with elements $x = (\xi_1, \ldots, \xi_m)$. The subset of points $r = (\rho_1, \ldots, \rho_n, 0, 0, \ldots 0)$ (where of course $n < m$) equipped with the restriction of the Euclidean metric is essentially† the Euclidean n-space E^n, and for brevity we shall refer to the set of points of form r as points of E^n. We wish to show that E^n *is a closed subspace of E^m*. All we have to show is that if $x \in E^m$ is an accumulation point of E^n, then it actually lies in E^n. Suppose x is an accumulation point. Then there must exist at least one $r \in E^n$ such that $d^2(x, r) < \epsilon$, for *any* $\epsilon > 0$. Thus, we must have

$$\sum_{k=1}^{m} |\xi_k - \rho_k|^2 \equiv \sum_{k=1}^{n} |\xi_k - \rho_k|^2 + \sum_{k=n+1}^{m} |\xi_k|^2 < \epsilon.$$

Since both terms are positive, each must be arbitrary small, in particular, we must have $\sum_{k=n+1}^{m} |\xi_k|^2 < \eta$ for *any* $\eta > 0$. But this is possible only if each $\xi_k = 0$ for $k > n$, hence $x = (\xi_1, \ldots, \xi_n, 0, \ldots, 0)$, so $x \in E^n$, q.e.d. With a small refinement of the argument the reader will have no difficulty to show that *every Euclidean space E^n is a closed subspace of E^∞*.

PROBLEMS

5.3-1. Show that a set A is open iff it is a neighborhood of each of its points.

5.3-2. Show that a point x of a *set* A is an isolated point of A iff the singleton set $\{x\}$ is open in the sub*space* A (i.e. with respect to the relative topology of A).

---

†The term "essentially" means here "homeomorphic to," a concept of equivalence for topological spaces which will be discussed in Section 5.6. For our present purposes, this is not important.

5.3-3. Let A and B be subsets of a topological space (X, τ). Denote by A' and B' the set of all accumulation points of A and B, respectively. Show that the set of all accumulation points of $A \cup B$ is the set $A' \cup B'$.

5.3-4. A subset $A \subset X$ is said to be a *perfect set* if it coincides with the set of its accumulation points. Show that A is perfect iff it is closed and has no isolated points.

5.3-5. Consider the following subsets of the real line, equipped with the usual topology:
(a) all integers,
(b) all rationals,
(c) all irrationals,
(d) $[0, 1] \cup \{1, 2\}$, (i.e. $\{x \mid 0 \leqslant x < 1$ or $x = 1, x = 2\}$).
Determine in each case the isolated points and the accumulation points. Which of the sets are open? Which are closed? Which are perfect?

5.3-6. Prove that in the discrete metric space (where $d(x, y) = 0$ if $x = y$, $d(x, y) = 1$ if $x \neq y$), every subset is both open and closed.

5.3-7. Let E^{n+1} be the Euclidean $(n + 1)$-space and let S^n be the subset of E^{n+1} whose points all lie at unit distance from the zero element $o = (0, 0, \ldots, 0)$. Thus,

$$S^n = \{x \in E^{n+1} \mid d(x, o) = 1\}.$$

S^n is called the *unit n-sphere* in E^{n+1}. (Do not confuse S^n with a ball. Visualize S^0, S^1, and S^2.) Prove rigorously that S^n is a closed subset of E^{n+1}.

5.3a. Interior, Closure, Boundary

In this subsection we complete our insight regarding the "looks" of a topological space, by means of introducing three simple pictorial auxiliary concepts.

Let A be a *subset* of a topological space. We define the *interior* of A (usually denoted by $\mathring{A}$ or by Int A) as the largest open subset contained in A. Thus,

$$\mathring{A} = \text{union of all open subsets contained in } A.$$

From its definition it is obvious that $\mathring{A}$ is a subset of A and $\mathring{A}$ is an *open* set.

A point $x \in A$ belongs to $\mathring{A}$ iff it has a neighborhood that is entirely contained in A. It then follows that A *is an open set iff* $A = \mathring{A}$.

Interiors are easily visualized. For example, an interior of the semiclosed interval $(a, b]$ of the real numbers is the open interval (a, b). In a discrete space, every set A coincides with its interior. In contrast, in an indiscrete space, the interior of any proper subset A is empty (because the only nonempty open set is X itself).

We note that given an arbitrary set A, we can form from it an open set, by simply taking its interior.†

Next we discuss the antithesis of interiors. Let again A be a *subset* of a topological space. We define the *closure of* A (usually denoted by $\bar{A}$) as the smallest closed set that contains A. Thus,

$$\bar{A} = \text{intersection of all closed sets which contain } A.$$

From its definition it is obvious that A is a subset of $\bar{A}$ and $\bar{A}$ is a *closed set*.

A point $x \in X$ belongs to $\bar{A}$ iff all neighborhoods of x have a nonempty intersection with A. (The proof is left as a problem.) It then easily follows that $\bar{A}$ is the union of A and of all its accumulation points. Hence, A *is a closed set iff* $A = \bar{A}$.

We give some simple examples. The closure of any interval of the real line is the corresponding closed interval. The closure of the set Q of rationals on the real line is the entire real line R (because every irrational number has a neighborhood which contains rational numbers, or, if you like, because the set of all accumulation points of Q is R itself). In a discrete space, every set A coincides with its closure. In an indiscrete space, the closure of any set A is X itself.

We note that given an arbitrary set A, we can form from it a closed set, by simply taking its closure. The closure of a set is never empty.

A final auxiliary concept connects the notions of interior and closure. If A is a *subset*, we define its *boundary* (usually denoted by bA or Bound A) as

$$bA = \bar{A} - \mathring{A}.$$

This means that the boundary consists of all accumulation points which are not interior points, plus all isolated points which are not interior points. This in turn implies that $x \in X$ is a boundary point of A iff each neighborhood of x has a nonempty intersection with both A and its complement. Therefore, every point $x \in bA$ is both in $\bar{A}$ and in $\overline{A^c}$. In other words, an alternative definition for the boundary is

$$bA = \bar{A} \cap \overline{A^c}.$$

This clearly shows that bA *is a closed set.* Furthermore, A is a closed set iff it contains its boundary.

We leave the fabrication of examples for boundaries to the reader, but,

---

†Of course, it just may happen that $\mathring{A}$ so formed is empty, as in the above example.

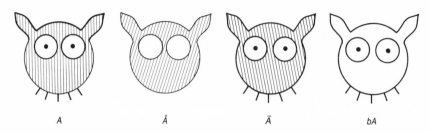

A   Å   Ā   bA

Fig. 5.5.

as an antidote for boredom often arising at this point, we illustrate all concepts discussed above for a particularly interesting set in Fig. 5.5.

PROBLEMS

5.3a-1. Show that if $A \subset B$, then $\mathring{A} \subset \mathring{B}$.

5.3a-2. Prove that for any set $A \subset X$, $\text{Int}(\text{Int}(A)) = \text{Int}(A)$ and also that $\bar{\bar{A}} = \bar{A}$.

5.3a-3. Show that the operation of "forming closures" has the following property: $\overline{A \cup B} = \bar{A} \cup \bar{B}$.

5.3a-4. In contrast to the result of the preceding problem, it is *not* true that $\text{Int}(A \cup B) = \text{Int } A \cup \text{Int } B$. Find an example of two subsets A and B of the real line (with the usual topology) to illustrate this statement.

5.3a-5. Prove the statement made in the text which claims that $x \in \bar{A}$ iff every neighborhood of x has a nonempty intersection with A.

5.3a-6. Show that $\text{Int}(A^c) = (\bar{A})^c$.

5.3a-7. Let V be an open set and show that V is disjoint from some given arbitrary set A iff V is disjoint from $\bar{A}$.

5.3a-8. Consider a discrete metric space and show that here the closure $\overline{S_r(x)}$ of an open ball $(r > 0)$ is not necessarily the corresponding closed ball, even though it is, naturally, a closed set. On the other hand, show that, for an arbitrary metric space, one always has $\overline{S_r(x)} \subset S_r[x]$.

5.3a-9. Show that $A \subset X$ is closed iff it contains its boundary. Show also that A is both open and closed iff it has an empty boundary.

5.3a-10. Show that in the metric space E^n the interior of any finite subset is empty. What is the closure and the boundary of such a set?

5.3a-11. Consider again the sets of Problem 5.3-5. Find for each set its interior, closure, and boundary.

5.4 CONVERGENCE

One of the major reasons for endowing a set with a topology is that we can define the concept of convergence. As everybody knows, this concept plays a crucial role not only in abstract analysis, but also in many branches of applied mathematics and science. We start with the general

DEFINITION 5.4(1). *A sequence* (x_n) *of points of a topological space is said to have a limit x, iff there exists a point* $x \in X$ *such that for each neighborhood* $N(x)$ *one can find a positive integer K such that* $x_n \in N(x)$ *provided* $n \geqslant K$. *If* (x_n) *has a limit, then the sequence is said to be convergent and we write* $\lim_{n \to \infty} x = x$ *or just* $\lim x_n = x$, *or sometimes* $x_n \to x$.

An informal way to paraphrase this definition is to say that, if (x_n) is convergent, then there exists some point x such that *any* arbitrary neighborhood $N(x)$ contains "almost all"† terms of the sequence. Of course, for different neighborhoods the critical number K will be different, but however we chose $N(x)$, we "lose" only a finite number of terms from the sequence. In passing we observe that a limit of a convergent sequence may or may not be a term of the sequence: the limit is defined as a point of X, not of $\{\ldots x_n \ldots\}$. (See, however, Problem 5.4-2.)

Before proceeding, the reader should make it absolutely clear in his mind that the limit of a convergent sequence must not be confused with the accumulation point of a subset. For example, in E, the sequence $(1, 1, \ldots, 1, \ldots)$ is convergent and has the limit 1 (because every neighborhood of 1 contains actually all terms of the sequence); but the set of the terms of the sequence (namely the singleton set $\{1\}$) has no accumulation point (because there is no number $x \neq 1$ such that every neighborhood of it would contain 1). Conversely, the sequence $(1, 2, 1/2, 3/2, \ldots, 1 - 1/n, 2 - 1/n, \ldots)$ is not convergent (no number captures in its neighborhoods "almost all" terms), but the set of terms has actually two accumulation points, namely the numbers 1 and 2.

As a matter of fact, there *is* a relation between limits of convergent sequences and accumulation points, which is this: *If a sequence which contains infinitely many distinct terms*‡ *is convergent, then a limit is an accumulation point of the set consisting of the terms of the sequence.*

---

†"Almost all" is used here in the sense of "all with the exception of finitely many."

‡This means that, even though there may be some terms which are repeated infinitely many times, there are still infinitely many *different* terms in the sequence too.

The definition of convergence simplifies considerably for the particular case of a *metric space*. From the basic definition and from the nature of neighborhoods in metric spaces it follows with ease that *a sequence* (x_n) *of a metric space* (X, d) *has a limit* x *and* $x_n \to x$ *iff for any given* $\epsilon > 0$ *there exist a positive integer* N *such that* $d(x_n, x) < \epsilon$ *provided* $n \geq N$. This can be also expressed by saying that (x_n) has a limit x iff the sequence $d(x_n, x)$ of nonnegative real numbers converges to the number zero,† i.e. in customary symbolism, iff $d(x_n, x) \to 0$ when $n \to \infty$.

The relation between limits and accumulation points for metric spaces is considerably stronger than what we pointed out for the general case. A moment's reflection shows the validity of the following

THEOREM 5.4(1).   *In a metric space* x *is an accumulation point of a subset* A *iff there exists a convergent sequence* (x_n) *consisting of distinct points of* A *which has* x *as its limit.*‡

We now give a brief assortment of examples for convergent sequences.

Example α.   Let (X, τ) be a discrete topological space. Recall that here for any point x, *every* subset which contains x is a neighborhood of x. Therefore, if (x_n) is a sequence which has one term, say $x_k \equiv \hat{x}$ which is repeated infinitely many times while all other terms occur only finitely many times, then this sequence is convergent and has as its limit the element $\hat{x}$. This is so because any neighborhood of $\hat{x}$ will contain all terms of (x_n) with the possible exception of a finite number of terms. It should also be clear that no other type of sequence can be convergent in this space.

Example β.   Let (x_k) be a sequence in E^n, with points $x_k = (\xi_1^{(k)}, \xi_2^{(k)}, \ldots, \xi_i^{(k)}, \ldots, \xi_n^{(k)})$. Suppose that for each fixed $i = 1, 2, \ldots, n$ the sequence $\xi_i^{(k)}$ of real numbers is convergent and $\lim_{k \to \infty} \xi_i^{(k)} = \xi_i$ (where, of course, ξ_i is a real number, and convergence is meant in the sense of sequences of real numbers). Thus, $|\xi_i^{(k)} - \xi_i| < \epsilon$ for each $i = 1, 2, \ldots, n$ and any $\epsilon > 0$ if only $k \geq N_i$. Then the sequence (x_k) of "vectors"§ converges in the usual topology of E^n and $\lim x_k = x$, where x

†This is so because for the real numbers with the usual metric, $\alpha_n \to \alpha$ iff $|\alpha_n - \alpha| < \epsilon$ for $n \geq N$.

‡This is the reason why in many texts the term "limit point" is used instead of the more correct term "accumulation point."

§We use here the term "vector" only for brevity, since the linear space structure of the underlying set is completely irrelevant.

is the "vector" $x = (\xi_1, \ldots, \xi_i, \ldots, \xi_n)$. The proof is simple. We have $d^2(x_k, x) = \sum_{i=1}^n |\xi_i^{(k)} - \xi_i|^2$ and if we want this to be less than any prescribed $\rho > 0$ when $k \geq N$, we only have to take each term sufficiently small, which in turn is possible by choosing for each i a suitable N_i. It is not difficult to see also the converse: a sequence (x_k) in E^n can be convergent only if for each i the sequence $(\xi_i^{(k)})$ is convergent when $k \to \infty$. Thus, the necessary and sufficient condition of convergence in E^n is the "pointwise convergence" of the components. For example, the sequence of vectors

$$x_1 = (1, 2, \ldots, n)$$

$$x_2 = \left(\frac{1}{2}, \frac{2}{2}, \ldots, \frac{n}{2}\right)$$

$$\vdots$$

$$x_k = \left(\frac{1}{k}, \frac{2}{k}, \ldots, \frac{n}{k}\right)$$

$$\vdots$$

is convergent and has the limit $x = (0, 0, \ldots, 0)$. The sequence with $x_k = (k^2, 0, \ldots, 0)$ is not convergent.

Example γ.   In the metric function space $C[0, 1]$ of all continuous† real functions on the interval $0 \leq x \leq 1$ (cf. Example ν in Section 5.1) the sequence (f_n) given by $f_n(x) = x/n$ is convergent and its limit is the constant function f given by $f(x) = 0$. Indeed,‡ $\sup_x |f_n(x) - f(x)| = 1/n$, so $d(f_n, f) < \epsilon$ provided $n > N = 1/\epsilon$. At this point we take the opportunity to point out that convergence in the metric of $C[0, 1]$ (or of more general function spaces) must not be confused with what is called, in elementary calculus, *pointwise convergence* of a sequence of functions. The letter term means that there exists a function f such that $|f_n(x) - f(x)| < \epsilon$ for any point x in the domain provided $n \geq K$, *where K may depend on x.* It should be obvious that this condition is less stringent than, for example, convergence in the metric of $C[a, b]$ which demands that $\sup_{a \leq x \leq b} |f_n(x) - f(x)| < \epsilon$ provided $n \geq N$. For example, consider on $[0, 1]$ the sequence f_n given by $f_n(x) = x^n$. The pointwise limit of this sequence is the function f given by

$$f(x) = \begin{cases} 0 & \text{if} \quad x \neq 1 \\ 1 & \text{if} \quad x = 1, \end{cases}$$

†Once again we anticipate that the reader is familiar with continuous real valued functions.

‡Draw a picture to visualize this.

as a graph easily reveals.† However, the sequence is not convergent in the metric topology of $C[0, 1]$ because for an arbitrarily large n, $\sup_x |f_n(x) - f|$ will be still equal to one. On the other hand, if the sequence f_n is *uniformly convergent* in the sense of elementary calculus, i.e. if $|f_n(x) - f(x)| < \epsilon$ whenever $n \geq K$ with K independent of x, then convergence in the $C[a, b]$ metric topology follows. For this reason, convergence in the metric space $C[a, b]$ is often referred to as *uniform convergence*.

Example δ. Let X be an indiscrete topological space. Here the following surprising situation is found: in this space *every* possible sequence is convergent and *every* point of the space is a limit of a given sequence. The reason for this phenomenon is that in X, any point x has only one neighborhood, namely the entire space X, so that in this unique neighborhood of *any* point we will find *all* points of *any* sequence.

The last example reveals that, contrary to nursery school prejudices, a limit of a convergent sequence is not necessarily unique. This is why we spoke so far about "a limit" of a sequence rather than "*the* limit." Obviously, in spaces where limits are not unique, convergence has only a restricted usefulness. We therefore wish to discover types of spaces where we can be assured that a convergent sequence has a unique limit. Inspection of Example δ reveals that the reason for lack of uniqueness is that neighborhoods cannot be "well enough separated." Accordingly, we are searching for some suitable "axiom of separation." In this spirit, we introduce the following

DEFINITION 5.4(2).   *Let (X, τ) be a topological space with the property that for every pair x and y of distinct points there exists a pair of nonintersecting neighborhoods. Then we call the space a Hausdorff space.*

In other words, if $x \neq y$, in a Hausdorff space one can always find respective neighborhoods $N(x)$ and $N(y)$ such that $N(x) \cap N(y) = \emptyset$.

The usefulness of this concept is borne out by the following

THEOREM 5.4(2).   *In a Hausdorff space every convergent sequence has a* UNIQUE *limit.*

Proof. Assuming that there were two different limits x and y, we would have $x_n \in N(x)$ provided $n \geq K_1$, and $x_n \in N(y)$ provided $n \geq K_2$,

---

†It is true that f is not continuous, but this is another matter.

for *any* pair of neighborhoods. Hence $N(x) \cap N(y) \neq \emptyset$, contrary to the definition of a Hausdorff space.

We give some examples.

Example ϵ. Every metric space is a Hausdorff space. Indeed, if $x \neq y$, then $d(x, y) \equiv \alpha > 0$, so choose an open ball $S_{\alpha/4}(x)$ about x and an open ball $S_{\alpha/4}(y)$ about y. There can be no common point in them, as an elementary use of the triangle inequality will demonstrate. Hence, we found a pair of nonintersecting neighborhoods. The importance of this result is that *in a metric space every convergent sequence is guaranteed to have a unique limit.*

Example ζ. A discrete topological space is Hausdorff. If $x \neq y$, the pair of neighborhoods $N(x) = \{x\}$ and $N(y) = \{y\}$ will do to illustrate the validity of the criterion.

Example η. Let $X = \mathbf{R}$ be the real line equipped with the following topology: $V \subset \mathbf{R}$ is open iff it is an open set in the usual topology less an arbitrary countable set. It is easy to verify that this is a topology. We observe that all usual open sets are also open in the new topology, but there are many more open sets. It is then clear that the space is Hausdorff, because even from the subcollection of "old" open sets we can choose nonintersecting neighborhoods for distinct pairs x and y.

It is fairly easy to see that *every metrizable space is a Hausdorff space.*† However, the converse is not true: *there are many nonmetrizable Hausdorff spaces.* Example η is a case in point. That the space in question is not metrizable, may be seen as follows. Let V be the set of all reals x such that $a < x < b$. This is open. Hence, if the space were metrizable, this interval ought to be a union of open balls. From this follows that every open ball used for covering the interval must contain at least one real number, distinct from its center. On the other hand, consider now the set W which consists of all irrational numbers between a and b. In the given topology, this is also open. But it cannot be a union of open balls because it does not contain a single rational point, contrary to the fact that open balls centered anywhere in the interval contain at least one rational point. Thus, no metric can reproduce the given topology, so that the space is not metrizable.

The importance of Hausdorff spaces lies precisely in the fact that, while they are considerably more general structures than metric spaces and

†This observation may serve as a *necessary* criterion for *metrizability.*

even than metrizable spaces, yet they share with these simpler structures the vital property of uniqueness of limits. In conclusion we mention that there exist spaces more general† than Hausdorff spaces where uniqueness of limits is still guaranteed. But these spaces play a less important role in applications.

PROBLEMS

5.4-1. Let (x_n) be a convergent sequence in an arbitrary topological space. Show that every (infinite) subsequence is also convergent.

5.4-2. Let (X, τ) be an arbitrary topological space and let (x_n) be a sequence whose terms are contained in some subset $A \subset X$. Suppose $x_n \to x$, and show that $x \in \bar{A}$.

5.4-3. Let $X = \{a, b, c\}$ be equipped with the topology $\tau = \{\emptyset, \{a\}, \{a, b\}, X\}$. Determine all possible convergent sequences in this space and find their limits.

5.4-4. Let (X, d) be a metric space. Call *two arbitrary sequences* (x_n) and (y_n) equivalent iff $\lim_{n \to \infty} d(x_n, y_n) = 0$. (The limit is here, of course, the limit of a sequence of nonnegative real numbers, in the usual sense.) Show that this definition leads to an equivalence relation on the set of all sequences in (X, d). Next show that if (x_n) and (y_n) are equivalent and if one of them is convergent, then the other is also convergent and they have the same limit.

5.4-5. Let $x_n \to x$ and $y_n \to y$ be two convergent sequences in (X, d). Show that the set of real numbers $d(x_n, y_n)$ is convergent and has the limit $d(x, y)$.

5.4-6. Show directly (without reference to Hausdorff spaces) that in a metric space every convergent sequence has a unique limit.

5.4-7. Show that every subspace of a Hausdorff space is a Hausdorff space.

5.4-8. Prove that the topological product of two Hausdorff spaces is a Hausdorff space.

5.4-9. Show that in a Hausdorff space every subset consisting of a single point is a closed set. (*Remarks*: (a) A space where every one-point set $\{x\}$ is closed is often called a T_1 space. Thus, you proved that every Hausdorff space is a T_1 space. (b) Incidentally, Hausdorff spaces are sometimes called T_2 spaces. (c) The converse of the theorem is not true: there exist T_1 spaces which are not T_2 spaces. Try to cook up an example.)

5.5 CONTINUITY

One of the most important properties which a function can have is that of being continuous. In this section we shall give a firm foundation for this

---

†By this we mean that less stringent "separation axioms of points" are satisfied.

notion, transferring the concept from its elementary setting in basic calculus to arbitrary topological spaces.

DEFINITION 5.5(1).   *A function $f: (X, \tau) \to (X', \tau')$ from one topological space to another is said to be continuous iff the inverse image $f^{-1}(V')$ of every open set $V' \subset X'$ is an open set V of X.*

Remarks: (1) It is clear that continuity can be defined only if both the domain and the codomain are *topological* spaces.

(2) Continuity of a given map $f: X \to X'$ between *sets* depends critically on the *topologies* of both X and X'. (See Problem 5.5-1.)

(3) From complementation it follows that in the above definition the term "open" may be replaced by "closed." Thus, f is continuous iff $C' \subset X'$ closed implies $f^{-1}(C')$ closed.

(4) The reader should well bear in mind that the definition of continuity goes sort of "backward" and the image of an open (or closed) set of X need by no means be an open (or closed) set of X'. We shall have to say more about this question in Section 5.6.

It is an easy exercise to show that the definition of continuity leads to the following alternative, and rather descriptive formulation:

THEOREM 5.5(1).   *If $f: (X, \tau) \to (X', \tau')$ is a continuous function, then for each given point $x \in X$ and each neighborhood $N'(f(x))$ of its image, there exists a neighborhood $N(x)$ of x such that $f(N(x)) \subset N'(f(x))$.*

This theorem is visualized in Fig. 5.6. Once again, bear well in mind "in which way" the statement goes.

Some simple examples follow.

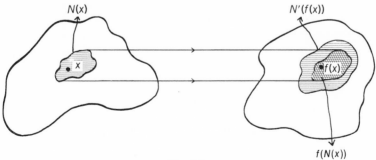

Fig. 5.6.

Example α.   Let X and X' be arbitrary topological spaces and let f be a constant function, i.e. $f(x) = c \in X'$ for every $x \in X$. If V' is an open set of X' which does not contain c, then $f^{-1}(V') = \emptyset$; and if V' contains c, then $f^{-1}(V') = X$. Since both $\emptyset$ or X are open, f is a continuous map.†

Example β.   Let $f: (X, \tau) \to (X, \tau)$ be the identity map $f = 1_X$ on an arbitrary space. Then, if $V \subset X$ is open, $f^{-1}(V) = V$ is open, hence f is continuous.

Example γ.   Let f be an *arbitrary* map from an *arbitrary* space X into an *indiscrete* space X'. Then f is continuous because the only open sets of X' are X' and $\emptyset$ and both $f^{-1}(X') = X$ and $f^{-1}(\emptyset) = \emptyset$ are open.

In the following, we will discuss the relation between continuity and convergence.

Suppose $f: (X, \tau) \to (X', \tau')$ is a continuous function and suppose (x_n) is a convergent sequence in X, so that $x_n \to x$. Then, using Theorem 5.5(1), it easily follows that the sequence $(f(x_n))$ of the images is convergent in (X', τ') and in fact $f(x_n) \to f(x)$. Thus, *continuity implies convergence for the image of a convergent sequence.* Unfortunately, the converse is not necessarily true. For example, let (X, τ) be the set of reals equipped with the finite complement topology (V is open iff V^c is a finite set or $V = \emptyset$). Let E be the set of reals, equipped with the usual topology. Let $f: (X, \tau) \to E$ be given by $f(x) = x$ for each real number. This is *not* a continuous function, because the inverse image of the open interval (a, b) of E is the set $a < x < b$ of (X, τ), which is not open in the given topology (its complement is infinite). On the other hand, the sequence $(x, x, \ldots, x, \ldots)$ (with arbitrary $x \in \mathbf{R}$) is clearly convergent in both spaces, and f carries it precisely from one space to the other. Hence, we have preserved convergence by a discontinuous function.

We would like to have a criterion for ensuring that continuity and preservation of convergence become equivalent. We precede this by a definition. A space is said to be *first countable* if for each point $x \in X$ there exists a *countable* class $\mathcal{B}_x$ of neighborhoods $N_k(x)$ ($k = 1, 2, \ldots$) such that *any* neighborhood $N(x)$ of x can be represented as a union of members from $\mathcal{B}_x$. (We say for short that "every point has a countable local base.") Now, we can state the following

THEOREM 5.5(2).   *Let f be a map from a first countable space into an*

---

†Since $\{c\} \in X'$ is not necessarily open, this is a good illustration to show that the image of an open set need not be open.

arbitrary space. Then f is continuous iff $x_n \to x$ implies $f(x_n) \to f(x)$ for all convergent sequences (x_n) of the domain space.

The reader surely realizes that *every metric space is first countable.*† Hence, for maps defined on metric spaces the "convergence test" is a necessary *and* sufficient criterion for continuity. The test becomes particularly simple if the codomain is also a metric space. Thus, *if $f: (X, d) \to (X', d')$ is a map between two metric spaces then f is continuous iff $d(x_n, x) \to 0$ implies $d'(f(x_n), f(x)) \to 0$ for all convergent sequences (x_n) in (X, d).* This ties up the general notion of continuity with the one of basic calculus, when f is a real valued function on a set of the reals.

So far we have investigated continuity from a global viewpoint. A closely related concept is that of *continuity at a point.* A map $f: (X, \tau) \to (X', \tau')$ is said to be continuous at $x_0 \in X$ iff for each neighborhood $N'(f(x_0))$ in X' there exists a neighborhood $N(x_0)$ in X such that $f(N(x_0)) \subset N'(f(x_0))$. Clearly, f is continuous on (X, τ) iff it is continuous at each point x of X.

We conclude this section with some comments on the restriction of functions, in relation to the property of continuity.

Let $f: (X, \tau) \to (X', \tau')$ be an arbitrary map and let f_A be the restriction of f to some subset A of X. We say that f *is continuous on A* iff $f_A: (A, \tau_A) \to (X', \tau')$ is continuous, where τ_A denotes the relative topology on the sub*space* (A, τ_A), and of course continuity is decided using *this* topology.

The reader will readily prove the following

THEOREM 5.5(3).   *If $f: (X, \tau) \to (X', \tau')$ is a continuous function and $A \subset X$, then f is continuous‡ on A.*

The converse, of course, does not hold: Continuity of f_A will not, in general imply the continuity of an extension to X. Equally important is the more subtle observation that, even though f may be a discontinuous function on X, a restriction of f to some subset may very well be continuous. For example, let $f: E \to E$ be given by

$$f(x) = \begin{cases} 0 & \text{if } x \text{ is rational,} \\ 1 & \text{if } x \text{ is irrational.} \end{cases}$$

†Think of open balls with rational radii.
‡That is, $f_A: (A, \tau_A) \to (X', \tau')$ is continuous.

This map is as discontinuous† as can be: it is actually discontinuous at each point of its domain. On the other hand, its restriction f_Q to the rationals Q is continuous. Furthermore, the restriction f_I to the set I of all irrational numbers is also continuous.‡

PROBLEMS

5.5-1. Let (X, τ) be a discrete space and (X', τ') an arbitrary topological space. Which functions $f: X \to X'$ are continuous? Compare your result with Example γ in the text. (*Remark*: It is not difficult to see that the larger the topology on the domain and the smaller the topology on the codomain, the more functions will become continuous. Thus, by suitably changing the topologies, one may enforce desired classes of functions to become continuous.)

5.5-2. Let $f: (X, \tau) \to (X', \tau')$ and $g: (X', \tau') \to (X'', \tau'')$ be continuous maps. Show that $g \circ f: (X, \tau) \to (X'', \tau'')$ is continuous. (*Hint*: Note that

$$(g \circ f)^{-1}(A) = f^{-1}(g^{-1}(A))$$

for any $A \subset X''$.)

5.5-3. Let f be a map between two metric spaces. Show that f is continuous iff for every point $x \in X$ and every open ball $S'_\varepsilon(f(x))$ in X' there exists an open ball $S_\delta(x)$ in X such that $f(S_\delta(x)) \subset S_\varepsilon(f(x))$. Visualize the result by a figure.

5.5-4. Let (X, d) be a metric space and let $x_0 \in X$. Define the function $f: X \to R$ by setting $f(x) = d(x, x_0)$. Show that f is continuous.

5.5-5. Let (X, τ) be an arbitrary topological space and let (Y, d) be a metric space. Show that $f: X \to Y$ is continuous at x_0 iff for each open ball $S_r(f(x_0))$ there exists a neighborhood $N(x_0)$ in X such that $f(N(x_0)) \subset S_r(f(x_0))$.

5.5-6. Let f be a map between two metric spaces and show that f is continuous at x_0 iff $x_n \to x_0$ implies $f(x_n) \to f(x_0)$.

5.5-7. Let $f: (X, \tau) \to (X', \tau')$ be continuous. Let A be a subspace of X (with the relative topology τ_A). Show that the restriction of f to A is continuous, as was claimed in the text.

5.6 HOMEOMORPHISM AND ISOMETRY

In this section we shall introduce the fundamental concept of homeomorphism which, for topological structures, plays a role completely analogous to that of isomorphism for algebraic structures.

What we are after is a class of maps which provide a one-to-one

---

†If $0 \in (a, b)$ but $1 \notin (a, b)$, then $f^{-1}(a, b)$ is the set of rationals; if $1 \in (a, b)$ but $0 \notin (a, b)$, then $f^{-1}(a, b)$ is the set of irrationals; both these sets are not open in E.

‡This example is truly remarkable, because $R = Q \cup I$.

correspondence not only between the underlying sets, but also between the topologies with which they are equipped. Now, continuous maps will not do because, apart from the fact that they need not be bijective, they do not preserve open (or closed) sets. So we look in the other direction and make the following

DEFINITION 5.6(1).   *A function* $f: (X, \tau) \to (X', \tau')$ *is said to be an open map iff the image* $f(V)$ *of every open set* V *of* X *is an open set* V' *of* X'.

(A *closed map* is defined in an analogous way: it maps closed sets onto closed sets.)

It is clear that an open (or a closed) map need not be continuous. For example, any function from an indiscrete space into a discrete space is open (as well as closed), but, unless it is a constant map, it cannot be continuous.

Let now $f: (X, \tau) \to (X', \tau')$ be a *bijective* map, so that the inverse function $f^{-1}: (X', \tau') \to (X, \tau)$ is guaranteed to exist. Suppose furthermore that f is continuous. Does this imply that f is open? Certainly not: all we can say is that its inverse f^{-1} is open. On the other hand, let f be bijective and suppose f^{-1} is open. Will f be continuous? The answer is positive.

These considerations lead us naturally to the following basic

DEFINITION 5.6(2).   *A map* $f: (X, \tau) \to (X', \tau')$ *which is bijective, continuous, and open is said to be a homeomorphism*† *from* (X, τ) *onto* (X', τ').

Since a bijective continuous map is open iff it is closed, in the above definition the term "open" may be replaced by "closed," so that *f is a homeomorphism if it is bijective, continuous, and closed.*

There is yet another equivalent and useful form of defining a homeomorphism. Since, as we saw above, the openness of a bijective map implies that f^{-1} is continuous, we may say that *a map f:* $(X, \tau) \to (X', \tau')$ *is a homeomorphism iff it is bijective, continuous, and its inverse is also continuous.*‡

The importance of homeomorphisms should now be obvious. *We shall call two topological spaces homeomorphic if there exists a homeomor*

---

†A homeomorphism is sometimes also called a *topological mapping.*

‡A continuous map which has a continuous inverse function is sometimes said to be a *bicontinuous function.* Thus, "homeomorphism" and "bicontinuous map" are synonymous.

phism from one to the other. The identity map is obviously a homeomorphism of X onto X, so each space is homeomorphic to itself. The inverse of a homeomorphism is a homeomorphism (this follows from bicontinuity) so that if f yields a homeomorphism from X to Y, then f^{-1} yields a homeomorphism from Y to X. Finally, the composite $g \circ f$ of two homeomorphic maps is homeomorphic,† so that if X is homeomorphic to Y and Y is homeomorphic to Z, then X is homeomorphic to Z. In summary, *homeomorphism is an equivalence relation on the class of all topological spaces.* We shall denote this equivalence relation by writing

$$(X, \tau) \underset{h}{\approx} (X', \tau') \qquad \text{or simply} \qquad X \underset{h}{\approx} X'.$$

If $X \underset{h}{\approx} X'$, then, as sets, they are in a one-two-one correspondence (i.e. we have a set isomorphism), but beyond that, the two topologies, i.e. the two classes of open sets, are also in a one-to-one correspondence. Homeomorphic spaces are indistinguishable from one another both in their set-theoretic and in their topological properties. They may differ only in the concrete nature of their elements and in notation. Homeomorphic spaces are equivalent and may be "identified." It is clear that homeomorphism plays in topology a role analogous to that of isomorphism in algebra.‡

Any property which holds for all homeomorphic images of a given topological space is called a *topological property.* We shall meet many examples later on (see, to start with, Problems 5.6-4 and 5.6-5) and mention now only that this definition is the analog of an "algebraic property." As a matter of fact, one may think of topology as the study of topological properties. This leads to visualizing topology as some kind of "rubber sheet geometry." A space is some configuration (a set of points) on a rubber sheet, a homeomorphism is a deformation (stretching, bending, folding, etc.) of the sheet which is reversible and does not tear the sheet. The properties of the configuration that do not change under such deformations are the topological properties of the space.

It is time for a mixed bag of examples of homeomorphisms.

Example α. Any parallelogram, considered as a subspace of E^3 is homeomorphic to any polygon, any disk (and many other shapes) in E^3

---

†This follows from the easily established fact that the composite of two continuous functions is continuous.

‡There is also a somewhat weaker analog for arbitrary morphisms of algebraic systems, cf. Subsection 5.6a.

(again with the relative topology). However, the parallelogram is not homeomorphic to the cylinder mantle.

Example β. The closed interval $[-\pi/2, +\pi/2]$ of the real line is homeomorphic to the closed interval $[-1, 1]$ (both equipped with the relative topology), since the function $f(x) = \sin x$, for example, furnishes a homeomorphism.

Example γ. Every (nonempty) bounded open interval (a, b) of the real line E (with the relative topology) is homeomorphic to any other (nonempty) open interval (c, d). Indeed, the map $f: (a, b) \to (c, d)$ given by

$$f(x) = \frac{d-c}{b-a}(x - a) + c$$

is easily seen to be bicontinuous. Geometrically, it corresponds to a translation, followed by a stretching and another translation.

Example δ. Every bounded open interval of E (as a subspace) is homeomorphic to all of E. To prove this, first take the open interval $(-1, +1)$; then the map $f: (-1, +1) \to E$ given by

$$f(x) = \frac{1}{1 - |x|}$$

is easily seen to be bicontinuous. (It is simply an infinite stretching.) Thus, $(-1, +1) \underset{h}{\approx} E$. But the preceding example tells us that $(a, b) \underset{h}{\approx} (-1, +1)$ for any given (a, b). Hence, by transitivity, our present statement is proven.†
It is not difficult to see that, more generally, every open rectangle in E^n is homeomorphic to all of E^n.

Example ε. Let X be the set $\{a, b, c, d\}$. Let $\tau = \{\{d\}, \{d, c\}, \{d, b\}, \{d, b, c\}, \emptyset, X\}$. Let $\tau' = \{\{a\}, \{a, b\}, \{a, c\}, \{a, b, c\}, \emptyset, X\}$. The two spaces (X, τ) and (X, τ') are homeomorphic, because the map f given by

$$d \mapsto a, \quad b \mapsto c, \quad c \mapsto b, \quad a \mapsto d$$

is bijective, open, and continuous. (For example, the image of the open set $\{d, b\}$ is the open set $\{a, c\}$, or the inverse image of the open set $\{a, b, c\}$ is the open set $\{d, c, b\}$ and so on.)

Example ζ. An interesting counterexample is the following. Let E be the real line with the usual topology and let (X, τ) be an indiscrete space.

---

†Why can not one omit the term "open" from the formulation of the statement?

These cannot be homeomorphic. For, suppose f is bijective, then $f^{-1}(X) = E$ and $f^{-1}(\emptyset) = \emptyset$. Then, if V is a proper open subset of E, $f(V)$ is a proper set of X, hence it cannot be open.

As an interesting application of homeomorphic maps, we digress for a moment and introduce the following

DEFINITION 5.6(3). *A topological space is said to be a homogeneous space iff for any given pair x, y of points there exists some homeomorphism h of the space onto itself such that $h(x) = y$.*

One may express this informally by saying that a homogeneous space "looks the same" when viewed from any of its points.† We give a few examples.

Example η. The real line E is a homogeneous topological space. Consider the map $h_\alpha : E \to E$ given by $x \mapsto x + \alpha$, where α is an arbitrary real number. This map is obviously a homeomorphism. Now let u, v be given arbitrary points. Then $h_{v-u}(u) = u + v - u = v$, as desired.

Example ϑ. The sphere ‡ S^2 of E^3 is a homogeneous space, since any given point x can be carried over into any other given point y by a suitable rotation of the sphere, and it is easy to see geometrically that this transformation is a homeomorphism. The result obviously generalizes to S^n of E^{n+1}.

Example κ. Any open interval of E is a homogeneous space. This follows easily from Example β. On the other hand, a closed interval is not a homogeneous space. The reader will easily verify that, for example, the point 0 of the closed interval $[0, 1]$ cannot be taken to the point $\frac{1}{2}$ by any homeomorphism.

In passing we observe that homogeneity is a topological property. Let X be a homogeneous space and let $f: X \to Y$ be a homeomorphism. Let y_1 and y_2 be two points of Y and let $h: X \to X$ be the homeomorphism which carries $f^{-1}(y_1) \equiv x_1$ to $f^{-1}(y_2) \equiv x_2$. Then, evidently, $f \circ h \circ f^{-1}$ is the desired homeomorphism which carries y_1 to y_2.

We now turn our interest specifically to metric spaces and consider a special, very important type of homeomorphism for such spaces.

---

†We recall that, at the end of Subsection 4.1a we introduced the algebraic concept of homogeneity of a set under the action of a group. Despite the similarity, the two concepts must be distinguished.

‡For the definition of n-spheres S^n, see Problem 5.3-7.

Definition 5.6(4).   *A surjective map* $f \colon (X, d) \to (X', d')$ *between two metric spaces is called an isometry iff for each pair* $x, y \in X$ *we have* $d'(f(x), f(y)) = d(x, y)$.

Thus, the distinctive feature of an isometry is that it preserves distances.

We now proceed to show that *any isometry* as defined above *is a homeomorphism*. We already know that, by definition, an isometry is surjective. To show that it is injective, note only that if we had $f(x) = f(y)$ for $x \neq y$, then $d'(f(x), f(y)) = 0$, which is contradictory since $d'(f(x), f(y)) = d(x, y) \neq 0$. Thus, f is bijective. Now let $x_n \to x$; then $d(x_n, x) \equiv \alpha < \epsilon$ for $n \geq N$, so that $d'(f(x_n), f(x)) = \alpha < \epsilon$. This means that $f(x_n) \to f(x)$, which, by Theorem 5.5(2) means that f is continuous. The continuity of f^{-1} is proven similarly: writing $x'_n = f(x_n)$ and $x' = f(x)$, the relation $d'(x'_n, x') < \epsilon$ implies $d(f^{-1}(x'_n), f^{-1}(x')) < \epsilon$. In summary, f is bicontinuous, hence it is a homeomorphism.

We give now an illustration, and show that C^n *is isometric to* E^{2n}. Let $z \in C^n$ be given as

$$z = (\zeta_1, \ldots, \zeta_k, \ldots, \zeta_n), \qquad \text{with} \qquad \zeta_k = \xi_k + i\eta_k.$$

Let $f \colon C^n \to E^{2n}$ be defined by

$$f(z) = (\xi_1, \eta_1, \ldots, \xi_k, \eta_k, \ldots, \xi_n, \eta_n).$$

This map is clearly surjective and

$$d^2(z, z') = \sum_{k=1}^{n} |\zeta_k - \zeta'_k|^2 = \sum_{k=1}^{n} |(\xi_k - \xi'_k) + i(\eta_k - \eta'_k)|^2$$

$$= \sum_{k=1}^{n} \{|\xi_k - \xi'_k|^2 + |\eta_k - \eta'_k|^2\} = d^2(f(z), f(z')).$$

Hence, the isometry is proven. In particular, we note that the space C of complex numbers is isometric to the real plane E^2. Since every isometry is a homeomorphism, we also know now that C^n and E^{2n} are homeomorphic, i.e. topologically equivalent.†

It goes without saying that, while every isometry is a homeomorphism of metric spaces, the converse is not true: metric spaces have many more homeomorphisms. For example, the map $f \colon E \to E$ given by $f(x) = \alpha x$

---

†This is interesting because we recall that these two spaces considered as *algebraic systems* with a linear space structure are not isomorphic.

with $\alpha \neq 0$ and $\alpha \neq 1$ real, is a homeomorphism (it is just a uniform stretching), but it is not an isometry, since

$$d(f(x), f(y)) = |\alpha x - \alpha y| = |\alpha| |x - y| = |\alpha| d(x, y) \neq d(x, y).$$

On the other hand, it is easy to see that isometry is an equivalence relation on the class of all metric spaces. Since an isometry is always a homeomorphism, we see that isometric metric spaces are indistinguishable in their set-theoretic, topological, *and* metric properties.

The term "isometry" is often used in a somewhat relaxed sense. Suppose $f: (X, d) \to (X', d')$ is a map which is not surjective although it preserves distances. Then we say that f is an isometry from (X, d) *into* (X', d'). In this case, clearly, (X, d) is isometric, not to (X', d'), but only to $(f(X), d')$ which is a metric subspace of (X', d').

We give an interesting example. Let (X, d) be the subset of the real line $a \leq x \leq b$, with the usual metric. Let $(X', d') = C[a, b]$ be the space of continuous real functions $F, G, \ldots$ on $[a, b]$, equipped with the usual uniform metric $d'(F, G) = \sup_{y \in [a,b]} |F(y) - G(y)|$. Consider now the map

$$f: (X, d) \to (X', d') \quad \text{given by} \quad x \mapsto F_x \quad \text{where} \quad F_x \in C[a, b]$$
$$\text{is defined by} \quad F_x(y) = x \quad \text{for all} \quad a \leq y \leq b.$$

(Thus, f is a map from $[a, b]$ *into* the class of all constant functions on $[a, b]$, which class itself is a subset of $C[a, b]$.) The map is obviously not surjective, but it preserves distances. Indeed,

$$d'(f(x), f(z)) = d'(F_x, F_z) = \sup_{y \in [a,b]} |F_x(y) - F_z(y)|$$
$$= \sup_{y \in [a,b]} |x - z| = |x - z| = d(x, z).$$

Thus, $[a, b]$ is isometric to the subspace of all constant functions in $C[a, b]$ which have their constant value between a and b.

As an exercise for isometries and homeomorphisms, we conclude this section with a sufficient criterion to recognize metrizability.†

THEOREM 5.6(1). *If a topological space (X, τ) is homeomorphic to a metric space (Y, d), then it is metrizable.*

Proof. Let $f: (X, \tau) \to (Y, d)$ be a homeomorphism. Construct from the underlying set X the metric space (X, e), where the metric e is defined

---

†In Section 5.4 we have seen that the Hausdorff property is a necessary, but not sufficient criterion for metrizability.

by

$$e(x, y) \equiv d(f(x), f(y)) \quad \text{for all} \quad x, y \in X. \tag{5.2}$$

This is a valid definition of a metric, since nonnegativeness, symmetry, and the triangular property are clear consequences of the metric d on Y, and if $x \neq y$, then $e(x, y) \neq 0$, because by the bijectiveness of f, $f(x) \neq f(y)$. Now, by Eq. (5.2), (X, e) and (Y, d) are isometric, hence also homeomorphic. But, by assumption, $(Y, d) \underset{h}{\approx} (X, \tau)$ hence, by transitivity, $(X, e) \underset{h}{\approx} (X, \tau)$. Therefore, the open sets of the metric space (X, e) (as defined via the metric) are precisely the open sets of (X, τ) (as originally given). Hence, (X, τ) has been "metrized" and in fact Eq. (5.2) exhibits an appropriate metric.

An almost trivial illustration of our theorem is provided by a discrete topological space (X, τ_d). Consider the discrete metric space (X, d_d) over the same underlying set X. The map $f: (X, \tau_d) \to (X, d_d)$ given by $f(x) = x$ is a homeomorphism, because it is bijective, open, and continuous.† Hence, (X, τ_d) is metrizable and Eq. (5.2) actually tells us that $e(x, y) = d_d(x, y)$.

PROBLEMS

5.6-1. Let $f: (X, \tau) \to (X', \tau')$ and $g: (X', \tau') \to (X'', \tau'')$. Suppose $g \circ f$ is continuous and that one of the two maps is both continuous and open. Show that then the other map is also continuous. Can you omit the premise on openness?

5.6-2. Let $(X \times Y, \tau_X \times \tau_Y)$ be a product space. Show that the projections $p: X \times Y \to X$ and $g: X \times Y \to Y$ are open functions.

5.6-3. Let X be an arbitrary set, τ_d the discrete topology, τ_i the indiscrete topology. Give an example of a bijection $f: (X, \tau_d) \to (X, \tau_i)$ which is continuous but which is not a homeomorphism. Can you make a general statement about this situation?

5.6-4. Show that the homeomorphic image of a Hausdorff space is a Hausdorff space, i.e. that "being Hausdorff" is a topological property.

5.6-5. Show that each of the following properties is a topological property:
(a) The space has infinitely many points,
(b) The space has exactly n points,
(c) The space is discrete,
(d) The space has an infinite number of closed sets.

5.6-6. Show that boundedness of a metric space is not a topological property. (*Hint*: Consider an unbounded metric space (such as E) and show that one can

---

†Recall that all sets of (X, τ_d) are open and so are those of (X, d_d).

construct an equivalent metric, i.e. one which gives the same topology, but for which the space is bounded.)

5.6-7. Let (X, τ) be a Hausdorff space which has the property that each point x has some neighborhood $N(x)$ which is homeomorphic to the Euclidean n-space E^n. Then (X, τ) is called a *topological manifold*,† more precisely, a topological n-manifold. (E^n itself is, by this definition, a manifold with $N(x) = E^n$ for each $x \in E^n$.)

Show that any sphere‡ S^n is an n-manifold. (If an analytic proof is too difficult, devise one using a geometrical construction.) Show that the torus (doughnut surface) is a 2-manifold. (It may help to realize that the torus is homeomorphic to $S^1 \times S^1$.) Give an example of a space which is not a manifold.

5.6-8. Show that if (X, τ) is homeomorphic to a topological manifold, then it is also a manifold. ("Being a manifold" is a topological property.)

5.6-9. Show that each topological manifold is homeomorphic to a subspace of some Euclidean E^n space.

5.6-10. Show that the union of the two axes of the Euclidean plane is not a homogeneous space.

5.6-11. Let $X = \{1, 2, \ldots, n\}$ and let X be equipped with the discrete metric. Let $C(X, \mathbf{R})$ be the set of all continuous bounded real valued functions on X, equipped with the usual metric $d_c(f, g) = \sup |f(k) - g(k)|$ $(k = 1, 2, \ldots, n)$. Let E^n be the real Euclidean n-space. Make $C(X, \mathbf{R})$ a linear space (by defining, as usual, the pointwise sum and scalar product of functions). Make also E^n a linear space (by defining the usual sum and scalar product of vectors). Show that $C(X, \mathbf{R})$ and E^n are isomorphic as linear spaces. Show that this (algebraic) isomorphism is also a (topological) homeomorphism. However, show that this homeomorphism is not an isometry.

5.6-12. Let E^n be the n-dimensional Euclidean space. Define the map $r: E^n \to E^n$ by setting $(\xi_1, \ldots, \xi_n) \to (\alpha_{1k}\xi_k, \ldots, \alpha_{nk}\xi_k)$, where the shorthand $\alpha_{ik}\xi_k \equiv \Sigma_{k=1}^n \alpha_{ik}\xi_k$ has been used, and where the n^2 real numbers α_{ik} obey the following condition:

$$\alpha_{ik}\alpha_{ij} = \delta_{kj}$$

(where δ_{kj} is the Kronecker delta, and the dummy index summation convention has been used again). Show that the map r is an isometry of E^n onto itself. (*Remark*: Looking at E^n as a linear space $\mathbf{R}^n$, the map r is also easily seen to be an isomorphism of the space onto itself, i.e. an automorphism of $\mathbf{R}^n$. Geometrically, r is a rotation or a reflection or a combination of these, i.e. an "orthogonal transformation" on $\mathbf{R}^n$.) Consider now the map $t: E^n \to E^n$ which is defined by $(\xi_1, \ldots, \xi_n) \to (\xi_1 + t_1, \ldots, \xi_n + t_n)$, where the t_k are arbitrary real numbers. Show that t is an isometry but not an automorphism. (Geometrically, it represents a translation.)

---

†This concept must not be confused with the algebraic concept of a (linear) manifold of a linear space.

‡For the definition of an n-sphere, see Problem 5.3-7.

5.6-13. Let (X, d) be a metric space and show that the set of all isometries of this space onto itself forms a group (with the group composition being given by the usual composition of functions).

5.6-14. Show that the set of all homeomorphisms of an arbitrary topological space onto itself is a group, and that (for metric spaces) the group of isometries (discussed in the preceding problem) is a subgroup of this group.

5.6a. Quotient Topology; Homeomorphism Theorem

In the study of algebraic systems, we often managed to endow certain quotient sets with an algebraic structure inherited from the original system. In this way, we produced quotient systems (quotient groups, rings, spaces, algebras). Using these structures, we arrived at important isomorphism theorems. In this section we study the analogous constructions for topological systems.

To introduce the discussion, suppose that E is some equivalence relation on some set X. We recall that the quotient set X/E consists of all distinct equivalence classes and that the surjective canonical map $c: X \to X/E$ is defined by the assignment $x \mapsto [x]$. Suppose now that X is actually a topological space, with some topology τ. We wish to introduce on X/E some "useful" topology, closely related to that on X. To this end, we define on X/E the class σ of subsets $Q \subset X/E$ in the following way:·

$$\sigma = \{Q \subset X/E \,|\, c^{-1}(Q) \in \tau\}. \tag{5.3}$$

That is, σ consists of all subsets of X/E whose inverse image under the canonical map is an open set of X. We claim that σ is a topology for X/E. This follows simply by observing that†

$$c^{-1}(\bigcup_{\alpha} Q_{\alpha}) = \bigcup_{\alpha} c^{-1}(Q_{\alpha}) \equiv \bigcup_{\alpha} V_{\alpha}$$

and

$$c^{-1}\left(\bigcap_{k=1}^{n} Q_k\right) = \bigcap_{k=1}^{n} c^{-1}(Q_k) = \bigcap_{k=1}^{n} V_k,$$

and we know that the r.h.s. of either equation is an open set of X. Thus, σ is a topology. It has two interesting properties. First, *the canonical map*

$$c: (X, \tau) \to (X/E, \sigma) \tag{5.4}$$

is a continuous function. This follows trivially from the definition of open sets on X/E. Second, σ *is the largest topology for which c can be continuous.* For indeed, if $\sigma' \supset \sigma$ were another topology, then it would

---

†Here $Q_{\alpha}(Q_k)$ denote members of an arbitrary (finite) collection of sets in σ and $V_{\alpha}(V_k)$ denote similar collections of open sets in X.

contain a set R of X/E which is not in σ, hence $c^{-1}(R)$ could not be open in X, so that c could not be continuous.

The topology σ defined by Eq. (5.3) is called the *quotient topology* on X/E, and it can be characterized as the largest (strongest) topology on X/E for which the canonical map c is continuous.†

The usefulness of the quotient topology reveals itself in a *decomposition theorem for continuous functions.*

Recall that if $f\colon X \to Y$ is an arbitrary function, then ker (f), defined by "x_1 and x_2 are equivalent elements of X iff $f(x_1) = f(x_2)$" is an equivalence relation on X. Recall also the decomposition Theorem 2.2(3) and the subsequent discussion which taught us that we have the unique decomposition chain

$$f = i \circ b \circ c,$$

where $c\colon X \to X/\mathrm{ker}\,(f)$ is the *surjective* canonical map given by $x \mapsto [x]$; $b\colon X/\mathrm{ker}\,(f) \to \mathrm{Im}\,(f)$ is the *bijective* induced map defined by $[x] \mapsto f(x)$; and $i\colon \mathrm{Im}\,(f) \to Y$ is the trivial *injective* insertion.

Now suppose that (X, τ) and (Y, ρ) are actually topological spaces and f is a map from the first to the second. Endow the quotient space $X/\mathrm{ker}\,(f)$ with the quotient topology introduced above. For brevity, denote this new topological space $(X/\mathrm{ker}\,(f), \sigma)$ by the simple symbol X/f. With this notation, the decomposition theorem for the present case of the three topological spaces is visualized by the following commutative diagram:‡

$$(X, \tau) \xrightarrow{\ c\ } X/f \xrightarrow{\ b\ } \mathrm{Im}\,(f) \xrightarrow{\ i\ } (Y, \rho)$$
$$\underbrace{\hspace{6cm}}_{f}$$

In particular, suppose that f is a continuous map. Then we have the remarkable

THEOREM 5.6a(1). *All factors in the decomposition $f = i \circ b \circ c$ of a continuous function f are themselves continuous functions.*

Proof. The continuity of i is trivial and that of c follows from the definition of the quotient topology. We only have to prove the crucial fact that *the induced map b of a continuous function is continuous.* From

---

†It is very easy, but rather useless, to define smaller topologies on X/E for which c is continuous. For example, if we endowed X/E with the indiscrete topology, then, as we know, *any* function into X/E would be continuous.

‡The topological space $\mathrm{Im}\,(f)$ is, of course, endowed with the relative topology inherited from ρ.

$f = i \circ b \circ c$ it easily follows for any set $U \subset \text{Im}(f)$ that

$$f^{-1}(U) = c^{-1}[f^{-1}(U)]. \tag{5.5}$$

Let now, in particular, U be an open set of $\text{Im}(f)$, i.e. $U = \text{Im}(f) \cap W$, where W is some open set of Y. Then $f^{-1}(U) = f^{-1}(\text{Im}(f)) \cap f^{-1}(W) = X \cap V = V$, where V is some open set† of X. Thus, $f^{-1}(U)$ is open. Because of the definition of the quotient topology, Eq. (5.5) now tells us that $b^{-1}(U)$ is open in X/f whenever U is open in $\text{Im}(f)$; hence, as claimed, b is a continuous map.

We are now prepared to formulate a homeomorphism theorem which, for topological structures, plays a role analogous to the typical isomorphism theorems we met for algebraic systems:

THEOREM 5.6a(2). *Let $f(X, \tau) \to (Y, \rho)$ be an open and continuous function between two arbitrary topological spaces. Then the induced map $b: X/f \to \text{Im}(f)$ is a homeomorphism, i.e. $X/f \underset{h}{\approx} \text{Im}(f)$.*

Proof. We already know that b is bijective and continuous, so that we must show only that it is open. Let $Q \subset X/f$. Then, from $f = i \circ b \circ c$ we find that

$$b(Q) = i^{-1}[f(c^{-1}(Q))]. \tag{5.6}$$

Suppose Q is open in X/f. Then $c^{-1}(Q)$ is open in X, hence $f(c^{-1}(Q)) \equiv W$ is open in Y (because by assumption, f is an open map). But i is continuous, so that $i^{-1}(W)$ is an open set in $\text{Im}(f)$. Hence, by Eq. (5.6), the image under b of an open set $Q \subset X/f$ is an open set of $\text{Im}(f)$, i.e. b is an open map, q.e.d.

We observe that, if in our theorem, f is a surjective map, then, of course, $X/f \underset{h}{\approx} Y$.

We now point out the analogies between algebraic and topological systems we referred to earlier. Clearly,

arbitrary morphism	corresponds to	open and continuous map
homomorphism	corresponds to	surjective open and continuous map
isomorphism	corresponds to	homeomorphism
$X/\text{Ker}(f)$	corresponds to	X/f

†This is so because f is continuous by assumption.

To conclude this topic, we add some remarks to Theorem 5.6a(2).

Remarks: (1) The term "open" may be replaced by "closed," because a bijection is open iff it is closed.

(2) If f is an open (or closed) continuous map (so that b is a homeomorphism), Im (f) has the same topology as X/f. Therefore, we can say that the image of an arbitrary topological space under a continuous and open (or closed) map has the quotient topology.

We now discuss an example which will illustrate the concepts introduced in this subsection.

Let $I = [0, 1]$ be the closed unit interval of the real line, equipped with the usual relative topology. Let S^1 be the "one-sphere" in E^2, i.e. the unit circle in the plane, centered at the origin.† Let $f: I \rightarrow S^1$ be the map which "wrappes" I around S^1 and joins the endpoints. This map can be given in analytic form by the assignment

$$x \mapsto (\cos 2\pi x, \sin 2\pi x) \quad \text{for all} \quad x \in I.$$

Note that 0 and 1 of I are both mapped on the point $(1, 0)$ of the circle, see Fig. 5.7. The map f is obviously surjective, and from the continuity of sin

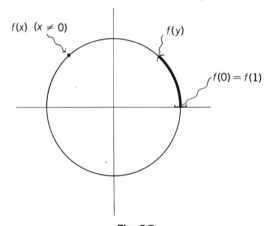

Fig. 5.7.

and cos it is not difficult to show that f is continuous. But f is not an open map. This can be seen if one considers the set $[0, y)$ (which is open in I) and locates on Fig. 5.7 its image (which is not open). On the other hand, f

---

†The definition of n-spheres was given in Problem 5.3-7.

is a closed map, as the reader may verify by a similar consideration.†
Finally, we point out that even though f is both continuous and closed,
it is not a homeomorphism, because it fails to be injective.

Now let us construct I/f. To start with, if $x \neq 0$ or 1, then $z \neq x$ implies
$f(z) \neq f(x)$, so that $[x] = \{x\}$ for all $x \neq 0$ or 1. But $x = 0$ and $x = 1$ are in
the same class (which contains nothing else), so that $[0] = [1] = \{0, 1\}$.
Thus, $X/f = \{$all singletons $\{x\}$ with $x \neq 0$, 1, and $\{0, 1\}\}$. We may visualize
this space as shown in Fig. 5.8. We may say that the canonical map
$c : I \to I/f$ given by

$$x \mapsto \{x\} \qquad \text{if } x \neq 0, 1,$$
$$x \mapsto \{0, 1\} \qquad \text{if } x = 0,$$
$$x \mapsto \{0, 1\} \qquad \text{if } x = 1,$$

"identifies" the endpoints of I, or collapses them into a single element of I/f.

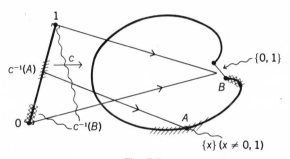

Fig. 5.8.

Next we wish to establish the quotient topology. Let A be a set of I/f
which does not contain $\{0, 1\}$. (Such a set is shown in Fig. 5.8.) Then A is
the union of singletons $\{x\}$, hence $c^{-1}(A)$ is a certain set of points amongst
which 0 and 1 do not occur. Therefore, by the definition of the quotient
topology, such an A is open in I/f if $c^{-1}(A)$ is some open *interior* set of I.
Now let $B \subset I/f$ contain $\{0, 1\}$ (cf. Fig. 5.8). Then $c^{-1}(B)$ will consist of
0, 1, and some interior set N of I. Hence, B can be open in the quotient
topology only if N is a neighborhood of both 0 and 1. Thus, for example,
the set B shown in Fig. 5.8 is not open in the quotient space. (Its inverse
image is an open interval starting at 0 *plus* the isolated point 1.) On the
other hand, Fig. 5.9 illustrates an open set of I/f which contains $\{0, 1\}$.

†The exact proof may be based on the observation that if C is closed in I, then $f(C)$ is
closed in E^2 and then $f(C) \cap S^1$ is closed in S^1.

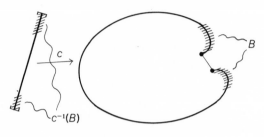

Fig. 5.9.

The induced map $b: I/f \to S^1$ is given by $[x] \mapsto f(x)$, i.e. by

$$\{x\} \mapsto (\cos 2\pi x, \sin 2\pi x), \qquad \text{where} \quad x = 0, 1,$$
$$\{0, 1\} \mapsto (1, 0).$$

This is illustrated in Fig. 5.10. The continuity of b, as claimed by Theorem 5.6a(1), can be visualized by examining the sets drawn in the figure. Finally, by contemplating the inverse map b^{-1}, it will be easily seen that b

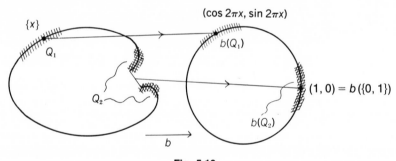

Fig. 5.10.

is actually a homeomorphism, as it should be on account of Theorem 5.6a(2) and the subsequent Remark 1 (since we know that f is continuous and closed). The reader is urged to produce formal proofs of these observations.

In summary, we see that I/f is homeomorphic to the circle. Therefore, we may simply "identify" in our mind I/f and the circle. Intuitively, this may have been expected from the fact that the quotient map c identified the endpoints of I.

PROBLEMS

5.6a-1. The n-cube I^n is defined as that subspace of E^n which consists of all points $x = (\xi_1, \ldots, \xi_n)$ such that $0 \leq \xi_k \leq 1$ for $k = 1, 2, \ldots, n$. (I^n is obviously the generalization of the closed unit interval on the line, etc.) Consider the map $f: I^n \to S^n$ (where S^n is the n-sphere) which consists of wrapping I^n over S^n and "sewing together to a point all loose edges." Show that the quotient space I^n/f is homeomorphic to S^n. (*Hint*: Try to visualize things for $n = 2$, and be content with qualitative considerations, relying on Theorem 5.6a(2).)

5.6a-2. Let $C[a, b]$ be the metric space of continuous real valued functions on the closed and bounded interval $[a, b]$ of the reals. Let t be a fixed but arbitrary real number in $[a, b]$ and define the map

$$e_t: C[a, b] \to E \quad \text{given by} \quad e_t(f) = f(t) \quad \text{for all} \quad f \in C[a, b],$$

where E denotes the space of real numbers with the usual topology. Show that the quotient space C/e_t is homeomorphic to E. Visualize the mappings involved in pictorial terms as well.

5.6a-3. Let R be the equivalence relation on E^2 which is given by the agreement "$x \sim y$ if $x = y$ or if both belong to the closed unit ball $S_1[o]$." Show that the quotient space E^2/R is homeomorphic to E^2. Visualize all details.

6

Topological Spaces
with Special Properties

Topological spaces play an important role in a variety of applications if they (or at least some of their interesting subspaces) exhibit certain special structural properties. Connectedness, separability, compactness, and (for metric spaces) completeness are the most important special properties. In this chapter we shall study these questions in turn.

6.1 CONNECTED SPACES

Perhaps the simplest and yet most important topological property which plays a great role both in geometry and in analysis is that of connectivity. Intuitively, we would say that a space is connected if it consists of "one piece." This notion is formalized by the following

DEFINITION 6.1(1). *A topological space* (X, τ) *is said to be disconnected if one can find two (nonempty, proper) open disjoint subsets A and B such that* $X = A \cup B$. *Conversely, if it is impossible to represent X as the union of a pair of (nonempty, proper) open disjoint sets, then* (X, τ) *is connected.*

We remark here that if X is disconnected, then, in general, there are several representations $X = A \cup B$ with open disjoint sets. Each such decomposition is called "a disconnection of X." A good way to visualize connected spaces is to observe that the definition is tantamount to saying that the only possible partitions of X consisting of a class of *open* sets are $\{X\}$ and $\{\emptyset, X\}$.

In Definition 6.1(1) *the term "open" can be replaced by "closed."* This is so because if $X = A \cup B$ with both A and B open and $A \cap B = \emptyset$, then $A^c = B$ and $B^c = A$, hence B and A are both closed.

The same observation leads to the following interesting theorem:

THEOREM 6.1(1). *If* (X, τ) *is connected, then the only subsets of* X *which are simultaneously open and closed are* $\emptyset$ *and* X.

Proof. If A were a proper subset which is both open and closed. then we could write the disconnection $X = A \cup A^c$.

This theorem may serve as an alternative definition of connectedness.

We wish to give yet another equivalent characterization of connectedness. To this end we pose the following

DEFINITION 6.1(2). *Two sets A and B are said to be separated iff*

$$A \cap \bar{B} = \emptyset \quad and \quad \bar{A} \cap B = \emptyset.$$

Observe that separation of sets is a much stronger requirement than disjointness. For example, on the real line, $[a, b]$ and (b, c) (with $a < b < c$) are disjoint but not separated, because $\overline{(b, c)} = [b, c]$ and $[a, b] \cap [b, c] \neq \emptyset$.

In terms of separated sets, we can now establish the following

THEOREM 6.1(2). *A topological space is connected iff it cannot be represented as the union of two (nonempty) separated sets.*

The *proof* follows directly from the basic definition and from the theorem expressed in Problem 5.3a-7. The usefulness of the theorem is that it provides us with a relatively easy criterion of connectedness.

It is high time for examples.

Example α. The real line is connected: the only open sets of E are unions of open intervals and it is impossible to partition all of E in terms of nonintersecting proper open intervals.

Example β. Let X be an *infinite* set and τ the finite complement topology. This space is connected because if there existed a disconnection $X = A \cup B$, then B and A ought to be finite to have closed complements, but X cannot be the union of two finite sets.

Example γ. The space $\mathbf{Q}$ of rational numbers as a subspace of the reals is *disconnected*, because if $A = \{$all rationals $<y\}$ where y is an irrational number, then $\mathbf{Q} = A \cup A^c$, with both A and A^c open (since $A = (-\infty, y) \cap \mathbf{Q}$ is open in the relative topology). This is rather remarkable because $\mathbf{Q}$ is not a discrete space.

As the last example suggests, the concept of connectivity can be applied to subsets of a space. *If $Y \subset (X, \tau)$, then we say that Y is a connected subset of X if, as a subspace (Y, τ_Y), it is a connected space.* In particular, this means that Y is a connected subset iff it does not contain any proper nonempty subset which is both open and closed *in Y* (not in X). We continue with examples.

Example δ. Any singleton subset $\{x\}$ of an arbitrary space is (fortunately!) connected because it does not contain *any* proper subset, so that the condition on connectedness is vacuously fulfilled.

Example ε. The subset $S = \{x \in \mathbf{R} | a < x < b$ or $c < x < d$, where $a < b < c < d\}$ of the space E of the reals is disconnected. Note that $S = (a, b) \cup (c, d)$ and $\overline{(a, b)} \cap (c, d) = \emptyset$, $(a, b) \cap \overline{(c, d)} = \emptyset$. The subset $(a, b) \subset S$ is both open and closed: it is open because of the definition of the relative topology and it is closed because it is the complement in S of (c, d) which is also open. Likewise, (c, d) is also both open and closed.

Example ζ. Every interval (a, b), $(a, b]$, $[a, b)$, or $[a, b]$ of the real line is a connected subset of the reals. Conversely, the only connected sets in E are the various types of intervals. (These include singletons, because $\{x\}$ can be characterized as the closed interval $[x, x]$.) Even though these observations look very plausible, their proof is surprisingly technical (cf. Simmons[36], p. 143).

Example η. The Euclidean 3-space E^3 is connected.† However, if we "cut out of it" the coordinate plane $X_1 X_2$, then the remaining subset P of E^3 is disconnected. Indeed, if A and B are the "upper" and "lower" halves of the space, we see that $\bar{A} \cap B = \emptyset$ and $A \cap \bar{B} = \emptyset$, so that A and B are separated sets, whereas $P = A \cup B$. On the other hand, if we remove from E^3 just one point (say the origin o), we have a connected space Y. To see this, note that, since E^3 is connected, any decomposition $E^3 = A \cup B$ is such that either $A \cap \bar{B} \neq \emptyset$, or $\bar{A} \cap B \neq \emptyset$, or both statements hold true. Now our space $Y = \{o\}^c$, so that any splitting of Y is obtained from some above described splitting of E^3 by omitting $\{o\}$ from

---

†This is not easy to prove. We shall come back to the connectedness of E^n a little later.

one of the two sets A or B. Suppose we take the case that $Y = A' \cup B \equiv (A - \{o\}) \cup B$. But $\overline{A'} = \overline{A}$ so that either $\overline{A'} \cap B \neq \emptyset$ or $A' \cap \overline{B} = (A \cap \overline{B}) \cup (\{o\} \cap \overline{B}) \neq \emptyset$, due to our previous observation on the splittings of E^3. It is not difficult to see that, even if we omit a whole line from E^3, the remaining set is still connected. These statements generalize easily to higher n.

After having familiarized ourselves with connected spaces and sets and explored the geometrical structure, we turn our interest to related topics of analysis. The central fact about functions defined on connected spaces is expressed by the following

THEOREM 6.1(3). *Let* $f: (X, \tau) \to (X', \tau')$ *be a continuous map whose domain* (X, τ) *is connected. Then the image* $f(X)$ *is a connected subspace of* (X', τ').

Proof. Suppose $f(X)$ was disconnected. Then we could write $f(X) = A' \cup B'$ with A', B' open in $f(X)$ and $A' \cap B' = \emptyset$. But then $f^{-1}(\emptyset) \equiv \emptyset = f^{-1}(A' \cap B') = f^{-1}(A') \cap f^{-1}(B')$, where, because of the continuity of f, both $f^{-1}(A')$ and $f^{-1}(B')$ are open in X. On the other hand, we also would have $f^{-1}(f(X)) \equiv X = f^{-1}(A' \cup B') = f^{-1}(A') \cup f^{-1}(B')$. Putting together these facts, we see that we would have a disconnection of X, contrary to the assumption that X is connected. Hence, the supposition that $f(X)$ is disconnected, is untenable, which proves the theorem.

The essence of our theorem is that continuous images of connected sets are connected. Since every homeomorphism is *eo ipso* a continuous map, we have also the (weaker) statement that *connectedness is a topological property.*

This theorem has many important consequences. First, it can be used to show that *the topological product of connected spaces is a connected space.*† In particular, we then see that E^n *for any n is connected* (because $E^n = E \times E \times \cdots \times E$). Furthermore, since C^n is homeomorphic to E^{2n} (see Section 5.6, p. 225), it follows from our present theorem that C^n *is connected.*

Another direct application of Theorem 6.1(3) is the generalization of the "intermediate value theorem" of elementary calculus. Let (X, τ) be an arbitrary connected space and let f be a real valued continuous map on X. Then the range of this map must be an interval of the reals, because (as stated in Example ζ) every connected set of the reals is some interval.

†See, for example, McCarty [28], p. 154.

So far we have looked at questions of connectivity from the global point of view. We may also adopt a local viewpoint.

DEFINITION 6.1(3). *A topological space which has the property that every neighborhood $N(x)$ of each point x of the space contains a connected subneighborhood $M(x)$, is said to be a locally connected space.*

It should be obvious that local connectivity does not imply connectivity. For example, the space consisting of the union of two disjoint intervals on the real axis is locally connected but not connected. More surprising is the fact that *connectedness does not necessarily imply local connectedness.* We give two typical examples.

Example ϑ. An indiscrete topological space is both connected and locally connected. The latter follows from the fact that the only neighborhood a point can have, is all of X.

Example κ. Let A be that subset of the plane which consists of the points of the "Y-axis" lying between -1 and $+1$, inclusive. Let B be the graph of the function $\sin 1/x$, for $0 < x \leqslant 1$ (cf. Fig. 6.1). Let X be $A \cup B$, with the relative topology of E^2. This space is connected. (The only suspicious decomposition is $X = A \cup B$ itself, but $\bar{B} \cap A = A \neq \emptyset$.) On the other hand, X is not locally connected. If $x \in A$, then all sufficiently small neighborhoods $N(x)$ are disconnected, since they can be decom-

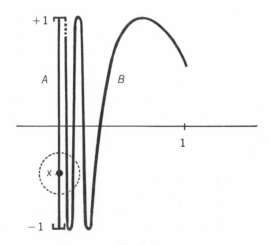

Fig. 6.1.

posed into an open set in B and a disjoint open set in $A \cup B$. (The reader should observe here what are the open sets in the relative topology.)

PROBLEMS

6.1-1. Let A be the open interval $(0, 1)$, B be the open interval $(1, 2)$, and C be the closed interval $[1, 2]$. Which of these subsets of the real line are separated from the others?

6.1-2. Consider the union of two open discs in E^2, which are externally tangent to each other. Is this subspace of E^2 connected? Let now one of the discs be open and the other closed. Finally, let both be closed. Reconsider the situation in each case.

6.1-3. Consider a discrete topological space. Is it connected? What about an indiscrete topological space?

6.1-4. Show that in any *metric* space, any countable subset consisting of more than one point is disconnected. Construct a simple example of a nonmetric topological space where the above statement is false.

6.1-5. Show that any n-sphere S^n is connected. (*Hint*: Recall that $E^{n+1} - \{o\}$ is connected (o is here the element $x = (0, \ldots, 0)$) and consider the map $E^{n+1} - \{o\}$ given by $f(x) = x/d(x, o)$.)

6.1-6. Show that the union of two not separated connected subsets is always a connected subset.

6.1-7. Show that if a connected space is the domain of some real valued nonconstant continuous function, then it must contain uncountably many points.

6.1-8. Show that a topological space is disconnected iff there exists a continuous mapping from it onto the discrete two-point space $\{0, 1\}$. (*Hint*: You will need a function which has the value $f(x) = 0$ for all points x of some particular set, and the value $f(x) = 1$ for all points x of some other particular set.)

6.1-9. Show that any discrete space is *locally* connected.

6.1-10. Let x be that subspace of E^2 which arises if one omits from E^2 a set of "vertical lines," which are intersecting the "absciss" at the points $\xi_1 = 1, 1/2, \ldots, 1/n, \ldots$ and for each of which $-1 \le \xi_2 \le +1$ (cf. Fig. 6.2). Show that this space is connected but not locally connected.

6.1-11. Show that the continuous image of a locally connected space need not be locally connected, but if the map is both continuous and open, the image will be also locally connected.

6.1-12. A subset A of a space (X, τ) is called a *component of X* iff it is a connected set which is not contained in any other connected subset of X. Show the following:
(a) Each component of X is a closed set.
(b) Each connected set is contained in some component of the space.

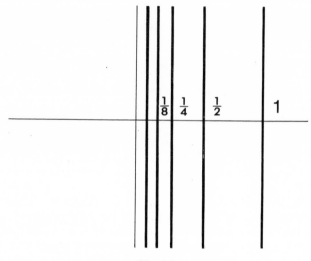

$$\frac{1}{8} \quad \frac{1}{4} \quad \frac{1}{2} \quad 1$$

Fig. 6.2.

(c) The class of components is a partition of the space.
(d) A space is connected iff it has one component.

6.1-13. A topological space is said to be *totally disconnected* if every singleton set is a component (cf. preceding problem) of the space. Show that the following are totally disconnected spaces:
(a) Any discrete space,
(b) The subspace of E consisting of all irrational numbers,
(c) Any countable subspace of a *metric* space.

6.1a. Path Connectivity; Homotopy

There exists a rather pictorial concept, path connectedness, which is closely related to the notion of connectedness studied above and which plays an important role in geometry and in the theory of topological groups. We briefly review here this topic.

Let (X, τ) be a topological space and let $a, b \in X$. *A path from a to b is any continuous map φ from the closed unit interval $I = [0, 1]$ of the reals into the space (X, τ) which has the property that $\varphi(0) = a$ and $\varphi(1) = b$.* It is customary to denote the range (image) of φ (i.e. $\varphi(I)$) by the symbol $\varphi_{a \to b}$ and refer to $\varphi_{a \to b}$ as "a path which connects a to b." The reader is warned to carefully distinguish between the map

$$\varphi : I \to X$$

and the "actual path" $\varphi_{a \to b}$ (which is a subset of X). One may remember this by saying that the "path function φ generates the path $\varphi_{a \to b}$." The concept of path is visualized in Fig. 6.3. The entire heavy curve shown is the path $\varphi_{a \to b}$, and a particular point, which corresponds to the real number $t \in [0, 1]$, is labeled by $\varphi(t)$.

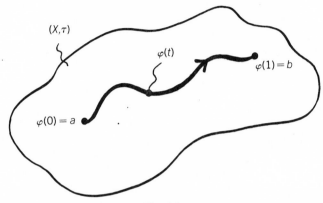

Fig. 6.3.

If, for a given pair a, b of points in X it is possible to find a path from a to b, then we say that a and b can be connected by a path.† Intuitively speaking, two points can be connected by a path if there exists some one-parameter family taken from X along which one can "proceed continuously" from a to b.

To give an example, consider the space E^n and single out the points $o = (0, \dots, 0)$ and $x = (\xi_1, \dots, \xi_n)$. These can be connected by a path, because, for example, the map $\varphi: I \to E^n$ given by $\varphi(t) = tx \equiv (t\xi_1, \dots, t\xi_n)$ for all $t \in [0, 1]$ is clearly continuous and $\varphi(0) = o$, $\varphi(1) = x$. The image (range) of the path is the set

$$\varphi_{o \to x} = \{(t\xi_1, t\xi_2, \dots, t\xi_n) | 0 \le t \le 1\}.$$

We now make the following

DEFINITION 6.1a(1).   *A topological space (X, τ) is said to be path-connected iff any pair of its points can be connected by a path*, i.e. *iff for*

†Obviously, there may exist pairs of points that cannot be connected by a path. For example, in the space consisting of two disjoint intervals of the real line, there is no path from a to b if the first and second point lies in the first and second interval, respectively.

any $x, y \in X$ *there exists at least one continuous map* $\varphi : [0, 1] \to X$ *such that* $\varphi(0) = x$, $\varphi(1) = y$.

For example, the space E^n is path-connected, because given $x, y \in E^n$, the map defined by

$$\varphi(t) = (1 - t)x + ty \quad \text{for all} \quad 0 \le t \le 1$$

generates a path which connects x to y.

The reader may now wonder, what is the relationship between path-connectedness and connectedness as previously defined. The answer is given by the following

THEOREM 6.1a(1).   *Every path-connected space is also connected.*

The *proof* follows from the observation that the path $\varphi_{x \to y}$, being the continuous image of a connected set of the reals, is a connected set of (X, τ), and from the theorem expressed in Problem 6.1-6.

This theorem is quite useful inasmuch as visualization of path-connectedness is often much easier than that of connectedness, so that we have at least a good intuitive guide for establishing connectedness. The reader should reconsider in this light previously discussed proofs of connectedness. For example, the connectedness of E^n follows now directly from the observation that E^n is path-connected, which (apart from the explicit proof given above) is intuitively obvious.

On the other hand, Theorem 6.1a(1) has no converse: there exist connected spaces which are not path connected. For example, the space described in Example κ on p. 240 was seen to be connected, but it is rather obvious that it is not path-connected: it is impossible to reach continuously a point in A from a point in B, since the curve does not lead into A (the origin is not in B). Thus, path-connectedness is a more restrictive concept than connectedness.

In the following we shall consider certain classes of paths in a given topological space (X, τ) and endow these classes of "topological objects" with an algebraic structure.†

If $\varphi_{a \to b}$ and $\psi_{b \to c}$ are two paths in (X, τ) generated by φ and ψ respectively, then the function $\chi : I \to X$, given by

---

†These considerations are a very simple example of that branch of mathematics which is called *algebraic topology*.

$$\chi(t) = \begin{cases} \varphi(2t) & \text{if } 0 \leq t \leq \frac{1}{2} \\ \psi(2t - 1) & \text{if } \frac{1}{2} \leq t \leq 1 \end{cases}$$

is easily seen to be continuous and to obey the boundary conditions $\chi(0) = a$, $\chi(1) = c$. Hence, χ generates a path from a to c. This tells us that we may define a composition law for paths. The path $\chi_{a \to c} \equiv \varphi_{a \to b} \psi_{b \to c}$ is called the composite of the two given paths.†

Next we consider the function $\epsilon: I \to X$, given by $\epsilon(t) = a$ for all $t \in [0, 1]$, where a is some fixed element of X. The image of I in X is simply the point a, so we can look upon $\epsilon_{a \to a} \equiv \epsilon_a$ as the *null path* at a.

Finally, given a path φ, the map $\varphi^{-1}: I \to X$ defined by $\varphi^{-1}(t) = \varphi(1 - t)$ is also continuous and, if φ generated the path $\varphi_{a \to b}$, the function φ^{-1} generates the path from b to a, to be denoted by $\varphi^{-1}_{b \to a}$, because $\varphi^{-1}(0) = \varphi(1) = b$, $\varphi^{-1}(1) = \varphi(0) = a$. The image $\varphi^{-1}_{b \to a}$ is called *the inverse*‡ *of the path* $\varphi_{a \to b}$. We observe that the inverse path $\varphi^{-1}_{b \to a}$ consists of the same points of X as $\varphi_{a \to b}$, but they are "run through" in the opposite sense.

Next, we define an equivalence relation between paths. Let a and b be two fixed points of X and let φ and ψ define paths such that both connect a to b. Suppose one can "continuously deform" $\varphi_{a \to b}$ into $\psi_{a \to b}$, while leaving the endpoints fixed. The intuitive meaning of this term is illustrated in Fig. 6.4, and the concept is rigorously formulated in Problem 6.1a-3. If the two paths have the stated property, then we say that they are *homotopic*. It is quite easy to convince oneself that *homotopy* is an *equivalence relation* for the collection of paths with specified common beginning- and endpoints. Each corresponding equivalence class $[\varphi]$ is called a *connection* between a and b. In Fig. 6.5, the paths φ and φ' are in the same class, but the path ψ belongs to a different class (the hole in the middle is not in the space).

Further insight into the structures under consideration is obtained if we consider closed paths or loops. Let $x_0 \in X$ be a fixed, arbitrary point. A path φ which connects x_0 to itself is called a *loop* at x_0. Since we shall keep x_0 fixed in the following discussion, we shall denote loops at x_0 simply by symbols φ, ψ, χ, etc. rather than referring to their images $\varphi_{x_0 \to x_0}$, etc. and we shall speak of "loops" without referring to x_0. Just as we defined equivalent (homotopic) paths with distinct endpoints a and b, we

†Strictly speaking, the composition is that of the path *functions* φ and ψ, giving the function χ.

‡The notation is somewhat misleading, because φ^{-1} is not the inverse function of φ: as a matter of fact, since $\varphi: I \to X$ is not injective, no inverse function exists.

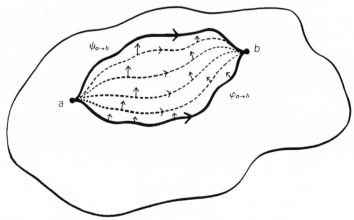

Fig. 6.4.

of course can define homotopy for the collection of loops at x_0. Thus, two loops are in the same class (i.e. are homotopic loops) if they can be "continuously deformed" into each other. For example, the loops φ and φ' in Fig. 6.6 are homotopic, but ψ is in another class.

The null path ϵ_{x_0} at x_0 is obviously a special case of a loop at x_0. Any loop φ which is homotopic to† ϵ is called a *contractable loop*, i.e. a contractable loop is one which can be continuously "shrunk to the point x_0." Clearly, the contractable loops form one homotopy class, which can be symbolized by

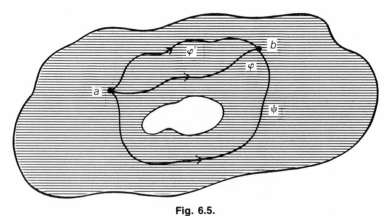

Fig. 6.5.

†In accord with our above statement on notation, we omit the subscript x_0 on ϵ.

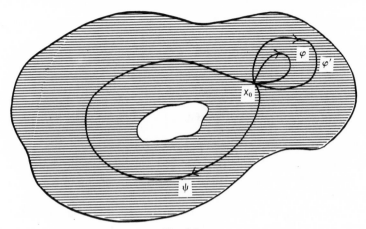

Fig. 6.6.

$[\epsilon]$; all other loops must belong to different classes. We remark that a loop and its inverse loop may or may not belong to the same homotopy class. For example, on the sphere S^2 every loop is homotopic to its inverse (because a loop may be "slipped around" on the sphere surface, keeping one of its points fixed). On the torus, in contrast, loops that wind around the "dough-nut hole" are not homotopic to their inverses, but all other loops are. In general, all contractable loops are homotopic to their inverses. This can be seen by imagining the loop to be contracted to ϵ and then "expanding" ϵ to φ^{-1}.

Let us now consider the collection Ω_{x_0} of all loops at x_0. The homotopy equivalence relation H leads to a partition of Ω_{x_0} into classes $[\epsilon]$, $[\varphi]$, $[\psi]$, ... and we are led to the quotient set Ω_{x_0}/H. This set (consisting of distinct classes of equivalent loops) *may be endowed with a group struc-ture.* We define the product of $[\varphi]$ and $[\psi]$ as the class $[\varphi\psi]$ which is generated by the composite loop $\varphi\psi$, where the composition of loops is the specialization of composition of arbitrary paths as discussed before. It can be shown† that the composition law is independent of the choice of representatives. It is rather obvious that associativity $[\varphi][\psi\chi] = [\varphi\psi][\chi]$ holds. The contractable class $[\epsilon]$ is clearly the identity element, since $[\epsilon][\varphi] = [\epsilon\varphi] = [\varphi]$. It is less easy to see that every element $[\varphi]$ has an inverse. Nevertheless, it is true that $[\varphi]^{-1} = [\varphi^{-1}]$, because $[\varphi][\varphi]^{-1} = [\varphi][\varphi^{-1}] = [\varphi\varphi^{-1}]$ can be shown to be the same class as $[\epsilon]$, i.e. one can

†For detailed discussion of homotopy topics see, for example, McCarty[28] and the classical text by Pontrjagin[32].

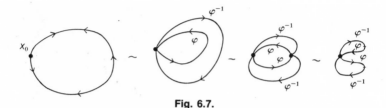

Fig. 6.7.

see that $\varphi\varphi^{-1}$ is contractable.† Instead of a proof, we refer to Fig. 6.7 where we first replace (in the sense of equivalence) the double loop by two closely adjacent ones, then join them again at one point, finally contract each loop-part to x_0.

Thus, the set Ω_{x_0}/H has now an algebraic structure, that of a group. The elements are the loop classes $[\varphi]$, the identity is the contractable class $[\epsilon]$. We denote this group by $\pi(X, x_0)$. It can be shown that if X is a path-connected space, then the construction $\pi(X, x_0)$ is essentially independent of the choice of x_0, in the sense that the groups $\pi(X, x_0)$ and $\pi(X, x_1)$, with arbitrary focal points x_0 and x_1, are isomorphic and can be "identified." In that sense, we may the speak of "the" *homotopy group* (or *fundamental group*) $\pi(X)$ of the path-connected space X. This group plays a very important role in the theory of topological groups.

In the subsequent discussion we assume that (X, τ) is path-connected (hence connected) and we shall give a simple application of the preceding ideas to the question of different "degrees" of connectedness.

A (path) connected space is said to be *simply connected* if all loops at all points are contractable. In view of our previous results, this can be expressed by saying that, for a simply connected space, the fundamental group $\pi(X)$ consists of one element, namely $[\epsilon]$. Similarly, a connected space is called *n-fold connected* if each point has exactly n connections to itself, i.e. if at each point the loop system has n distinct equivalence classes. Yet in other words, the group $\pi(X)$ has n elements. It may happen that the group $\pi(X)$ has infinitely many elements; then the space is said to be infinitely connected.

For establishing the degree of connectivity, the following theorems are useful:

(1) If (X, τ) and (Y, ρ) are homeomorphic (path-connected) spaces, then $\pi(X)$ and $\pi(Y)$ are isomorphic. Hence, the two spaces have the same degree of connectedness.

(2) If the spaces (X, τ) and (Y, ρ) have fundamental groups $\pi(X)$ and

---

†Also, $[\varphi]^{-1}[\varphi] = [\varphi^{-1}\varphi] = [\epsilon]$.

$\pi(Y)$, then the product space $(X \times Y, \tau \times \rho)$ has a fundamental group $\pi(X \times Y)$ which is isomorphic to the direct product group $\pi(X) \times \pi(Y)$. Hence, the degree of connectedness of a product space is the product of that of its factors.

(3) If (X, τ) is a homogeneous (and path-connected) space, then the number of connections of a point x_0 to itself is the same as the number of connections between x_0 and any other given point y. This often simplifies the counting of equivalence classes.

We now give a few examples.

Example α. The real line E is simply connected: obviously, there is only one connection between a pair x, y of points (which actually consists of a single path).

Example β. All E^n spaces are simply connected, because $E^n = E \times \cdots \times E$, each component in the product being simply connected. Since $C^n \approx_h E^{2n}$, the C^n spaces are also simply connected.

Example γ. A surprise: the circle S^1 in the plane is infinitely connected. It is not easy to give an exact proof. However, intuition convinces one of the following. The contractable class $[\epsilon]$ consists, for a given point x_0, of the loops that go from x_0 to any other point of the circle and back to x_0, without circumnavigating the circle completely. Another class has as its representative element the loop that connects x_0 to itself by one full clockwise wounding round the circle. Distinct classes are obtained by 2, 3, ... clockwise trips; and also, new distinct classes arise by 1, 2, 3, ... counterclockwise rounds. Thus, the fundamental group $\pi(S^1)$ is seen to be isomorphic to the additive group $\mathbf{Z}$ of all integers.

Example ϑ. Lest you think that the situation generalizes to arbitrary n-spheres, here is the pleasant countertheorem: Any sphere S^n with $n \geqslant 2$ is *simply* connected. You may easily visualize this for S^2: any loop is contractable. Moreover, even if a point p (say the North Pole) is missing, $S^n - \{p\}$ is still simply connected.

PROBLEMS

6.1a-1. Give a detailed proof that the space E^∞ is path-connected (hence, connected).

6.1a-2. Show that path-connectedness is preserved by continuous maps (even though it is a more specific property than connectedness).

6.1a-3. Make the pictorial definition of homotopic paths (given in the text) mathematically precise as follows: Two paths $\varphi_{a \to b}$ and $\psi_{a \to b}$ are homotopic if there exists a continuous function $g: I \times I \to X$ such that

$$g(0, t) = \varphi(t),$$
$$g(1, t) = \psi(t),$$
$$g(s, 0) = \varphi(0) = \psi(0) = a,$$
$$g(s, 1) = \varphi(1) = \psi(1) = b.$$

(Draw a figure to illustrate homotopy.) Using this definition, prove rigorously that homotopy is an equivalence relation, as was claimed in the text.

6.1a-4. Give a precise mathematical definition for the homotopy of a loop system at x_0.

6.1a-5. Show that the torus is infinitely connected and determine its fundamental group. (*Hint*: Recall that the torus is homeomorphic to $S^1 \times S^1$.)

6.1a-6. Show that the projective line **R*** (cf. Problem 1.2a-4) is infinitely connected, in contrast to the usual real line.

6.1a-7. Let (X, τ) be the real plane E^2 from which the origin (or even a finite disk centered at the origin) has been deleted. Show that this space is infinitely connected. (*Hint*: By realizing that *every* point x is uniquely determined by an angle and a distance, show first that the space is homomorphic to $S^1 \times E$.) What happens if you delete a hole from E^3?

6.2 SEPARABLE SPACES

Separability of topological spaces is a rather technical concept, but it plays an important role in the theory of Hilbert spaces, so we give a brief review.

We start with a simple

DEFINITION 6.2(1).   *A subset A of a topological space (X, τ) is said to be dense iff $\bar{A} = X$.*

Thus, if A is dense in X, then every point $x \in X$ is either a point of A or at least an accumulation point of A. In this sense, a dense set "almost fills" the whole space. For example, the rationals are dense in E, because every irrational number is an accumulation point (in fact, it is the limit of some rational sequence).

For *metric* spaces, the definition of denseness implies that A is dense iff for *every* $x \in X$ and any given $\epsilon > 0$, there exists an element $a \in A$ such that $d(x, a) < \epsilon$. In other words, in a metric space, *every* point $x \in X$ lies arbitrarily close to *some* point a of a given dense set A.

Quite generally, a *proper* dense set A of a topological space cannot be closed (because $\bar{A} = A$). In fact, a dense set A cannot be even contained in a closed proper subset of X.

Now we are prepared to make the following

DEFINITION 6.2(2).   *A topological space (X, τ) is called a separable space† iff it contains a countable dense subset.*

We give some of the most frequently occurring examples of separable spaces.

Example α.   The real line E is separable because, as pointed out above, the countable subset of rationals is a dense subset.

Example β.   The Euclidean space E^n is separable. Consider the subset A with elements $a = (r_1, r_2, \ldots, r_n)$, where all components r_k are rational numbers. This subset A is obviously countable ($n\aleph_0 = \aleph_0$). Now let $x = (\xi_1, \ldots, \xi_n)$ be an arbitrary point of E^n. Given any $\epsilon > 0$, for a suitable choice of rationals $r_1, \ldots, r_n$ we surely have $d^2(x, a) = \Sigma_{k=1}^n (\xi_k - r_k)^2 < \epsilon$, because, by the denseness of the rationals in E, each term in the sum can be made as small as desired. Hence the countable set A is dense, so that E^n is separable.

Example γ.   The E^∞ space is also separable. Consider the subset A which consists of the elements

$$a_1 = (r_{11}, 0, 0, \ldots)$$
$$a_2 = (r_{21}, r_{22}, 0, \ldots)$$
$$\cdot$$
$$\cdot$$
$$\cdot$$
$$a_N = (r_{N1}, r_{N2}, \ldots, r_{NN}, 0, \ldots)$$
$$\cdot$$
$$\cdot$$
$$\cdot$$

where all r_{kl} components are rational numbers. This set A is countable ($\aleph_0 \aleph_0 N = \aleph_0$). To see that it is dense, let $x = (\xi_1, \ldots, \xi_k, \ldots)$ be an arbitrary element and let $y = (\xi_1, \ldots, \xi_N, 0, \ldots)$ be a corresponding truncated element. We have $d(x, y) = (\Sigma_{k=N+1}^\infty \xi_k^2)^{1/2}$, and since $\Sigma_{k=1}^\infty \xi_k^2$ is

†The reader should not be bothered by the fact that terms resembling the same word occur in other contexts too. We talked about "separation of points" (in connection with Hausdorff spaces) and "separated sets" (in connection with connectivity). The present concept must not be confused with either of these.

convergent, we can choose N sufficiently large so as to have $d(x, y) <$ $\epsilon/2$. We also can choose from the set A some element a_N such that

$$d(y, a_N) = \left(\sum_{k=1}^{N} (\xi_k - r_{Nk})^2 \right)^{1/2} < \frac{\epsilon}{2},$$

because the rationals are dense in E. Then, using the triangle inequality, $d(x, a_N) \le d(x, y) + d(y, a_N) < \epsilon$, so that the countable set A is dense in E^∞, q.e.d. In a similar manner one can show that all E_p^n spaces and E_p^∞ are also separable.

Example δ. The space $C[a, b]$ of continuous real functions on the closed and bounded interval $[a, b]$ is separable. We only sketch the proof. Divide $[a, b]$ into $2n - 1$ equal parts. Define the *countable* set of functions f_k^n by

$$f_k^n(x) = r_k \quad \text{for} \quad x_{2k} < x < x_{2k+1} \qquad (k = 0, 1, \ldots, n - 1)$$

where the r_k are rational numbers, and interpolate linearly in between. Such a function is shown in Fig. 6.8. This set is *dense*, because it is not

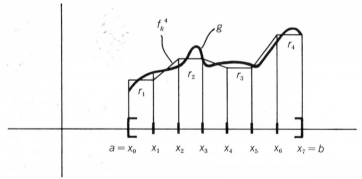

Fig. 6.8.

difficult to show that for any given continuous function $g(x)$, one may find a large enough n and a suitably chosen set of rationals r_k such that $\sup_x |f_k^n(x) - g(x)|$ is as small as desired.

We now give a theorem which can be used as a good criterion for establishing separability.

THEOREM 6.2(1). *If (X, τ) is a topological space which possesses a countable base, then it is separable.*

Proof. Let $\mathcal{B} = \{\ldots B_n \ldots\}$ be the countable base and construct the countable set $D = \{x_1, x_2, \ldots, x_n, \ldots\}$, where $x_n \in B_n$. This set is also *dense* because every neighborhood $N(x)$ of any point $x \in X$ can be represented as some union of sets from the collection $\mathcal{B}$, so $N(x)$ contains at least one point of D, hence x is an accumulation point of D.

The separability of E^n (which we proved above directly) is a good illustration of the theorem. Actually, for *metric spaces* the converse also holds true: *If a metric space is separable, then it possesses a countable base.*

We add here some remarks on a closely related topic. Suppose (X, τ) is a topological space and $\mathcal{O} = \{\ldots V_\alpha \ldots\}$ is a cover† of X which consists of *open* sets V_α. We call $\mathcal{O}$ an *open cover* of (X, τ). Any subclass of $\mathcal{O}$ which is a cover itself, is called an *open subcover*. The following definition is frequently met:

DEFINITION 6.2(3). *A topological space which has the property that every open cover of it contains a countable open subcover, is called a Lindelöf space.*

It can be shown that *if (X, τ) has a countable base, then it is a Lindelöf space.* However, even though the same criterion leads to separable spaces, it is *not* true that every Lindelöf space is separable, nor does separability imply the Lindelöf property. On the other hand, because of our previous remark on metric spaces, we certainly have the following

THEOREM 6.2(2). *Every separable metric space is a Lindelöf space*, i.e. *from each open cover we can extract a countable subcover.*

We conclude this section with the following

THEOREM 6.2(3). *Separability is a topological property*, i.e. *the homeomorphic image of a separable space is separable.*

As a matter of fact, even (surjective) continuous maps preserve separability.

A consequence of Theorem 6.2(3) is that C^n, C^∞, $C_p{}^n$, $C_p{}^\infty$ are separable spaces.

---

†We remind the reader that a cover $\mathcal{O} = \{\ldots A_\alpha \ldots\}$ of the space X is a class of sets of X such that $\bigcup_\alpha A_\alpha = X$. Similarly, a cover of a subset $S \subset X$ is a class $\mathcal{O}$ such that $S \subset \bigcup_\alpha A_\alpha$.

PROBLEMS

6.2-1. Show that a set D is dense iff it has a nonempty intersection with every (nonempty) open set. (*Hint*: Recall that D cannot be contained in any closed proper subset of the space and that every open set is the complement of some closed set.)

6.2-2. Let (X, τ) be the arbitrary space and let (X', τ') be a Hausdorff space. Let D be a dense subset of X, and let $f_D: (D, \tau_D) \to (X', \tau')$ be a continuous function. Show that f_D has at most one continuous extension $f: (X, \tau) \to (X', \tau')$. (That is, show that if it is possible at all to extend f_D to a continuous function on (X, τ), then this extension is unique.)

6.2-3. Show that a discrete metric space is separable iff it consists of not more than countably many points.

6.2-4. Prove that any subspace of a separable *metric* space is also separable. (*Hint*: Recall that in a metric space the limit of a sequence has "almost all" members of the sequence arbitrarily close to itself. Use also the triangular inequality.)

6.2-5. Let X be an uncountable set and let p be some specified point of X. Define on X the topology τ as follows: $V \subset X$ is open iff either $V = \emptyset$ or $p \in V$. Show that (X, τ) is a separable space which possesses a nonseparable subspace. (*Remark*: This example illustrates that the theorem of the preceding problem does not hold for arbitrary topological spaces.)

6.2-6. Show that contrary to the disappointing result of the preceding problem, a somewhat weaker result is true: Every *open* subspace† of an arbitrary separable topological space is separable.

6.3. COMPACT SPACES

Obviously, the simplest spaces are those that have only a finite number of points. A rather natural generalization would be a space which, roughly speaking, can be at least covered with a finite number of simple sets. Compact spaces have essentially this feature. Their major importance is connected with the favorable behavior of sequences and functions in such spaces. We shall discuss many properties of compact spaces, but, because of their highly technical character, we shall omit most proofs. The interested reader is referred to any textbook on classical topology.

We start right away with the fundamental

DEFINITION 6.3(1).   *A topological space is said to be compact iff every open cover‡ contains a finite open subcover.*

†That is, a subspace which, considered as a subset, is an open set of X.

‡The definition of an open cover and subcover was given at the end of Section 6.2.

We observe that compact spaces are a kind of improved Lindelöf spaces, where only the extraction of a *countable* open subcover was guaranteed.

Before giving examples, we wish to make it clear that (as was in the case of connectedness and separability) a subset A of a space is said to be compact if, as a subspace with the relative topology, (A, τ_A) is a compact space.

Example α.   Any finite space is compact, because if we have an open cover, it already consists of only finitely many sets. By the same token, any finite subset of an arbitrary space is compact.

Example β.   Let (X, τ) be an infinite space with the finite complement topology. This space is compact, because any open set omits only finitely many points, so from a given open cover you may omit all members except one, say the set V_0, and then the points that are left uncovered are $x_1, \ldots, x_n$. Now you take the open sets $V_1, V_2, \ldots, V_n$ with each V_k being in the cover and such that $x_k \in V_k$. Obviously, $\{V_0, V_1, \ldots, V_n\}$ is a finite open subcover.

Example γ.   It is a well-known, although not quite easily proved, theorem of classical analysis that *every closed and bounded interval of the real line is compact.* Moreover, as a generalization, the celebrated Heine–Borel theorem tells us that every closed and bounded *subset* of E is compact.

The following two theorems relate closedness and compactness, and furnish useful criteria to discover compact sets.

THEOREM 6.3(1).   *If (X, τ) is compact, then every closed subset is automatically compact.*

One application of this theorem is the proof of the Heine–Borel theorem from the simpler statement that closed and bounded *intervals* are compact. Since any bounded *set* is a subset of a closed interval, the Heine–Borel theorem follows directly. Another application is the sad result that *the real line*, as a whole, *is not compact.* Indeed, it possesses closed subsets that are not compact. For example, any countable and unbounded set of points is closed but not compact, because we cannot extract a finite open subcover from any open cover. In a similar manner one sees that E^n is not compact.

The converse of Theorem 6.3(1) is, in general, not true: A space, whether compact or not, may have compact sets that are not closed.

However, for Hausdorff spaces a rather strong statement holds:

THEOREM 6.3(2).   *In an arbitrary Hausdorff space, every compact sub-set must be closed.*

An application of this theorem is that on the real line or in E^n, etc., no open set can be compact.

The next theorem shows one of the strong constraints that compactness imposes upon a space. The theorem can be also used as a workable necessary criterion of compactness:

THEOREM 6.3(3).   *In a compact space every infinite subset has an accumulation point.*†

It is a tradition to say that a space where every infinite subset has an accumulation point, has the *Bolzano–Weierstrass property*. Thus, Theorem 6.3(3) tells us that every compact space has the Bolzano–Weierstrass property.‡

We may say that Theorem 6.3(3) is the reason why we use the terminology "compact" space. In such a space, each infinite subset is so strongly "compacted" that it cannot fail to have an accumulation point.

On the real line, the set $0, 1, 2, \ldots$ has no accumulation point. This illustrates the already known fact that E is not compact.

A more sophisticated example for the use of Theorem 6.3(3) is the following. Let $S_1[o]$ be the closed unit ball centered at $o = (0, 0, \ldots, 0, \ldots)$ of the E^∞ space. Thus, $x \in S_1[o]$ if $\Sigma_{k=1}^\infty \xi_k^2 \leq 1$. We wish to show that $S_1[o]$ is not a compact space. Consider the subset

$$x_1 = (1, 0, \ldots, 0, \ldots)$$
$$x_2 = (0, 1, \ldots, 0, \ldots)$$
$$\vdots$$
$$x_n = (0, 0, \ldots, 1, 0, \ldots)$$
$$\vdots$$

---

†It should be noted that the accumulation point x_0 of the infinite subset A whose existence is claimed by theorem, may or may not belong to A.

‡Sometimes the term "countably compact space" is applied to spaces with the Bolzano–Weierstrass property. Since the terminology is ambiguous, we shall avoid it.

For any two distinct points in this set, $d(x_i, x_j) = \sqrt{2}$. Therefore, this infinite set cannot have an accumulation point, since, if y was one, then $d(x_i, y) < \epsilon$, $d(x_j, y) < \epsilon$ provided $i, j \geqslant N$, which is impossible because $d(x_i, y) + d(y, x_j) \geqslant d(x_i, x_j) = \sqrt{2}$. Hence, by Theorem 6.3(3), *the closed and bounded subspace $S_1[o]$ of E^∞ is not compact.* This is surprising, because in E^n every closed and bounded subset is compact.† Incidentally, Theorem 6.3(1) tells us that E^∞ cannot be compact. (We may have anticipated this on simpler grounds, of course.)

Yet another useful theorem for recognizing compact spaces is known as *Tychonoff's theorem*:

THEOREM 6.3(4).   *The topological product of compact spaces is compact.*

An important application is the *generalized Heine–Borel theorem*. Since we know that every closed and bounded subset of the real line is compact, it immediately follows that *every closed and bounded subset of E^n is compact.*

The next three theorems refer specifically to *metric spaces*. For these, especially strong statements follow from compactness and they display clearly the geometrical significance of compactness.

The first statement claims that, for metric spaces the converse of Theorem 6.3(3) holds (whereas, for arbitrary spaces this is not true):

THEOREM 6.3(5).   *If a metric space has the Bolzano–Weierstrass property, then it is compact.*‡

The following theorem is a consequence of the Bolzano–Weierstrass property, but usually easier to apply and more useful, too:

THEOREM 6.3(6).   *A metric space is compact iff every sequence in it contains a convergent subsequence.*

The property that from every (infinite) sequence one can extract a convergent subsequence is often referred to as *sequential compactness*.

---

†So far we know the validity of this statement for E. The result for arbitrary n will be discussed presently.

‡Putting this together with Theorem 6.3(3), we may say that, for metric spaces, the Bolzano–Weierstrass property is a necessary *and* sufficient criterion of compactness.

Thus, we may say that a metric space is compact iff it is sequentially compact. For variety's sake, we prove this theorem.

(a) Assume that (X, d) is sequentially compact. Let A be an infinite subset. Then we can extract from A an infinite sequence (x_n) which consists of *distinct* points. By the assumed sequential compactness, there exists a convergent subsequence so that $x_{n_k} \to x$. But we recall that if a convergent sequence has distinct terms, then its limit is an accumulation point. Hence x is an accumulation point of the *set* $\{\ldots x_{n_k} \ldots\} \subset A$, therefore it is an accumulation point of A. Hence, X has the Bolzano–Weierstrass property, so that by Theorem 6.3(5) it is compact.

(b) Assume now that (X, d) is compact. Let (x_n) be an arbitrary sequence in X. If (x_n) has a point x_k which is infinitely many times repeated, then the subsequence (x_k) is obviously convergent, so there is nothing to prove. Therefore, we may assume that (x_n) does not have any of its terms infinitely many times repeated. Consequently, the *set* $\{\ldots x_n \ldots\}$ is an *infinite* subset of X. Then, because of the assumed compactness of X and the concomitant Bolzano–Weierstrass property, $\{\ldots x_n \ldots\}$ has an accumulation point x. Since we are in a metric space, this x is the limit of some subsequence of (x_n), q.e.d.

We give a highly nontrivial illustration of Theorem 6.3(6). Let K be the subspace of E^∞ (or C^∞) whose elements $x = (\xi_1, \ldots, \xi_k, \ldots)$ have the property that $|\xi_k| \leq 1/k$. This subspace is called the *Hilbert cube* of E^∞ (or C^∞). We shall show (with a lot of work) that K is compact.† Let (x_n) be some sequence in K. If some term x_m is infinitely many times repeated, then (x_n) is trivially convergent so that we may restrict ourselves to the case when no term occurs infinitely many times. Let A^m be the subset of E^m defined by

$$A^m = \left\{ y = (\eta_1, \ldots, \eta_m) \middle| |\eta_k| \leq 1/k \right\}.$$

Because of the generalized Heine–Borel theorem, A^m (being closed and bounded in E^m) is obviously compact. Define, for each m, the "projection map" $\tau_m : K \to A^m$ by setting

$$\tau_m(z) = (\xi_1, \ldots, \xi_m) \quad \text{for all} \quad z = (\xi_1, \ldots, \xi_m, \ldots) \quad \text{of} \quad K.$$

If (x_n) is a sequence in K, then $(\tau_m(x_n))$ is a sequence in A^m, and since A^m is compact, we can find a convergent subsequence in A^m, so that (for any

---

†The reader who wishes to make fast progress may, at the present moment, omit this lengthy proof, without endangering his understanding of subsequent developments.

m) there is a subsequence of (x_n) whose projection onto A^m is convergent in A^m.

Let now $\epsilon_1 = 1$, $\epsilon_2 = 1/2^2, \ldots, \epsilon_j = 1/j^2, \ldots$. Since $\Sigma_{k=1}^{\infty} 1/k^2$ is convergent, one can certainly choose a sequence of associated integers $N_1, N_2, \ldots, N_j, \ldots$ such that

$$\sum_{k=N_j+1}^{\infty} \frac{1}{k^2} < \left(\frac{\epsilon_j}{3}\right)^2. \tag{α}$$

Consider now the above discussed subsequence of (x_n) whose projection converges in A^{N_1}. Because of convergence, we can discard a finite number of terms from the beginning so that each remaining term in the projected subsequence is within a distance $\epsilon_1/3$ from any other term. Denote the so defined subsequence of (x_n) by

$$(z_{1n}) \equiv z_{11}, z_{12}, \ldots, z_{1n}, \ldots .$$

Now we may start all over: we can take this (z_{1n}) as an "original" sequence of K and extract a subsequence such that its projection onto A^{N_2} converges, then we throw away a finite number of terms from the beginning so that the remaining projected terms are all within the distance $\epsilon_2/3$ from each other. The subsequence of (z_{1n}) (hence of (x_n)) which is so defined we denote by

$$(z_{2n}) \equiv z_{21}, z_{22}, \ldots, z_{2n}, \ldots .$$

Repeating these steps, we are led to a sequence of subsequences

$$z_{11}, z_{12}, \ldots, z_{1n}, \ldots$$
$$z_{21}, z_{22}, \ldots, z_{2n}, \ldots$$
$$\cdot$$
$$\cdot$$
$$\cdot$$
$$z_{l1}, z_{l2}, \ldots, z_{ln}, \ldots \tag{β}$$
$$\cdot$$
$$\cdot$$
$$\cdot$$

where each $z_{lm} = x_n$ for some n and

$$d_{N_j}\left(\tau_{N_j}(z_{lr}), \tau_{N_j}(z_{ls})\right) < \frac{\epsilon_j}{3} \tag{γ}$$

for any r, s and any $l \geq j$.

Let now

$$\lim_{n\to\infty} \tau_{N_1}(z_{1n}) = a_1 = (\xi_1, \ldots, \xi_{N_1}),$$

.
.
.

$$\lim_{n\to\infty} \tau_{N_j}(z_{jn}) = a_j = (\xi_1, \ldots, \xi_{N_1}, \ldots, \xi_{N_j}),$$

.
.
.

and set

$$z = (\xi_1, \xi_2, \ldots, \xi_k, \ldots),$$

where ξ_k is the kth component of every a_j that has at least k components. Obviously, $z \in K$. We now claim that *the subsequence of* (x_n), *defined by*

$$(u_n) \equiv z_{11}, z_{22}, z_{33}, \ldots, z_{nn}, \ldots$$

which is diagonally chosen from the array (β), *converges to* $z \in K$. To see this, choose an arbitrary $\epsilon > 0$. Find an integer j such that $\epsilon_j < \epsilon$. Then, taking $n \geq j$ and using the notation

$$u_n \equiv z_{nn} = (\xi_1^{\,n}, \xi_2^{\,n}, \ldots, \xi_k^{\,n}, \ldots),$$

we have

$$d(u_n, z)^2 = \sum_{k=1}^{\infty} |\xi_k^{\,n} - \xi_k|^2 = \sum_{k=1}^{N_j} |\xi_k^{\,n} - \xi_k|^2 + \sum_{k=N_j+1}^{\infty} |\xi_k^{\,n} - \xi_k|^2. \qquad (\delta)$$

Setting, for short, $\tau_{N_j}(u_n) \equiv \tau_{N_j}(z_{nn}) = \bar{z}_{nn}$ and $\tau_{N_j}(z) = \bar{z}$, the first term of Eq. (δ) can be written as

$$\sum_{k=1}^{N_j} |\xi_k^{\,n} - \xi_k|^2 = d_{N_j}(\bar{z}_{nn}, \bar{z})^2. \qquad (\epsilon)$$

Noting that $|\xi_k^{\,n} - \xi_k| \leq |\xi_k^{\,n}| + |\xi_k| \leq 2/k$, we then have, from Eqs. (δ) and (ϵ),

$$d(u_n, z)^2 \leq d_{N_j}(\bar{z}_{nn}, \bar{z})^2 + 4 \sum_{k=N_j+1}^{\infty} \frac{1}{k^2}.$$

Using then Eqs. (γ) and (α), we see that

$$d(u_n, z)^2 < \frac{5\epsilon_j^2}{9} < \epsilon_j^2 < \epsilon^2.$$

Thus, the laboriously constructed subsequence (u_n) of (x_n) is convergent (with limit $z \in K$) as claimed. Hence, Theorem 6.3(6) assures us that the

Hilbert cube is compact, q.e.d. This result is quite interesting because, as we saw above, simple bounded and closed subspaces of E^∞ (like $S_1[\text{o}]$) need not be compact. The Hilbert cube is one of the simplest compact subspaces of E^∞.

Before proceeding, it will be useful to summarize what we learned so far about metric spaces. In view of Theorems 6.3(3), 6.3(5), and 6.3(6), we can say that for *metric spaces* the following statements are equivalent:

(a) (X, d) is compact,
(b) (X, d) has the Bolzano–Weierstrass property,
(c) (X, d) is sequentially compact.

The next theorem relates compactness and boundedness.†

THEOREM 6.3(7). *In an arbitrary metric space, every compact subset must be bounded.*‡

The *proof* of this theorem is based on the following interesting

LEMMA. *If A is a compact set of a metric space, then, for any given $r > 0$, there exists a finite set of points $x_1, x_2, \ldots, x_N$ such that the open balls $S_r(x_1), S_r(x_2), \ldots, S_r(x_N)$ cover A.*

Proof of Lemma. Suppose the statement was not true. Take a point p_1. Since $S_r(p_1)$ does not cover A, there exists a $p_2 \in A$ such that $p_2 \notin S_r(p_1)$, and hence $d(p_1, p_2) \geq r$. Consider the two balls $S_r(p_1), S_r(p_2)$. Since we do not believe that they cover A, we can find a p_3 such that $p_3 \notin S_r(p_1)$, $p_3 \notin S_r(p_2)$, and therefore $d(p_3, p_1) \geq r$, $d(p_3, p_2) \geq r$. By induction, for any n we could have a set of points $p_1, \ldots, p_n$ such that $d(p_i, p_k) \geq r$ ($i \neq k$). Since A is compact, the (infinite) sequence (p_n) must have a convergent subsequence with some limit p. This implies that for some i and some k, $d(p, p_i) < r/2$ and $d(p, p_k) < r/2$. But then $d(p_i, p_k) \leq d(p_i, p) + d(p, p_k) < r$, which contradicts our "result" that $d(p_i, p_k) \geq r$. Hence, the denial of the lemma's statement leads to a contradiction, which proves the lemma.

Proof of Theorem 6.3(7). To prove that the compact set A is bounded, we must show that $d(x, y)$ has a supremum when x, y range over all of A.

†We remind the reader that the exact definition of boundedness of an arbitrary metric space was given in Problem 5.1-6.

‡Naturally, this implies that a compact metric space must be bounded. This may be considered to be the reason why, for example, E^n cannot be compact.

Taking some $r > 0$, the lemma assures us that a certain finite collection $\{S_r(x_1), \ldots, S_r(x_N)\}$ of balls covers A. Thus, for any $x, y \in A$, there exists at least one x_i and one x_j such that $x \in S_r(x_i)$ and $y \in S_r(x_j)$. Now, $d(x, y) \leq d(x, x_i) + d(x_i, y) \leq d(x, x_i) + d(x_i, x_j) + d(x_j, y) < 2r + d(x_i, x_j)$, because x and y lie in the open balls $S_r(x_i)$ and $S_r(x_j)$, respectively. Since $\sup\{d(x_i, x_j)\}$ exists (because there are only finitely many points $x_1, \ldots, x_N$), the inequality assures us that $\sup\{d(x, y)\}$ exists, q.e.d.

Unfortunately, the converse of Theorem 6.3(7) does not hold in general: Boundedness does not necessarily imply compactness. For example, a discrete metric space which has an infinite number of points is surely bounded ($\sup\{d(x, y)\} = 1$) but cannot be compact because a sequence of distinct elements cannot have any convergent subsequence. For our purposes, the most important class of metric spaces where a partial converse of Theorem 6.3(7) is valid, is the class of Euclidean spaces E^n: a bounded and *closed* subset of E^n is compact. This follows from the generalized Heine–Borel theorem stated after the Tychonoff Theorem 6.3(4).

We summarize our results concerning metric and in particular Euclidean spaces:

(A) *In a metric space, any compact set must be both bounded and closed.*†

(B) *In a Euclidean n-space, a set is compact iff it is both bounded and closed.*‡

We add one more theorem on metric spaces, which can be used as a criterion for separability:

THEOREM 6.3(8).   *Every compact metric space is separable.*

The *proof* may be based on the lemma used above and on Theorem 6.2(1).

This theorem illustrates the not very surprising fact that compactness is a more restrictive property than separability.

In the preceding discussions, we surveyed the major intrinsic properties of compact spaces and sets. We now turn our attention to the

---

†This follows from Theorems 6.3(7) and 6.3(2).

‡This follows from (A) and from the generalized Heine–Borel theorem.

behavior of functions defined on compact spaces. The central fact is that *compactness is a topological property,* i.e. the homeomorphic image of a compact space is compact. Actually, a more general statement holds:

THEOREM 6.3(9). *Let $f:(X, \tau) \to (X', \tau')$ be a continuous map whose domain (X, τ) is compact. Then the image $f(X)$ is a compact subspace of (X', τ').*

Proof. Let $\mathcal{O} = \{\ldots V_\alpha \ldots\}$ be an open cover of $f(X)$. Then, because of continuity, $f^{-1}(V_\alpha)$ is an open set of X. Furthermore, $\cup_\alpha f^{-1}(V_\alpha) = f^{-1}(\cup_\alpha V_\alpha) = X$, so that $\{\ldots f^{-1}(V_\alpha) \ldots\}$ is an open cover of X. Since X is compact, there exists a finite open subcover $\{f^{-1}(V_1), \ldots, f^{-1}(V_n)\}$. Since $x \in f^{-1}(V_k)$ implies $f(x) \in V_k$, and since each $x \in X$ belongs to at least one $f^{-1}(V_k)$, the surjectiveness of $f: X \to f(X)$ leads to the statement that $\{V_1, \ldots, V_n\}$ covers $f(X)$. Thus, we found a finite subcover of $\mathcal{O}$, hence $f(X)$ is compact.

This theorem has many important consequences.† For example, from the compactness properties of subsets of E^n follow analogous properties for C^n, because of the homeomorphism $C^n \underset{h}{\approx} E^{2n}$. In particular, in the unitary space C^n, a set is compact iff it is both bounded and closed, just as we found for E^n.

More important for the use in analysis is the following result:

THEOREM 6.3(10). *If f is a continuous function from a compact space (X, τ) into a metric space (Y, d), then $f(X)$ is a bounded subset of (Y, d).*

Proof. Since $f(X)$ is compact and (Y, d) is a metric space, Theorem 6.3(7) assures us that $f(X)$ is bounded.

COROLLARY 1. *If f is a continuous real or complex valued function with an arbitrary compact space (X, τ) as its domain, then f is a bounded function,* i.e. *there exists a nonnegative real number K such that $|f(x)| \leq K$ for all $x \in X$.*

This is obvious because bounded sets in E or C are characterized precisely by the criterion that the magnitude of numbers in such a set is bounded by a constant. A familiar special case of Corollary 1 is

---

†Actually, one popular proof of Tychonoff's theorem (cf. McCarty [28], p. 155) is based on the preservation of compactness under a homeomorphism. But there are also different proofs (cf. Simmons [36], p. 119).

Weierstrass' theorem of elementary calculus: *Any continuous real function on a closed and bounded interval is automatically bounded.* We see here the role of compactness: the function $f(x) = 1/x$ defined on $0 < x \leq 1$ is continuous but not bounded. Similarly, the function $f(x) = e^x$ defined on the whole real line is continuous but not bounded.

A sharpening of these elementary observations is expressed by

COROLLARY 2. *If f is a continuous* REAL *valued function with an arbitrary compact space (X, τ) as its domain, then f attains both a minimum and a maximum in $f(X)$, i.e. there exists an x_m and an x_M in X such that $f(x_m) = \inf_{x \in X} \{\ldots f(x) \ldots\}$ and $f(x_M) = \sup_{x \in X} \{\ldots f(x) \ldots\}$.*

The *proof* follows from the observation that a compact subset of the reals is also a closed set and therefore the inf and sup of such a set belongs to the set.

Corollary 2 is important, because if X is not compact, one cannot conclude the existence of a maximum or minimum, *not even for continuous and bounded functions.* For example, let f be the map defined for all *nonzero* x that obey the relation $-1 \leq x \leq 1$ and where $f(x) = x^2$. This is both a continuous and a bounded function but has no minimum since $\inf\{\ldots f(x) \ldots\} = 0$ and 0 does not belong to the domain.

So far we considered compactness from the global point of view. There is also a local equivalent, similarly as for connectedness:

DEFINITION 6.3(2). *A topological space is said to be locally compact iff every point x has some neighborhood $N(x)$ such that its closure $\overline{N(x)}$ is compact.*†

It is clear that local compactness does not imply compactness. For example, E^n is not compact, but it *is* locally compact, since for any point x, the closure $\overline{S_r(x)}$ of an open ball is both closed and bounded, and hence compact. The same holds for C^n. On the other hand, *every compact space is locally compact.*‡ Indeed, every point x has X as a neighborhood and $\overline{X} = X$ is, by assumption, compact. Needless to say, there are many

†Some authors use the broader definition: A space X is locally compact if every neighborhood of each point is contained in some compact set of X.

‡Contrast this with the situation concerning the concepts of connectedness and local connectedness.

important spaces which are neither compact nor locally compact. Example: the space of all rational numbers.

PROBLEMS

6.3-1. Is an indiscrete space compact? Is a discrete space that has an infinite number of points, compact? Using the indiscrete space as an example, illustrate that a compact subspace need not be closed.

6.3-2. Construct an open cover of the real line which has no finite subcover (i.e. illustrate that E is not compact).

6.3-3. Let (X, τ) be an arbitrary topological space. Let $\{A_1, \ldots, A_n\}$ be a finite class of compact subspaces, and show that $\cup_{k=1}^{n} A_k$ is also compact.

6.3-4. Let (X, τ) be the subset $0 < x < 1$ of the real line, with the relative topology. Show that X does not have the Bolzano–Weierstrass property (hence, as already known to you, X cannot be compact).

6.3-5. Let V_n be the set of real numbers $\{x \,|\, n - 1 < x < n\}$ (where n is an integer). Let X be the union of all such sets V_n. Let τ be the topology generated by the family V_n. Prove that (X, τ) has the Bolzano–Weierstrass property but (X, τ) is not compact.

6.3-6. Using the generalized Heine–Borel theorem, prove the following: If A is a bounded infinite subset of E^n, then it is possible to extract from A a convergent sequence (x_n) whose terms are all distinct, but the limit x need not belong to A.

6.3-7. Using some results of Problems 6.3-1 and 6.2-3, give an example of a metric space which is separable but not compact. (*Remark*: This shows that the converse of our Theorem 6.3(8) is not true.)

6.3-8. Let $f : (X, \tau) \rightarrow (H, \rho)$ be a continuous map from an arbitrary compact space to a Hausdorff space. Show that f is a closed map. (*Hint*: Recall Theorems 6.3(1), 6.3(2), and 6.3(9).) Use this result to show that Im(f) (i.e. the subspace $f(X)$ of (H, ρ)) "has the quotient topology." (*Hint*: See the remarks in Subsection 5.6a, following Theorem 5.6a(2).)

6.3-9. Use the first part of the preceding problem to prove the following theorem: If $f : (X, \tau) \rightarrow (H, \rho)$ is a bijective continuous function from an arbitrary compact space onto a Hausdorff space, then f is a homeomorphism.

6.3-10. Let $f : (X, d) \rightarrow (X', d')$ be a map from a metric space to another with the property that for every given $\epsilon > 0$ there exists a $\delta > 0$ (independent of x) such that $d(x, y) < \delta$ implies $d'(f(x), f(y)) < \epsilon$ for *all* $x \in X$. Then f is said to be *uniformly continuous*. (Note that this implies continuity, but not conversely.) Show that if f is continuous and (X, d) is compact, then f is guaranteed to be uniformly continuous. (*Remark*: A familiar application is the theorem of real analysis which tells us that every continuous real function defined on a closed and bounded interval is uniformly continuous.)

6.3-11. Show that any discrete space is locally compact. By recalling the result of Problem 5.5-1 conclude that the continuous image of locally compact space is not necessarily locally compact.

6.3-12. Show that E^∞ (or C^∞) is not locally compact. (*Hint*: Recall what we showed for $S_1[o]$ in this space.)

6.3-13. Show that a Hausdorff space is locally compact iff each point of the space is an interior point of some compact set.

6.3a. Compactification

Since compact spaces have favorable properties concerning convergence and the behavior of functions defined on them, it is tempting to "modify" a given noncompact space so as to make it compact. A familiar, but rather special example is the construction of the "*extended real line* $\hat{\mathbf{R}}$." This consists of all real numbers to which two "ideal numbers" $+\infty$ and $-\infty$ have been adjoined. Then one defines the following topology: The basic open sets of $\hat{\mathbf{R}}$ are all usual basic open sets of $\mathbf{R}$, as well as sets that consist of elements x such that $a < x \leqslant +\infty$ or $-\infty \leqslant x < a$ (where a is a real number). Such sets are usually denoted by the symbols $(a, +\infty]$ and $[-\infty, a)$, respectively.† It is not difficult to convince oneself that $\hat{\mathbf{R}}$ is compact: the trouble of not being able to extract a finite subcover from an arbitrary open cover disappears because the "endsections" can now be covered by intervals $(a, +\infty]$, $[-\infty, b)$, which, in the new topology, are open. The advantage of this compactification is that now, for example, every infinite sequence has a convergent subsequence because if we have a positive (negative) unbounded sequence, we simply say that it converges to $+\infty$ $(-\infty)$. Indeed, "almost all" elements of such a sequence lie in a neighborhood $(a, +\infty]$ (or $[-\infty, a)$, respectively) of $+\infty$ $(-\infty)$, whatever a we choose. But the present compactification is merely an *ad hoc* convenience. (See also Problem 6.3a-1.)

We wish to show in the following that there exists a systematic procedure which allows us to *convert any topological space into a compact space by adjoining only one new point*. In order to simplify the discussion, we assume that the given space (X, τ) is a Hausdorff space.‡ Let p be a symbol ("object") which is not contained in X. Form the new set $X^* = X \cup \{p\}$. Define the topology τ^* on X^* as follows: V^* is open in X^* iff

---

†The order relation on $\hat{\mathbf{R}}$ is specified for the new elements by stipulating that for every $x \in \mathbf{R}$, $x < +\infty$, $x > -\infty$, and $-\infty < +\infty$.

‡The general case is left for the reader as Problem 6.3a-3.

(a) $V^* = X^*$,
(b) $V^* = V$, where V is open in (X, τ),
(c) $V^* = X^* - A$, where A is compact in (X, τ).

The last condition means that V^* is the complement in X^* of a compact subset A of X; hence open sets of type (c) may be also characterized by writing

$$V^* = (X \cup \{p\}) - A = (X - A) \cup \{p\}.$$

To start with, we must show that τ^* is indeed a topology for X^*. $\emptyset$ and X^* are trivially open. To show that arbitrary unions and finite intersections of various V^* sets are open, we recall that a compact subspace A of a Hausdorff space is closed, so that $X - A$ is open. Checking out then various possibilities with open V^* sets of the different types, it is a straightforward though somewhat tedious matter to verify the composition properties.†

Next we demonstrate that (X^*, τ^*) is compact. Let $\mathcal{O}^* = \{\ldots V_\alpha^* \ldots\}$ be an open cover of X^*. At least one member of $\mathcal{O}^*$, say $V_{\alpha_0}^*$, must contain the point p. This $V_{\alpha_0}^*$ is of type (c) above. Hence $V_{\alpha_0}^{*c}$ is a compact subset of X. Since X is a subset of X^*, $\mathcal{O}^*$ is an open cover of X and *eo ipso* of $V_{\alpha_0}^{*c} \subset X$; and since $V_{\alpha_0}^{*c}$ is compact, it can be covered by a finite subcollection $\{V_1^*, \ldots, V_n^*\}$ of $\mathcal{O}^*$, i.e. $V_{\alpha_0}^{*c} \subset \bigcup_{k=1}^n V_k^*$. Hence all of X^*, with the exception of points in $V_{\alpha_0}^*$, is covered by the subcollection $\{V_1^*, \ldots, V_n^*\}$. Consequently, *all* of X^* is covered by $\{V_1^*, \ldots, V_n^*, V_{\alpha_0}^*\}$. Thus, we succeeded in extracting from $\mathcal{O}^*$ a finite subcover for X^*, so we see that X^* is compact.

We observe that (X^*, τ^*) contains the original (X, τ) as a sub*space*, i.e. the old τ topology for the set X is the relative topology induced by τ^* on the sub*set* X. Indeed, $V \subset X$ is open in the old topology iff $V = V^* \cap X$, where V^* is a member of τ^*. This is seen by a glance at the three types of V^* sets.

Actually, the subset $X \subset X^*$ is dense in X^*, because the only point of X^* which is not in X, is the new point p which, however, is an accumulation point of X. This is so because all neighborhoods of p are open sets of type (c), hence they all contain some points‡ of X.

One final question that one may ask is this: will the compactification be a Hausdorff space? The answer is that this is true provided (X, τ) was not only Hausdorff but also locally compact.

†You will have to recall also Theorem 6.3(1) and the fact that the union of two compact sets is always compact.
‡Observe that, by assumption, X is not compact so that $X - A$ cannot ever be empty.

We discussed above in detail the general method of *one-point compactification*. We now illustrate the procedure in a direct manner on the example of the real line E. We adjoin one ideal element p, which specifically we denote by the symbol ∞ and call "the point at infinity."† The set $\mathbf{R} \cup \{\infty\}$ we denote by $\mathbf{R}^*$ and, according to the general method, the open sets of $\mathbf{R}^*$ are as follows:

(a) $\mathbf{R}^*$ itself,

(b) the old open sets of E (i.e. unions of open intervals of $\mathbf{R}$),

(c) complements in $\mathbf{R}^*$ of all compact subsets of E (i.e. sets which consist of ∞ and of $\mathbf{R}$ less any closed and bounded subset).

It is clear that the new open sets (type (c)) are of the form $\{\infty\} \cup$ (an open and both-directionally unbounded set of E). To put this in more exact language, any connected neighborhood of ∞ will be a set of the form $N(\infty) = \{x \in \mathbf{R} \mid x > r_1 \text{ or } x < r_2\}$, where r_1 and r_2 are some arbitrary given real numbers (cf. Fig. 6.9). In particular, this shows that any unbounded set will have ∞ as an accumulation point.

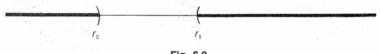

$$r_2 \qquad\qquad\qquad r_1$$

Fig. 6.9.

The one-point compactification $\mathbf{R}^*$ of the real line is called the *projective line*. This topological space may be very well visualized by noting that $\mathbf{R}^*$ is *homeomorphic to the circle* S^1 of the plane E^2. Instead of a formal proof, we refer to Fig. 6.10. The homeomorphism is "constructed" by projecting the points of the circle from the "north pole" n. The rule of this mapping $h: S^1 \to \mathbf{R}^*$ is: $h(s) = x$ if $s \neq n$ and $h(n) = \infty$.

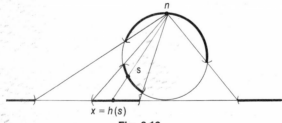

$$x = h(s)$$

Fig. 6.10.

†Note that ∞ "has no sign," there is only one ∞-point.

‡This must be well distinguished from the extended real line discussed earlier.

We also indicate in the figure two typical open sets of S^1 and their corresponding images. It should not be difficult now to see that h is indeed a homeomorphism. In passing we note that, since S^1 (being a closed and bounded subset of E^2) is compact, its homeomorphic image $\mathbf{R}^*$ is necessarily compact. This can be considered to be a direct proof of the compactness of $\mathbf{R}^*$. Similarly, it also follows that $\mathbf{R}^*$ is Hausdorff. Finally, we observe that the original line E is homeomorphic to the circle with the north pole n missing. "Plugging up" this hole, we achieve compactification.

A similar consideration applies to the one-point compactification of the complex space C. The "point at infinity" is added to the complex numbers, and a typical neighborhood of this is the set of complex numbers z such that $|z| > r$, where r is a nonnegative real number. This compactified space C^* is the one which is used in the theory of complex functions. It is usually called the complex plane of analysis. It is not difficult to see that C^* is homeomorphic to the sphere $S^2 \subset E^3$. The north pole corresponds to the point at infinity. The sphere S^2, considered as the homeomorphic image of C^*, is usually referred to as the Riemann sphere.

PROBLEMS

6.3a-1. Show that it is impossible to extend the metric of $\mathbf{R}$ to a metric on the extended real line $\hat{\mathbf{R}}$. (*Hint*: Recall the continuity of the distance function and consider $\lim_{n \to \infty} d(x, x_n)$, where x_n is an unbounded positive (or negative) sequence, i.e. $x_n \to +\infty$ (or $-\infty$) in the $\hat{\mathbf{R}}$ topology.)

6.3a-2. Show that the projective line $\mathbf{R}^*$ is metrizable. (*Hint*: Recall Theorem 5.6(1).)

6.3a-3. Discuss the one-point compactification of the space X, where X is that subspace of E^2 which consists of two parallel lines.

6.3a-4. The one-point compactification of Hausdorff spaces (which was discussed in the text) can be generalized to arbitrary topological spaces (X, τ) in the following manner. Let $X^* = X \cup \{p\}$ (where $p \notin X$) and define the topology τ^* as follows: (a) X^* is in τ^*, (b) all members of τ are in τ^*, (c) a set $V^* \subset X^*$ is in τ^* if $V^{*c} \equiv X^* - V^*$ is a *closed and compact* subset of X. Show that this is indeed a one-point compactification of (X, τ).

6.3a-5. If (X, τ) is a *compact* space, obviously it is still possible to adjoin one more element $p \notin X$ and introduce the topology as defined in the preceding problem. The new space (X^*, τ^*) is certainly compact. Show that if (X, τ) is compact, then (a) X is not dense in (X^*, τ^*); (b) the singleton set $\{p\}$ is a component of (X^*, τ^*) (cf. Problem 6.1-12), i.e. (X^*, τ^*) is not connected.

6.4   COMPLETE METRIC SPACES

The notion of completeness applies only to a class of *metric* spaces. Complete metric spaces have "many" convergent sequences and play a central role in functional analysis. We start the discussion with introducing an auxiliary concept:

DEFINITION 6.4(1).   *A sequence* (x_k) *in an arbitrary metric space* (X, d) *is called a Cauchy sequence† iff for any given* $\epsilon > 0$, $d(x_n, x_m) < \epsilon$ *whenever* $n, m \geq K$, *where* K *is some positive integer.*

For many applications it is helpful to realize that this definition is equivalent to the following. The sequence (x_k) is Cauchy if for *any fixed n* and for some fixed $m \geq M$, we have $d(x_{n+k}, x_m) < \epsilon$ whenever $k \geq N$.

Now we realize that in any metric space, *every convergent sequence is a Cauchy sequence.* Indeed, if $x_n \to x$, then $d(x_n, x_m) \leq d(x_n, x) + d(x, x_m) < \epsilon$ provided n and m are large enough. Loosely speaking, if a sequence is convergent, then its terms eventually come very close to one another.

However, in general the converse is not true: There are many spaces which have nonconvergent Cauchy sequences. For example, consider the subspace of E consisting of the real numbers $0 < x \leq 1$. The sequence $x_k = 1/k$ is Cauchy (since $|1/n - 1/m| \leq 1/n + 1/m < \epsilon$ if n, m is large enough) but clearly, (x_k) has no limit. (The point $x = 0$ is not in the space.) A more sophisticated familiar example is the space of all rationals (as a subspace of E). There are many Cauchy sequences that do not converge, for example, the sequence $0.1, 0.101, 0.101001, \ldots$. (The sequence "tends to" an unterminating decimal, which is actually an irrational number.)

We infer from these examples that a Cauchy sequence may fail to converge because, in some sense, the space is "incomplete." Adjoining a suitable set of new elements (namely the formal "limits" of all nonconvergent Cauchy sequences) we would have a "complete" space. We shall study this procedure in Subsection 6.4a, but right now we use these observations to motivate the following

DEFINITION 6.4(2).   *A metric space is called complete iff every Cauchy sequence is convergent.*

Example α.   The real line E is complete. Let (x_k) be Cauchy. Then $|x_n - x_m| < \delta$ provided $n, m \geq K$. Therefore, if m is fixed, then the set of

†The term "fundamental sequence" is also used.

all x_n with $n \geqslant K$ is in some open and bounded interval I which contains x_m. The closure $\bar{I}$ is closed and bounded, hence compact. Therefore, the infinite† set of points x_n will have an accumulation point x in $\bar{I} \subset E$. This x is then clearly the limit of the sequence (x_n). Hence, (x_k) is convergent. In a similar manner one shows that the complex space C is also complete.

Example β.   The space E^∞ or C^∞ is complete. Let

$$x_1 = (\xi_{11}, \xi_{12}, \ldots, \xi_{1k}, \ldots)$$
$$x_2 = (\xi_{21}, \xi_{22}, \ldots, \xi_{2k}, \ldots)$$
$$\vdots$$
$$x_n = (\xi_{n1}, \xi_{n2}, \ldots, \xi_{nk}, \ldots)$$
$$\vdots$$

be an arbitrary Cauchy sequence. Then, for $n, m \geqslant K$, we have

$$d^2(x_n, x_m) = \sum_{k=1}^{\infty} |\xi_{nk} - \xi_{mk}|^2 < \epsilon,$$

for any given $\epsilon > 0$. This tells us that the sequence (ξ_{nk}), $(n = 1, 2, \ldots)$ of numbers, for any fixed k, is a Cauchy sequence. (This is so because each term in the infinite series must be arbitrary small.) Since E (or C) is a complete space, it follows that (ξ_{nk}) is convergent so that the number $\xi_k \equiv \lim_{n \to \infty} \xi_{nk}$ exists, for any fixed k. Construct the element

$$x \equiv (\xi_1, \xi_2, \ldots, \xi_k, \ldots).$$

First, we wish to show that this x *is* an element of E^∞ (or C^∞). To this end, we must show that $\sum_{k=1}^{\infty} |\xi_k|^2$ is finite (i.e. the series converges). Now, writing $|\xi_k| \equiv |(\xi_{nk} - \xi_k) + \xi_{nk}|$, the Minkowski inequality (Appendix I) tells us that

$$\sum_{k=1}^{j} |\xi_k|^2 \leqslant \left\{ \left(\sum_{k=1}^{j} |\xi_{nk} - \xi_k|^2 \right)^{1/2} + \left(\sum_{k=1}^{j} |\xi_{nk}|^2 \right)^{1/2} \right\}^2. \qquad (\alpha)$$

Here

$$\sum_{k=1}^{j} |\xi_{nk} - \xi_k|^2 < \epsilon \qquad (\beta)$$

for any j whenever $n \geqslant K$, because, by definition, ξ_k is the limit of the sequence ξ_{nk}, so that each term in the sum can be made as small as

---

†We may assume that the Cauchy sequence (x_n) has infinitely many distinct terms, otherwise it would be trivially convergent, so that there would be nothing to prove.

desired. The second term on the r.h.s. of Eq. (α) is less than or equal to $(\Sigma_{k=1}^{\infty} |\xi_{nk}|^2)^{1/2}$, which is a finite number c because x_n is in the space E^{∞} (or C^{∞}). Hence, the inequality (α) leads to

$$\sum_{k=1}^{j} |\xi_k|^2 \leq \epsilon + c.$$

Since the r.h.s. is independent of j, the infinite series $\Sigma_{k=1}^{\infty} |\xi_k|^2$ must converge; so that x is in our space, as claimed. Finally, we show that the given Cauchy sequence (x_n) converges. Indeed, from Eq. (β) it is clear that

$$d^2(x_n, x) = \sum_{k=1}^{\infty} |\xi_{nk} - \xi_k|^2 < \epsilon$$

if $n \geq K$. This means that $x_n \to x$, so indeed (x_n) is convergent and actually, its limit is x as constructed above.

Example γ. In a similar manner one can show that E_p^{∞} (C_p^{∞}) is a complete space. With even less trouble one demonstrates that E^n, $E_p^{\ n}$ (or C^n, $C_p^{\ n}$) are complete spaces.

Example δ. The function space $C[a, b]$ of all real (or complex) valued continuous functions on the closed and bounded interval $[a, b]$ is complete with the uniform metric $\sup |f(x) - g(x)|$. Let (f_n) be a Cauchy sequence. Then $\sup_x |f_n(x) - f_m(x)| < \epsilon$ if $n, m \geq K$. This obviously implies that for *any fixed* x, the sequence $(f_n(x))$ of ordinary numbers is a Cauchy sequence in E(or C). Because of completeness of E (or C), the sequence is convergent, i.e. there exists a *number* $f(x)$ such that $f(x) = \lim_{n \to \infty} f_n(x)$. Since for any $x \in [a, b]$ we can construct in this way a value $f(x)$, we have here defined a map $f: [a, b] \to E$ (or C). We first show that this f is a continuous function, i.e. it belongs to our function space $C[a, b]$. Indeed, for any n,

$$|f(x) - f(x')| \leq |f(x) - f_n(x)| + |f_n(x) - f(x')|$$
$$\leq |f(x) - f_n(x)| + |f_n(x) - f_n(x')| + |f_n(x') - f(x')|.$$

If, in particular, we take n large enough, the first and third terms are as small as we wish, because by construction $f(x)$ is the limit of $f_n(x)$. If now $|x - x'|$ is taken sufficiently small, then the middle term is also arbitrarily small, because f_n is a continuous function on $[a, b]$. Thus,

$$|f(x) - f(x')| < \epsilon \quad \text{provided} \quad |x - x'| < \delta.$$

This shows that f is a continuous function on $[a, b]$ as claimed. Finally, we show that the given Cauchy sequence (f_n) converges. Indeed, as already stated, $|f_n(x) - f(x)| < \epsilon$ if $n \geq N$, hence†

$$d(f_n, f) = \sup_{x \in [a,b]} |f_n(x) - f(x)| < \epsilon,$$

which means that $f_n \to f$, so (f_n) is convergent and actually, its limit is the function f constructed above. In passing we note that the essence of our proof was to show that *in the $C[a, b]$ space*, pointwise convergence implies uniform convergence.

The reader is urged to discover the deep similarities between the proofs of completeness used in Examples β and δ. As a matter of fact, the method is rather typical and worthwhile remembering.

We now give a simple criterion for recognizing complete subspaces of a complete space:

THEOREM 6.4(1).   *A subspace of a complete metric space is complete iff it is closed.*

Proof. (a) Assume that (A, d_A) is a closed subspace (with d_A being the restriction of d) of the metric space (X, d). Since every Cauchy sequence in A is a Cauchy sequence in X, it certainly converges; and since A is closed, the limit must belong to A.

(b) Assume that (A, d_A) is complete. If $B \subset A$ and x is an accumulation point of B, then, given $\epsilon > 0$, $d(x, y_k) < \epsilon/2$ for an infinite subset $\{\ldots y_k \ldots\}$ of elements of A with x being the limit of (y_k). Hence, if $n \geq K$, $m \geq K$, we have

$$d(y_n, y_m) \leq d(y_n, x) + d(x, y_m) < \epsilon,$$

so that (y_k) is a Cauchy sequence. But since A is complete by assumption, $\lim_{k \to \infty} y_k = x$ belongs to A; thus A contains all its accumulation points, and is therefore closed, q.e.d.

A simple example of this theorem is the result that every n-sphere S^n is a complete subspace of E^{n+1}. To illustrate the converse part of the theorem, we observe that E^n must be a closed subspace of E^∞, because both are complete spaces.‡

Another, somewhat less useful criterion for completeness relates this property to compactness:

†Observe that, since f_n and f are continuous functions on the compact set $[a, b]$, they are bounded, and so is $f_n - f$.
‡Of course, we established the closedness of E^n previously by a direct calculation.

THEOREM 6.4(2). *Every compact metric space is complete.*

Proof. If (x_l) is an arbitrary Cauchy sequence, then, given $\epsilon > 0$, $d(x_k, x_m) < \epsilon$ for some fixed $m \geqslant M$ and sufficiently large k. Hence, all x_k (with $k \geqslant M$) will be contained in an open ball $S_\epsilon(x_m)$, and *eo ipso* in $\overline{S_\epsilon(x_m)}$. Since $\overline{S_\epsilon(x_m)}$ is closed and bounded (hence compact), the sequence (x_k) (with $k = M, M + 1, \ldots$) will contain a convergent subsequence whose limit x is in $\overline{S_\epsilon(x_k)}$. But if a subsequence (x_{k_j}) of the given Cauchy sequence (x_l) is convergent, then the sequence (x_l) itself converges, as a little consideration shows. Hence the space is complete.

A simple example is provided by a metric space which has a finite number of points.

The converse of Theorem 6.4(2) surely does not hold: E^n, for example, is complete but not compact.†

PROBLEMS

6.4-1. Show that a discrete metric space is complete.

6.4-2. Let (X, d) be the set of all real valued continuous functions on the closed unit interval $[0, 1]$, with the metric

$$d(f, g) = \int_0^1 |f(x) - g(x)| dx,$$

where the integral is the usual Riemann integral. Show that (X, d) is really a metric space. Show that this space is not complete, by considering the sequence (f_n) defined as follows (for $n \geqslant 2$):

$$f_n(x) = \begin{cases} 0 & \text{if } 0 \leqslant x \leqslant \dfrac{1}{2} - \dfrac{1}{n} \\[2mm] nx - \dfrac{n}{2} + 1 & \text{if } \dfrac{1}{2} - \dfrac{1}{n} \leqslant x \leqslant \dfrac{1}{2} \\[2mm] 1 & \text{if } \dfrac{1}{2} \leqslant x \leqslant 1 \end{cases}$$

(To facilitate calculations, draw some figures.) (*Remark*: Observe that the same set of functions, if equipped with the usual uniform metric $\sup |f(x) - g(x)|$, *is* a complete space, as we showed in the text.)

6.4-3. Show that if $C[a, b]$ is the complete metric space of all continuous real functions on $[a, b]$ with the usual uniform metric in this space, then the subset of functions f with the property that $c \leqslant f(x) \leqslant d$ for every $x \in [a, b]$ (and c, d any

†See, however, Problem 6.4-6.

fixed real number) is also a complete space. Show that this is no longer true if the above subset condition is replaced by the less restrictive one, $c < f(x) < d$ for all $x \in [a, b]$.

6.4-4. Let (X, d) be an arbitrary metric space and let $C(X, \mathbf{R})$ be the set of all bounded continuous real valued functions on X. Show that, with the usual metric $D(f, g) = \sup |f(x) - g(x)|$, the space (C, D) is complete. (*Hint*: Use the same pattern of argument as was used in the text to show that $C[a, b]$ is complete.)

6.4-5. Justify, *without any calculation*, the statement that an open ball in E^n is not complete but a closed ball is complete. By the same argument justify the statement that a subspace of a Euclidean space is complete if *and only if* it is compact. Can you justify the "if" part of the statement by some other theorem?

6.4-6. A metric space is said to be *totally bounded* if, given any $r > 0$, there exists a finite set $\{x_1, x_2, \ldots, x_N\}$ of points such that the open balls $S_r(x_1), S_r(x_2), \ldots, S_r(x_N)$ form an open cover of the space. Prove that a metric space which is both complete and totally bounded, is necessarily compact. (*Hint*: Recall that sequential compactness implies compactness and observe that, since the space is complete, the proof of sequential compactness requires only the demonstration that every sequence has a Cauchy subsequence.) (*Remark*: The theorem expressed in this problem is the partial converse of Theorem 6.4(2).)

6.4-7. Prove the following striking theorem: A closed subspace of a complete metric space is compact iff it is totally bounded. (*Hint*: Use the result of the preceding problem and recall also the lemma from Section 6.3 on p. 261.

6.4a. Completion

Since complete spaces have pleasant properties regarding convergence of sequences, one may desire to "modify" a given incomplete space so as to make it complete. If the given incomplete space is known to be a subspace of some complete space, then the task is easy: we simply have to take its closure. (That is, adjoin the limits of all nonconvergent Cauchy sequences.) This is how, for example, the space of rationals is completed: the limits of nonconvergent Cauchy sequences give all irrational numbers, so that the completion of $\mathbf{Q}$ is E. However, there exists a general method of completion which we shall study in this subsection. The considerations leading up to the final result are rather lengthy, and the reader who (maybe for lack of time) is not desirous to follow a beautiful exercise in structural analysis, may skip the details and be content with the statement of the final Theorem 6.4a(1).

To start with, let us recall (cf. Problem 5.4-4) that in an arbitrary metric space we may define an equivalence relation E for sequences:

$$(x_n) \sim (y_n) \quad \text{iff} \quad \lim_{n \to \infty} d(x_n, y_n) = 0. \tag{6.1}$$

Let now (X, d) be a given incomplete space and let *the collection of all Cauchy sequences in X be denoted by C*. We then introduce in C the equivalence relation E and are thus led to the quotient space C/E. The elements of C/E are the (distinct) equivalence classes of Cauchy sequences of (X, d). We shall denote the equivalence class with the representative element (x_n) by the symbol† $[x_n]$. Thus, $C = \cup [x_n]$.

We now wish to *introduce on C/E a metric*. If $[x_n] \in C/E$, $[y_n] \in C/E$, then we set

$$e([x_n], [y_n]) = \lim_{n \to \infty} d(x_n, y_n). \tag{6.2}$$

Since e is defined in terms of elements of the representative sequences of the classes, we first must show that the definition is independent of which representatives we take. This can be shown by the following

LEMMA. *If (x_n) and (y_n) are Cauchy sequences (i.e. elements of C), then $(a_n) \equiv (d(x_n, y_n))$ is a Cauchy sequence of real numbers. Furthermore, the value of $\lim_{n \to \infty} a_n$ is unchanged if we replace (x_n) and (y_n) by equivalent Cauchy sequences (x'_n) and (y'_n), respectively.*

Proof. It is not difficult to show that

$$|a_n - a_m| \equiv |d(x_n, y_n) - d(x_m, y_m)| \leq d(x_n, x_m) + d(y_n, y_m). \tag{6.3}$$

Therefore, given $\epsilon/2$, the r.h.s. is less then ϵ whenever $n, m \geq K$. Thus, (a_n) is Cauchy, hence it converges. Let $a_n \to a$. Now, if $(x_n) \sim (x'_n)$ and $(y_n) \sim (y'_n)$, then

$$|a_n - a'_n| \equiv |d(x_n, y_n) - d(x'_n, y'_n)| \leq d(x_n, x'_n) + d(y_n, y'_n),$$

where we used again Eq. (6.3); and, in view of Eq. (6.1), we see that $|a_n - a'_n|$ is as small as desired provided n is large enough. Hence $(a_n - a'_n) \to 0$, or $\lim a_n = \lim a'_n$, q.e.d.

This lemma then immediately assures us that, as anticipated, the r.h.s. of Eq. (6.2) does not depend on the representatives. Next we will show that e has all requisite properties of a metric.

(a) The nonnegativeness of e follows from that of d.

(b) If $[x_n] = [y_n]$, then $(x_n) \sim (y_n)$, therefore by Eqs. (6.1) and (6.2), $e([x_n], [y_n]) = 0$. The converse is seen in a similar manner.

---

†This is a somewhat misleading notation, and it would be more appropriate to write $[(x_n)]$. But we wish to avoid a proliferation of brackets.

(c) The symmetry of e follows from that of d.

(d) $e([x_n], [y_n])$ $=$ $\lim d(x_n, y_n)$ $\leqslant$ $\lim \{d(x_n, z_n) + d(z_n, y_n)\}$ $=$ $e([x_n], [z_n]) + e([z_n], [y_n])$, where we took (z_n) to be a Cauchy sequence. This proves the triangular inequality.

In summary, we now have a metric space $(C/E, e)$. Next we show that *the original space (X, d) is isometric to a certain subspace M of $(C/E, e)$.*

Let M be the collection of all equivalence classes $[x_n]$ which are generated by a Cauchy sequence (x_n) whose terms are all equal. To be clearer in our notation, denote the Cauchy sequence $(x) \equiv (x, x, \ldots, x, \ldots)$ by $(\bar{x})$ and the corresponding equivalence class by $[\bar{x}]$. Thus, M consists of all classes $[\bar{x}]$, where x is an arbitrary element of X. Observe that, by the definition (6.1), each class $[\bar{x}]$ contains all Cauchy sequences (y_n) which are such that $y_n \to x$. Define now the map

$$f: X \to M \quad \text{given by} \quad f(x) = [\bar{x}] \quad \text{for all} \quad x \in X. \tag{6.4}$$

This map is clearly surjective. It preserves distances, because

$$e(f(x), f(y)) = e([\bar{x}], [\bar{y}]) = \lim_{n \to \infty} d(x_n, y_n) = \lim_{n \to \infty} d(x, y) = d(x, y).$$

Thus, f is an isometry, and (X, d) is isometric to M, as proclaimed.

The next step in the program is to show that *M is a dense subset of C/E.*

Let $[x_n]$ be an arbitrary element of C/E. Then, given $\epsilon > 0$, $d(x_n, x_\nu) < \epsilon$ provided $n, \nu \geqslant K$, because (x_n) is Cauchy. Let $[\bar{x}_\nu]$ be the equivalence class which contains the sequence $(\bar{x}_\nu) \equiv (x_\nu, x_\nu, \ldots, x_\nu, \ldots)$. Clearly, $[\bar{x}_\nu] \in M$. Then $e([x_n], [\bar{x}_\nu]) = \lim_{n \to \infty} d(x_n, x_\nu) < \epsilon$ for any given ϵ provided ν is large enough, as follows from the above observation on $d(x_n, x_\nu)$. Thus, for any $[x_n] \in C/E$ there exists some $[\bar{x}_\nu] \in M$ in each neighborhood of $[x_n]$, so M is dense, q.e.d.

We are now prepared to show that $(C/E, e)$ *is a complete metric space.*
Let

$$([x_n^1], [x_n^2], \ldots, [x_n^k], \ldots) \equiv ([x_n^k]), \qquad k = 1, 2, \ldots$$

be some arbitrary Cauchy sequence in C/E; we shall show that it is convergent. To start with, we construct from $([x_n^k])$ a Cauchy sequence in the subspace M. To this end, take a fixed k and select an element† $[\bar{x}(k)]$ of M such that

$$e([x_n^k], [\bar{x}(k)]) < 1/k. \tag{6.5}$$

---

†The element $[\bar{x}(k)]$ of M consists of all Cauchy sequences of X that are equivalent to the Cauchy sequence $(\bar{x}(k)) = (x(k), x(k), \ldots, x(k), \ldots)$, where $x(k)$ is some element of X.

(Such an $[\bar{x}(k)] \in M$ surely exists because M is dense.) Do this for all $k = 1, 2, \ldots$. In this manner you get a sequence

$$([\bar{x}(1)], [\bar{x}(2)], \ldots, [\bar{x}(k)], \ldots) \equiv ([\bar{x}(k)]), \qquad k = 1, 2, \ldots$$

in the subspace M. Observe that

$$e([\bar{x}(p)], [\bar{x}(q)]) \leq e([\bar{x}(p)], [x_n^p]) + e([x_n^p], [\bar{x}(q)])$$
$$\leq e([\bar{x}(p)], [x_n^p]) + e([x_n^p], [x_n^q]) + e([x_n^q], [\bar{x}(q)]).$$

Because of Eq. (6.5), the first and last terms are less than $1/p$ and $1/q$, respectively. Since, by assumption, $([x_n^k])$ is a Cauchy sequence in C/E, the middle term becomes as small as desired provided both p and q are sufficiently large. Hence, given $\epsilon > 0$,

$$e([\bar{x}(p)], [\bar{x}(q)]) < \epsilon \quad \text{if } p, q \geq R$$

i.e. $([\bar{x}(k)])$ is a Cauchy sequence in M, as proclaimed.

Now we pass back for a moment from M to X and construct a Cauchy sequence in X. This is easy: since M and X are isometric, the sequence

$$(x(1), x(2), \ldots, x(k), \ldots) \equiv (x(k)), \qquad k = 1, 2, \ldots$$

is Cauchy. (Here $x(k) \in X$ is determined by $[\bar{x}(k)] = f(x(k))$, where f is the isometry described by Eq. (6.4).) Now, since $(x(k))$ is a Cauchy sequence of X, it belongs to C, and hence, there exists a unique equivalence class $[y_n]$ of C/E such that $(x(k))$ belongs to this class. By the definition (6.1) of equivalence, this means that

$$\lim_{k \to \infty} d(x(k), y_k) = 0. \tag{6.6}$$

We now claim that the given Cauchy sequence $([x_n^k])$ of C/E converges precisely to the element $[y_n]$ of C/E. Indeed,

$$e([x_n^k], [y_n]) \leq e([x_n^k], [\bar{x}(k)]) + e([\bar{x}(k)], [y_k]),$$

where, because of Eq. (6.5), the first term is less than $1/k$, and, because of Eq. (6.6), the second term can be made as small as desired provided we take k large enough. Hence, given any $\epsilon > 0$,

$$e([x_n^k], [y_n]) < \epsilon \quad \text{if } k \geq K.$$

Thus, the arbitrary Cauchy sequence $([x_n^k])$ is indeed *convergent* (and actually converges to $[y_n]$ as specified above). This completes the proof of the statement that C/E is a complete space.

We have now all information at our disposal and summarize our findings in the following

THEOREM 6.4a(1).   *Let (X, d) be an arbitrary incomplete metric space. Then there exists a complete metric space (Z, e) such that (X, d) is isometric to a dense subspace of Z.*

We actually know a little more. The complete space (Z, e) has as its underlying set $Z = C/E$, its elements are $z = [x_n]$, where (x_n) is a Cauchy sequence in X and $[x_n]$ is the equivalence class generated by (x_n). The distance in Z is given by $e(z, z') = \lim_{n \to \infty} d(x_n, x'_n)$.

The space (Z, e) so constructed is usually called the *sequential completion* of (X, d). It contains a certain subspace† (M, e_M), to which (X, d) is isometric. Because of this isometry, we may "identify" (X, d) and (M, e_M). With this in mind, we can describe the process of completion in rather simple terms: We first "replace" (X, d) by an isometric equivalent space (M, e_M) and then adjoin certain "ideal elements" so as to get a complete space. It should be obvious that the set of "ideal elements" is precisely $M^c \equiv C/E - M$, i.e. the ideal elements are equivalence classes of sequences which do *not* have the form $(x, x, \ldots, x, \ldots)$. (In contrast, M consists precisely of all equivalence classes which are generated by sequences which have one infinitely many times repeated term.) To visualize the whole construction, the reader should envisage the completion of the rationals to the space of reals, bearing in mind that every irrational number may be regarded as a sequence of rationals which has an infinite number of distinct terms.

The detailed procedure of sequential completion is not very practical. The above theorem serves mainly as an existence proof. In many cases, some *ad hoc* method is more feasible for practical use. Actually, there is a generalized concept of completion. Any complete space (Z, e) which contains (X, d) (or rather, an isometric image of (X, d)) as a dense set, is regarded as a completion of (X, d). As Problem 6.4a-1 shows, all such completions are essentially indistinguishable.

In conclusion we remark that the process of completion bears some resemblance to that of compactification.‡

---

†As always, e_M denotes the restriction of the metric e to the subset M.

‡The one-point compactification that we studied in detail may also be generalized. Any compact space that contains the original space (or rather a homeomorphic image of it) as a dense subspace is called a compactification.

PROBLEMS

6.4a-1. Let (X, d) be an incomplete metric space. Let (Z, ρ) and (W, σ) be two completions of (X, d). Show that they are isometric. (*Hint*: Recall that there exist the isometries $f: (X, d) \to (f(X), \rho)$ and $g: (X, d) \to (g(X), \sigma)$ such that $f(X)$ is dense in Z and $g(X)$ is dense in W. Observe that $gf^{-1}: (f(X), \rho) \to (g(X), \sigma)$ is an isometry. Extend now gf^{-1} from the dense subset $f(X)$ of Z to all of Z as follows: the mapping $h: (Z, \rho) \to (W, \sigma)$ is given by $h(z) = gf^{-1}(z)$ if $z \in f(X)$ and for any other $z \in Z$, $h(z) = \lim_{n \to \infty} h(z_n)$, where $z_n \in f(X)$ for all $n = 1, 2, \ldots$. Justify that this is really an extension to any z of Z, then show that h is an isometry.) (*Remark*: This theorem assures you that completion is essentially unique; in particular, if you find some *ad hoc* completion of (X, d), then this is isometric to the one which is given by the laborious method of sequential completion.)

6.4a-2. Consider the set of all real polynomials defined on a closed and bounded interval $[a, b]$ of the real axis, equipped with the metric $d(p^{(i)}, p^{(k)}) = \sup_{a \le x \le b} |p^{(i)}(x) - p^{(k)}(x)|$. Show that this metric space is not complete. Determine, without going through any elaborate construction, the completion of this space. (*Hint*: Recall the Weierstrass approximation theorem of classical analysis: any continuous (real) function f defined on $[a, b]$ can be approximated pointwise, with arbitrary accuracy, by some real polynomial p.)

6.4b. Contraction Mappings

The major purpose of this subsection is to illustrate that completeness is such a "strong" property of metric spaces that it enables one to give a complete solution to a large variety of difficult problems.

A frequent type of problem of analysis is the following. Let X be some set and let $T: X \to X$ be a given map from X into X. Consider the equation $T(x) = x$. Has this equation a solution? If the answer is "yes," is the solution unique? Can one give a method of constructing the solution? If the method of solution is based on a successive iteration process, can one estimate the accuracy of each step?

There are many cases where the "fixed point problem" described in the preceding paragraph appears in a veiled form. For instance, consider the algebraic equation

$$a_n x^n + \cdots + a_1 x + a_0 = 0, \qquad \text{where} \quad a_1 \ne 0.$$

If we define the map T by

$$T(x) = -\frac{a_n}{a_1} x^n - \cdots - \frac{a_2}{a_1} x^2 - \frac{a_0}{a_1},$$

then the above equation assumes the form $T(x) = x$, and solving the equation amounts to finding the fixed point of T. Another illustration arises in the realm of differential equations. For example, consider the differential equation

$$\frac{df}{dx} + \alpha(x)f = 0, \qquad \alpha(x) \quad \text{given function,}$$

and ask for a solution f which is an infinitely differentiable function. Defining T by

$$T(f) = \left(\frac{d}{dx} + \alpha(x) + 1\right)f,$$

the solution of the problem is reduced to finding the fixed point of T. Even more exciting examples arise in connection with integral equations (cf. toward the end of this subsection).

We wish to show that if X is a complete metric space and if T possesses a certain not too restrictive property, then the fixed point problem of T has a complete and very simple solution. The particular property that T must satisfy is that it be a contraction mapping. This concept is formalized by

DEFINITION 6.4b(1). *A map* $T: (X, d) \to (X, d)$ *is called a contraction mapping if, for every pair* $x, y \in X$, *we have*

$$d(T(x), T(y)) \leq \alpha d(x, y),$$

where $\alpha < 1$.

Informally speaking, a contraction map reduces distances.

We now formulate the very general and powerful *contraction mapping theorem*:

THEOREM 6.4b(1). *Let* (X, d) *be a complete metric space and let T be a contraction mapping on X. Then T has a fixed point, i.e. there exists a point $p \in X$ such that $T(p) = p$. The solution is unique. The solution p can be constructed by successive iteration as follows. Let x_0 be an arbitrary point of X and define*

$$X_1 = T(x_0), \qquad x_2 = T(x_1) = T(T(x_0)), \ldots,$$
$$x_n = T(x_{n-1}) = T(\ldots T(T(x_0)), \ldots \quad .$$

Then $p = \lim_{n \to \infty} x_n$. *Furthermore, the error of the nth iteration can be estimated by the formula*

$$d(p, x_n) \leq \frac{\alpha^n}{1 - \alpha} d(x_0, x_1).$$

Proof. Our strategy will be to show that the sequence (x_n) described in the theorem converges and that its limit satisfies the fixed point equation.

To show that (x_n) is convergent, we only have to prove that it is a Cauchy sequence. To start with, we observe that, by induction

$$d(x_1, x_2) = d(T(x_0), T(x_1)) \leq \alpha d(x_0, x_1),$$

$$d(x_2, x_3) = d(T(x_1), T(x_2)) \leq \alpha d(x_1, x_2) \leq \alpha^2 d(x_0, x_1),$$

$$\vdots$$

$$d(x_n, x_{n+1}) \leq \alpha^n d(x_0, x_1). \tag{6.7}$$

Therefore, by repeated use of the triangular inequality,

$$d(x_n, x_{n+k}) \leq d(x_n, x_{n+1}) + d(x_{n+1}, x_{n+k}) \leq \cdots$$

$$\leq d(x_n, x_{n+1}) + d(x_{n+1}, x_{n+2}) + \cdots + d(x_{n+k-1}, x_{n+k}),$$

and thus, with Eq. (6.7),

$$d(x_n, x_{n+k}) \leq (\alpha^n + \alpha^{n+1} + \cdots + \alpha^{n+k-1}) d(x_0, x_1),$$

or simply

$$d(x_n, x_{n+k}) \leq \frac{\alpha^n - \alpha^{n+k}}{1 - \alpha} d(x_0, x_1). \tag{6.8}$$

If we take an arbitrary fixed value for k and let n be sufficiently large, then the r.h.s. becomes as small as desired, because $\alpha < 1$. This shows that (x_n) is a Cauchy sequence, hence (x_n) is convergent, because the space is assumed to be complete. Thus, there exists a limit, and we set

$$\lim_{n \to \infty} x_n = p. \tag{6.9}$$

Now we observe that any contraction mapping is a continuous function. Indeed,

$$d(T(x), T(y)) \leq \alpha d(x, y)$$

implies that, given any $\epsilon > 0$,

$$d(T(x), T(y)) < \epsilon \quad \text{provided} \quad d(x, y) < \delta.$$

But, for functions on metric spaces, this is precisely the test of continuity.†

Because of continuity, we have, using Eq. (6.9),

$$\lim_{n\to\infty} T(x_n) = T(\lim_{n\to\infty} x_n) = T(p).$$

On the other hand, by the definition of the sequence (x_n), we can write

$$\lim_{n\to\infty} T(x_n) = \lim_{n\to\infty} x_{n+1} \equiv \lim_{n\to\infty} x_n = p.$$

Comparison of the last two equations gives the desired result $T(p) = p$, which was to be proven.

To show uniqueness of the solution, assume that there existed another solution p', i.e. that also $T(p') = p'$ held true. Then

$$d(p, p') = d(T(p), T(p')) \leqslant \alpha d(p, p').$$

Since, by assumption $p \neq p'$, we have $d(p, p') > 0$, so that upon dividing by $d(p, p')$ we would have $1 \leqslant \alpha$. But this is a contradiction since $\alpha < 1$. Hence, $p' = p$, i.e. there is only one fixed point.

Finally, to estimate the error of the nth iteration, we go back to Eq. (6.8), and taking n fixed, we let $k \to \infty$. The limit of the l.h.s. is obtained by referring to the continuity of the distance function (cf. Problem 5.5-4) and we get $\lim_{k\to\infty} d(x_n, x_{n+k}) = d(x_n, \lim_{k\to\infty} x_{n+k}) = d(x_n, p)$. On the r.h.s., $\alpha^{n+k} \to 0$ because $\alpha < 1$. Hence

$$d(x_n, p) \leqslant \frac{\alpha^n}{1 - \alpha} d(x_0, x_1),$$

as claimed in the theorem. This completes the proof.

The skillful application of the contraction mapping theorem is done in the following manner. Given the equation $T(x) = x$ on the set X, we search for a metric d on X such that (X, d) is a complete space and T is a contraction in this metric. If such a metric is found, the existence and uniqueness of a fixed point follows without any calculation. In order to actually construct a solution by iteration, we may start with any x_0 as the zeroth iteration, but a good guess (which makes $d(x_1, x_0)$ as small as possible) will favorably influence the rate of convergence of the procedure.

We illustrate the power of contraction mapping on the problem of solving a certain class of integral equations.

---

†We see that actually T is uniformly continuous.

Let $[a, b]$ be a closed and bounded interval of the reals. Let

$$f: [a, b] \to E$$

be a given real valued *continuous* function. Let

$$K: [a, b] \times [a, b] \to E$$

be another given real valued *continuous* function. (This is, $K(x, y)$ is a continuous real function in two variables, defined for $a \le x \le b, a \le y \le b$.) Let λ be a real number and consider the equation

$$\varphi(x) = f(x) + \lambda \int_a^b K(x, y)\varphi(y)\, dy. \qquad (6.10)$$

Here $\varphi: [a, b] \to \mathbf{R}$ is considered to be the "unknown function." Equation (6.10) is called an *inhomogeneous Fredholm integral equation of kind 2*, with the continuous kernel K and the continuous inhomogeneity term f, and parameter λ. We shall try to use the contraction mapping theorem to find a function φ which is the unique *continuous* solution of the equation.†

To start with, we reformulate the task as a fixed point problem. Consider the set of all real valued continuous functions ψ defined on $[a, b]$, and define on this set the map T by setting

$$(T\psi)(x) = \lambda \int_a^b K(x, y)\psi(y)\, dy + f(x) \qquad (6.11)$$

for each ψ. Here f and K are the above given entities, and, for simplifying the notation, we wrote‡

$$(T\psi)(x) \quad \text{for} \quad (T(\psi))(x).$$

With this definition of T, the Fredholm equation (6.10) assumes the form

$$T\psi = \psi \qquad (6.12)$$

and we are looking for a solution φ, i.e. we search for a particular continuous function φ in the set $\{\ldots \psi \ldots\}$ which satisfies Eq. (6.12).

In order to fulfill the criteria of the contraction mapping theorem, we

---

†For definiteness, we assumed that f, K, λ (and hence φ) are real. The entire discussion generalizes easily to the complex case.

‡In other words, $(T\psi)(x)$ denotes the value of the image of ψ under T, evaluated at $x \in [a, b]$. Thus, $T\psi$ is the image of ψ under T. From Eq. (6.11) it is obvious that $T\psi$ is a *continuous* function on $[a, b]$.

must endow the set ψ of real continuous functions on $[a, b]$ with a suitable metric. We suggest the usual uniform metric

$$d(\psi_1, \psi_2) = \sup_{x \in [a, b]} |\psi_1(x) - \psi_2(x)|,$$

i.e. we make the set $\{\ldots \psi \ldots\}$ become the function space $C[a, b]$. We recall that this space is *complete*; we must now check whether T is a contraction map. From Eq. (6-11) we see that

$$d(T\psi_1, T\psi_2) = \sup_{x \in [a, b]} |(T\psi_1)(x) - (T\psi_2)(x)|$$

$$= \sup_{x \in [a, b]} \left| \lambda \int_a^b K(x, y)(\psi_1(y) - \psi_2(y)) \, dy \right|. \qquad (6.13)$$

To estimate the r.h.s., we use Cauchy's integral estimate, according to which

$$\left| \int_a^b g(y) dy \right| \leq (b - a) \max_{y \in [a, b]} |g(y)|.$$

Setting then†

$$M \equiv \sup_{x \in [a, b]} \left(\max_{y \in [a, b]} |K(x, y)| \right), \qquad (6.14)$$

Eq. (6.13) gives

$$d(T\psi_1, T\psi_2) \leq |\lambda|(b - a)M \sup_{y \in [a, b]} |\psi_1(y) - \psi_2(y)|.$$

This can be rewritten as

$$d(T\psi_1, T\psi_2) \leq |\lambda|(b - a)M d(\psi_1, \psi_2),$$

or, introducing the notation

$$\alpha \equiv |\lambda|(b - a)M, \qquad (6.15)$$

we may write

$$d(T\psi_1, T\psi_2) \leq \alpha d(\psi_1, \psi_2).$$

Now, is T a contraction mapping or not? All depends on whether α (defined by Eq. (6.15)) is smaller than 1 or not. Since M, via Eq. (6.14), is determined by the kernel K, we see that the condition for T to be a contraction places a restriction on λ. Indeed, Eq. (6.15) tells us that we

---

†The existence of $\max_y |K(x, y)|$ is assured because for any given x, $K(x, y)$ is continuous on a closed and bounded interval of y.

have a contraction mapping if [†]

$$|\lambda| < \frac{1}{(b-a)M}. \tag{6.16}$$

In consequence of these considerations, we are now assured that, *for sufficiently small $|\lambda|$ parameters, the Fredholm equation* (6.10) *is guaranteed to have a unique continuous solution φ.*

We now proceed to actually construct the solution. Looking at Eq. (6.10) and considering λ to be "small" in some sense, a good guess appears to be to take for the zeroth iteration $\varphi_0 = f$. We then calculate the iterations as follows:

$$\varphi_0(x) = f(x),$$

$$\varphi_1(x) = (T\varphi_0)(x) = f(x) + \lambda \int_a^b K(x, y)f(y)\, dy,$$

$$\varphi_2(x) = (T\varphi_1)(x) = f(x) + \lambda \int_a^b K(x, y)f(y)\, dy$$

$$\vdots \qquad\qquad + \lambda^2 \int_a^b \int_a^b K(x, z)K(z, y)f(y)\, dz\, dy,$$

$$\varphi_n(x) = (T\varphi_{n-1})(x)$$

$$= f(x) + \sum_{k=1}^{n} \lambda^k \int_a^b \dots \int_a^b dx_1\, dx_2 \dots dx_k K(x, x_1)K(x_1, x_2) \dots K(x_{k-1}, x_k)f(x_k).$$

To simplify the notation, it is convenient to introduce the "*iterated kernels*" defined by

$$K_1(x, x_1) \equiv K(x, x_1),$$

$$K_2(x, x_2) \equiv \int_a^b K(x, x_1)K(x_1, x_2)\, dx_1,$$

$$\vdots$$

$$K_k(x, x_k) \equiv \int_a^b K_{k-1}(x, x_{k-1})K(x_{k-1}, x_k)\, dx_{k-1}.$$

[†]This condition on λ is surely sufficient but not necessary. The Cauchy estimate is rather crude and it can be shown that, for example, we have a contraction provided only

$$|\lambda| < \left(\int_a^b \int_a^b K^2(x, y)\, dx\, dy \right)^{-1/2}.$$

This restriction on λ is, in general, less stringent than the one given by Eq. (6.16). But it is possible that even larger λ values give a contraction.

We then have

$$\varphi_n(x) = f(x) + \sum_{k=1}^{n} \lambda^k \int_a^b K_k(x, x_k) f(x_k) \, dx_k. \tag{6.17}$$

The exact solution is obtained by taking the limit† when $n \to \infty$. We obtain

$$\varphi(x) = f(x) + \sum_{k=1}^{\infty} \lambda^k \int_a^b K_k(x, x_k) f(x_k) \, dx_k. \tag{6.18}$$

Equation (6.18) is called the Neumann–Liouville series solution of the Fredholm equation. It has been established laboriously and by an *ad hoc* procedure in the last century, long before contraction mappings (and topological spaces!) were known.‡

We may also easily estimate the accuracy of the nth iteration. We have, from the general formula and with α given by Eq. (6.16),

$$d(\varphi, \varphi_n) \leqslant \frac{|\lambda|^n (b-a)^n M^n}{1 - |\lambda|(b-a)M} \, d(f, \varphi_1). \tag{6.19}$$

Using φ_1 as given above, we calculate

$$d(f, \varphi_1) = \sup_x \left| \lambda \int_a^b K(x, y) f(y) \, dy \right| \leqslant |\lambda|(b-a)MN$$

where

$$N = \max_{x \in [a, b]} |f(x)| \tag{6.20}$$

and we again used the Cauchy estimate of integrals. Substituting back into Eq. (6.19) and rewriting the l.h.s. explicitly, we finally obtain

$$\max_{x \in [a, b]} |\varphi(x) - \varphi_n(x)| \leqslant \frac{|\lambda|^{n+1}(b-a)^{n+1} M^{n+1}}{1 - |\lambda|(b-a)M} \, N, \tag{6.21}$$

where M and N are given by Eqs. (6.14) and (6.20), respectively.

To conclude this topic, we remark that even if $|\lambda|$ is not "sufficiently small," the Fredholm equation may still have a unique and even continuous solution. But then the solution cannot be obtained with the help of the contraction mapping theorem.§ Furthermore, for any λ, there

†The existence of the limit, i.e. the convergence of the power series in λ, is assured precisely by the contraction mapping theorem, provided λ obeys the restriction (6.16).

‡It is amusing to note that the method was rediscovered by the famous physicist M. Born in 1926. In particular, the first two iterations φ_1 and φ_2 are commonly referred to in the physics literature as "the first and second Born approximation of scattering theory."

§Quite generally, the contraction mapping theorem gives only a sufficient but not necessary criterion for the existence of a fixed point.

may exist noncontinuous solutions, i.e. solutions that do not belong to the space $C[a, b]$. Finally, we did not even consider the case when f and/or K is discontinuous and when $[a, b]$ is not a bounded interval. In later chapters, equipped with more powerful methods, we shall touch upon some of these questions related to integral equations.

We now return to general questions and wish to indicate some generalizations of our results on fixed points.

First of all we remark that the premise of the fixed point theorem for complete metric spaces may be relaxed. Suppose (X, d) is a complete metric space and $T: (X, d) \to (X, d)$ is a map which has the property that

$$d(T^k(x), T^k(y)) \le \alpha d(x, y),$$

where $\alpha < 1$, k is some positive integer† and

$$T^k(x) = (T \circ T \circ \cdots \circ T)(x) = T(\cdots T(T(x))).$$

Then $T(x) = x$ still has a unique solution p and actually $p = \lim_{n \to \infty} T^n(x_0) = \lim_{n \to \infty} x_n$, as before, with $x_0 \in X$ being arbitrary. However, the estimate of the accuracy of the n th iteration is slightly different. We leave the detailed proof to the reader: it follows the same pattern as that of the original theorem.

Next we mention that the *existence* of a fixed point can be inferred from much weaker conditions. A typical, somewhat special yet very powerful fixed point theorem has been given by *Brouwer*:

THEOREM 6.4b(2). *Let E^n be the Euclidean n-space and consider the subspace consisting of the closed unit ball*

$$S_1[o] = \left\{ x \in E^n \,\middle|\, d(x, o) = \left(\sum_{k=1}^{n} \xi_k^2 \right)^{1/2} \le 1 \right\}.$$

Let $T: S_1[o] \to S_1[o]$ be an arbitrary continuous map on this space. Then T has a fixed point, i.e. *there exists a $p \in S_1[o]$ such that $T(p) = p$.*

This is a remarkable result because it tells us that in $S_1[o]$ continuity alone ‡ assures the existence of a fixed point. However, the theorem makes no statement about the uniqueness of the fixed point nor does it give a procedure to construct a fixed point. Figure 6.11 visualizes

---

†For $k = 1$, we have the previously discussed case.

‡Observe that while a contraction mapping is always continuous, there are many continuous maps that are not contractions. Thus, Brouwer's theorem is not valid for, say, arbitrary Euclidean spaces.

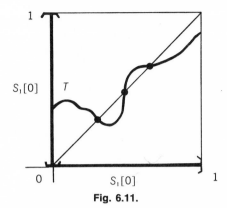

Fig. 6.11.

Brouwer's theorem for $n = 1$. (Note that there are several fixed points.) The proof for $n = 1$ is not too difficult. However, for $n \geqslant 2$, it is far from trivial.

There are several other fixed point theorems. Some of them apply even to certain nonmetric topological spaces, but these topics lie outside our introductory exposition.

PROBLEMS

6.4b-1. Use the contraction mapping theorem to prove that the equation $x^2 - 3x + 2 = 0$ has exactly one solution in the interval $0 \leqslant x \leqslant 1.4$.

6.4b-2. Let M be a real $n \times n$ matrix with the property that $\Sigma_{k=1}^{n} |M_{ik}| \leqslant \alpha < 1$ for all $i = 1, 2, \ldots, n$. Prove that the system of algebraic equations

$$\xi_i - \sum_{k=1}^{n} M_{ik}\xi_k = b_i, \qquad i = 1, 2, \ldots, n,$$

with $b_1, \ldots, b_n$ arbitrary, has exactly one solution. (*Hint*: Consider the space with elements $x = (\xi_1, \ldots, \xi_n)$, and with metric $d(x, y) = \max_{1 \leqslant i \leqslant n} |\xi_i - \eta_i|$. Show that this space is complete; show that all other conditions for a contraction mapping are satisfied.)

6.4b-3. Consider the algebraic equation system

$$\xi_1 = \tfrac{1}{3}\xi_1 - \tfrac{1}{4}\xi_2 + \tfrac{1}{4}\xi_3 - 1$$

$$\xi_2 = -\tfrac{1}{2}\xi_1 + \tfrac{1}{5}\xi_2 + \tfrac{1}{4}\xi_3 + 2$$

$$\xi_3 = \tfrac{1}{5}\xi_1 - \tfrac{1}{3}\xi_2 + \tfrac{1}{4}\xi_3 - 2$$

and show (by applying the result of the previous problem) that the system has exactly one solution. Find the solution by an iteration method, proceeding to a step n where the accuracy is such that $\max_{1 \leqslant i \leqslant 3} |\xi_i^{(n)} - \xi_i| \leqslant 1/100$. (Here ξ_i denotes the exact solution and $\xi_i^{(n)}$ is the nth approximation.)

6.4b-4. Consider the following integral equations:

(a) $\varphi(x) = x + \lambda \displaystyle\int_0^\pi \sin(x + y) \cdot \varphi(y)\, dy,$

(b) $\varphi(x) = x^2 + \lambda \displaystyle\int_0^1 xy\varphi(y)\, dy.$

Show, in each case that the contraction mapping theorem applies for a certain range of λ. Use the power series method to find the *exact* solutions, for the admissible range of λ.

6.4b-5. Show that the integral equation

$$\varphi(x) = 1 + \frac{a}{2} \int_0^1 (x + y + a)^{-1} \varphi(y)\, dy, \qquad a > 0,$$

can be solved by successive iteration. In this case, you will not be able to find a "closed form" for the exact solution. Calculate the first approximation; and estimate the error of the nth approximation.

IIC: Measure Structures

7

Measure Spaces

In our introduction to algebraic and to topological structures we noted that these highly abstract and general notions arise as natural generalizations of a child's first experiences when studying the system of real numbers. After he gains some maturity, the child will become aware of yet another structure on the real line. He will realize the necessity for a concept of "extent." He will first start talking about the "length" of an interval, even though he may be tempted to oversimplify this notion by regarding it as the "distance between the endpoints." However, when he proceeds in his geometrical education from the line to the plane and to the three-dimensional space, the new concepts of area and volume will reveal themselves to him as building blocks of a genuinely new pursuit. He may also discover that combining areas of arbitrary "shape" and arbitrary relative position is not a simple job. During his first experiences with material bodies, he will, furthermore, become conscious of another kind of extent or "measure," such as the mass of objects. He may even realize that combining areas or volumes is very similar to combining masses. When learning about electricity, he may become puzzled by noting that this "fluid" can have both a positive and a negative measure.

Coming back to pure mathematics, and having graduated to some institution of "higher education," our young fellow will learn that, apart from its basic role in geometry, area plays also a crucial role in analysis: the concept of the integral of a function will appear essentially as the "area under the graph of the function." As a matter of fact, measure theory originated historically from the attempt to generalize the classical notion of integration. However, the subject is now based on a much broader footing and once again we shall follow the modern "axiomatic" method of exploration.

As it should transpire from the above intuitive remarks, "measure" is some sort of numerical value associated with certain subsets of a universal set. The first logical step in developing the theory is therefore to endow a universal set with a structure which consists of a suitably chosen class of subsets, which are declared "measurable sets" of the "measurable space." Then, a definition for associating a numerical measure to each measurable set will have to be supplied. Measurable spaces equipped with a measure will be called "measure spaces." The rigorous establishment of these notions is both more subtle and more complicated than the foundation-laying of algebraic or of topological structures. We shall restrict ourselves to the essentials and frequently omit details and proofs.

7.1ˀ MEASURABLE SPACES

In this section we introduce the fundamental structure of measure theory, namely the concept of a measurable space. However, in contrast to the procedure followed in topology, we cannot start directly with the basic definition and must first study somewhat simpler (and less important) structures.

Let X be an arbitrary (universal) set. Let $\mathcal{R} = \{\ldots A_\alpha \ldots\}$ be any collection of subsets† of X which obeys the following axioms:

(i) $\emptyset$ belongs to $\mathcal{R}$,
(ii) If A_α and A_β belong to $\mathcal{R}$, then $A_\alpha \cup A_\beta$ belongs to $\mathcal{R}$,
(iii) If A_α and A_β belong to $\mathcal{R}$, then $A_\alpha - A_\beta$ belongs to $\mathcal{R}$.

Any collection $\mathcal{R}$ which has these properties is called a *ring $\mathcal{R}$ on X*. The essential point is that *a ring on X is a collection of subsets of X which is closed under formation of pairwise unions and differences of sets*.‡

From the definition it follows by trivial induction that *finite unions* $\bigcup_{k=1}^{n} A_k$ *of members of a ring also belong to the ring*. Similarly, noting that $A_k \cap A_l = A_k - (A_k - A_l)$ and using induction, it follows that *finite intersections* $\bigcap_{k=1}^{n} A_k$ *of members of a ring also belong to the ring*.

These observations give actually a justification of the nomenclature "ring" of sets. In Problem 4.2-3 it was shown that any class of subsets of X which obeys our present axioms (ii) and (iii) constitutes a ring in the

---

†That is, let $\mathcal{R}$ be a subset of the power set $\mathcal{P}(X)$.
‡Note that axiom (i) is, strictly speaking, redundant, because if $A \in \mathcal{R}$, then $A - A = \emptyset$.

algebraic sense with respect to symmetric difference and intersection as composition laws. However, this algebraic structure will not play an explicit role in the following.†
We now give a few examples.

Example α. Let X be an arbitrary set and define $\mathscr{R} = \mathscr{P}(X)$. This is clearly the biggest ring on X. Conversely, let $\mathscr{R} = \{\emptyset\}$. This is the smallest ring on X.

Example β. Let X be arbitrary and let A be an arbitrary fixed proper subset of X. Then $\mathscr{R} = \{\emptyset, A\}$ is a ring on X.

Example γ. Let X be an uncountable set. Then the collection $\mathscr{R}$ of all countable subsets of X is a ring.

Example δ. Let X be uncountable. The collection $\mathscr{R}$ of all finite subsets of X is a ring.

We now restrict our interest to a special type of rings which are characterized by the *additional* property that the union $\cup_{k=1}^{\infty} A_k$ of any *countable* ‡ collection of sets of $\mathscr{R}$ also belongs to $\mathscr{R}$. A ring with this property is called a *σ-ring on X* and will be denoted by the special symbol $\mathscr{S}$. Thus, in detail, a σ-ring $\mathscr{S}$ is a collection of subsets of X such that

   (i) $\emptyset \in \mathscr{S}$,
   (ii) If $A_k \in \mathscr{S}$ for $k = 1, 2, \ldots$, then $\cup_{k=1}^{\infty} A_k \in \mathscr{S}$,
   (iii) If $A_\alpha \in \mathscr{S}$ and $A_\beta \in \mathscr{S}$, then $A_\alpha - A_\beta \in \mathscr{S}$.

Observe that axiom (ii) trivially implies that finite unions, such as $A_\alpha \cup A_\beta$, belong to $\mathscr{S}$, because one may take $A_k = \emptyset$ for $k > 2$.

It is important to note that *countable intersections* $\cap_{k=1}^{\infty} A_k$ *of elements of a σ-ring also belong to the σ-ring.* Indeed, set $\cup_{k=1}^{\infty} A_k \equiv A$ and consider the simple identity

$$\bigcap_{k=1}^{\infty} A_k = A - \bigcup_{k=1}^{\infty} (A - A_k).$$

Axioms (ii) and (iii) lead then to the desired result.

It is easy to see that in the above Examples α through γ, the rings in question happen to be σ-rings. However, Example δ is not a σ-ring, because countable unions of finite sets need not be finite sets.

---

†Nevertheless, the reader is urged to restudy Problems 4.2-3 and 4.2-4.

‡The reader should note that, even though for an arbitrary ring, *finite* unions always belong to the ring, this does not in general imply that infinite unions belong to the ring, not even for the countable case.

We obtained the concept of a σ-ring from that of an arbitrary ring by demanding that countable unions belong to it. Another type of special rings is obtained if we demand that X *itself be an element of the ring*. A ring with this property is called an *algebra*† of sets on X. Putting it formally, an algebra $\mathscr{A}$ on X is a collection of subsets of X such that

   (i) $\emptyset \in \mathscr{A}$, and $X \in \mathscr{A}$,
   (ii) If $A_\alpha \in \mathscr{A}$ and $A_\beta \in \mathscr{A}$, then $A_\alpha \cup A_\beta \in \mathscr{A}$,
   (iii) If $A_\alpha \in \mathscr{A}$ and $A_\beta \in \mathscr{A}$, then $A_\alpha - A_\beta \in \mathscr{A}$.

Since any algebra is a ring, finite unions and finite intersections of its members belong to the algebra. More important is the following

LEMMA 7.1(1).   *If $\mathscr{A}$ is an algebra on X, then together with A also A^c belongs to $\mathscr{A}$. Conversely, if a ring has the property that together with an element A also A^c belongs to the ring, then the ring is actually an algebra.*

Proof. (a) If $A \in \mathscr{A}$, then $A^c \equiv X - A \in \mathscr{A}$, because of axioms (i) and (iii).

   (b) If $\mathscr{R}$ is a ring and $A \in \mathscr{R}$, $A^c \in \mathscr{R}$, then $X \equiv A \cup A^c \in \mathscr{R}$, on account of the first basic property of rings.

We now obtain the following

THEOREM 7.1(1).   *$\mathscr{A}$ is an algebra of sets on X iff*

   $(\alpha)\ A \in \mathscr{A}$ *and* $B \in \mathscr{A}$ *imply* $A \cup B \in \mathscr{A}$,
   $(\beta)\ A \in \mathscr{A}$ *implies* $A^c \in \mathscr{A}$.

Proof.   (a) Since $A - B = A \cap B^c = (A^c \cup B)^c$ and since $B \in \mathscr{A}$ by assumption while $A^c \in \mathscr{A}$ by (β), it follows from (α) that $(A^c \cup B)^c \in \mathscr{A}$, and hence $A - B \in \mathscr{A}$. Then, $\emptyset \equiv A - A \in \mathscr{A}$. Also, because of (α) and (β), $X \equiv A \cup A^c \in \mathscr{A}$. Thus, if properties (α) and (β) hold, then all algebra axioms (i), (ii), and (iii) set forth above, are satisfied.

   (b) The converse statement follows directly from the first part of Lemma 7.1(1).

The importance of Theorem 7.1(1) is that it can be used as an *alternative definition of an algebra. A class $\mathscr{A}$ of subsets of X is an algebra on X iff it is closed under the formation of complements and of pairwise unions.*

   For example, on an arbitrary set X the collection $\mathscr{A} = \{\emptyset, X\}$ is an algebra (actually the smallest one). Or, again for an arbitrary set X and a

---

†The terminology is not a particularly happy one.

fixed $A \subset X$, the collection $\mathcal{A} = \{\emptyset, A, A^c, X\}$ is an algebra. On the other hand, Example β given earlier is not an algebra ($A^c \notin \mathcal{A}$); nor is the ring of Example γ or δ an algebra (the complement of a countable or finite set is not countable).

We have introduced two types of special rings: σ-rings and algebras. Obviously, an even more restrictive type of structure is obtained if we combine the distinguishing features of these special rings. A σ-ring which is also an algebra is called, quite naturally, a σ-*algebra of sets on* X or, more frequently, a *Borel field* on X. Thus, we have

DEFINITION 7.1(1).   *A Borel field $\mathcal{B}$ on X is a collection of subsets of X obeying the following axioms*:

(i) *If $B_k \in \mathcal{B}$ for $k = 1, 2, \ldots$, then all finite unions $\cup_{k=1}^{n} B_k$ and even the countable union $\cup_{k=1}^{\infty} B_k$ belongs to $\mathcal{B}$,*
(ii) *If $B \in \mathcal{B}$, then $B^c \in \mathcal{B}$.*

Indeed, from axiom (i) it trivially follows that $B_\alpha \in \mathcal{B}$ and $B_\beta \in \mathcal{B}$ imply $B_\alpha \cup B_\beta$. Taking this together with axiom (ii) and recalling Theorem 7.1(1), we see that we have an algebra. Therefore, from the original definition of an algebra it follows that $B_\alpha - B_\beta \in \mathcal{B}$ and also $\emptyset \in \mathcal{B}$. Combining these statements with the fact, that by axiom (i), $\cup_{k=1}^{\infty} B_k \in \mathcal{B}$, we see that the axioms of a σ-ring (as given on p. 295) are satisfied. Thus, the Definition 7.1(1) of a Borel field describes precisely a system that is both a σ-ring and an algebra.

Since a Borel field $\mathcal{B}$ is an algebra, we know that X itself (as well as $\emptyset$) are elements of $\mathcal{B}$. Further, since $\mathcal{B}$ is a σ-algebra, we know that countable intersections of elements belong also to $\mathcal{B}$. It will be useful to summarize all properties of a Borel field (including redundant ones) in the following

THEOREM 7.1(2).   *A Borel field $\mathcal{B} = \{\ldots B_\alpha \ldots\}$ on X has the following properties*:

(α) *$\emptyset \in \mathcal{B}$ and $X \in \mathcal{B}$,*
(β) *Countable unions $\cup_{k=1}^{\infty} B_k$ belong to $\mathcal{B}$,*
(γ) *Countable intersections $\cap_{k=1}^{\infty} B_k$ belong to $\mathcal{B}$,*
(δ) *Complements B_α^c belong to $\mathcal{B}$,*
(ϵ) *Differences $B_\alpha - B_\beta$ belong to $\mathcal{B}$.*

Borel fields are, generally speaking, larger classes of sets than are algebras, σ-rings, or rings. This is so because we have a more restrictive structure and hence, more elements are needed to satisfy all criteria.

The members B_α of a Borel field $\mathcal{B}$ are called *Borel sets*. In a sense, Borel sets are the analogs of open sets and a Borel field is the analog of a topology. Actually, a set X on which a Borel field is defined, deserves a special name (similarly as a set X on which a topology is defined, has the name "topological space"). We adopt the following

DEFINITION 7.1(2).   *A* (*universal*) *set X equipped with a Borel field $\mathcal{B}$ is called a* BOREL SPACE *and is denoted by the ordered pair* $(X, \mathcal{B})$.

Thus, a Borel space is a system, or structure, very much like an algebraic system $(X, \square)$ or a topological space (X, τ). If it is clear from the context what the Borel sets are (i.e. what the Borel field on X is), we sometimes simply talk of the "Borel space X." Obviously, on a given set X one may define several (maybe even infinitely many) different Borel fields and *each* definition $\mathcal{B}$ gives rise, in general, to a *different* Borel space. Borel fields on a given set X, like topologies, may be compared: $\mathcal{B}_1$ is smaller than $\mathcal{B}_2$ iff $\mathcal{B}_1 \subset \mathcal{B}_2$.

Borel spaces will play a fundamental role in further developments. However, for some purposes, the Borel space structure is not sufficiently general. There exists a generalization, given by the following

DEFINITION 7.1(3).   *Let X be a* (*universal*) *set and let $\mathcal{S}$ be a σ-ring on X which has the property that X is a* (*not necessarily countable*) *union of sets taken from the collection $\mathcal{S}$. Then the ordered pair $(X, \mathcal{S})$ is called a measurable space. The members of $\mathcal{S}$ are referred to as the measurable sets of X.*

We may informally paraphrase this definition by saying that a measurable space is a σ-ring where the union of all measurable sets is the entire space. Yet in other words, in a measurable space every point $x \in X$ is contained in some measurable set. An example of a measurable space (which is *not* a Borel space) is the system $(X, \mathcal{S})$, where X is uncountable and $\mathcal{S}$ consists of all countable subsets of X. We saw previously that this system is a σ-ring (but not an algebra) and obviously, any x belongs to some countable set.

We observe that *every Borel space is a measurable space* since $\mathcal{B}$ is a σ-ring and $X \in \mathcal{B}$ (so that it is not even necessary to take any kind of union). For this reason, it is customary to call the Borel sets of a Borel space $(X, \mathscr{B})$ simply *measurable sets of X*. In the following, we shall restrict our interest mostly to the special but most important type of measurable spaces which are precisely the Borel spaces and the reader may visualize measurable sets by Borel sets.

It may be instructive at this point to compare Borel spaces and topological spaces. In a sense, a Borel space is both something less and something more than a topological space. A Borel space is *not*, in general, a topological space, because *arbitrary* unions of Borel sets do not belong necessarily to $\mathscr{B}$. On the other hand, the collection $\mathscr{B}$ has a property which, in general, is stronger than what holds for a topology τ: *countable* intersections of Borel sets do belong to $\mathscr{B}$. Furthermore, complements of Borel sets belong to $\mathscr{B}$, whereas, in general, complements of open sets do not belong to τ.

Nevertheless, there are quite strong relations between topological and Borel spaces, which we will have occasion to observe somewhat later.

We now proceed to discuss the most familiar and probably most important example of a Borel space.

Let X be the real line **R**. Let

$$[a, b) = \{x \,|\, a \leq x < b\}, \qquad \text{with } a, b \text{ real}, a \leq b,$$

be the standard symbol for a lower semiclosed bounded interval.† Define the collection $\mathscr{B}$ to be the smallest‡ Borel field that contains all these semiclosed intervals. It is easily seen that $\mathscr{B}$ will contain, in particular, all countable unions, countable intersections, and complements of these intervals.

Our collection generated by semiclosed intervals as defined above is called, for historical reasons, "THE" *class of Borel sets of the real line*, even though, naturally, many other Borel fields can be defined on **R**. If we talk about "the Borel space of the reals" we tacitly imply that the Borel sets are the "usual ones" as specified above.

The reader is urged to explicitly check that the additional derived properties of Borel fields (as given in Theorem 7.1(2)) hold also true for our Borel field on the reals. For example, $\mathbf{R} \in \mathscr{B}$ is shown directly by observing that the whole real line can be covered by countably many semiclosed intervals.

For many later applications, it is necessary that we be able to easily

---

†Note that if, in particular $a = b$, then the "semiclosed interval" $[a, a)$ is actually the *empty set*. This will prove to be a valuable observation in many subsequent discussions.
‡In Theorem 7.1(3) below it will be shown that such a smallest $\mathscr{B}$ exists and is unique.

recognize typical Borel sets of the reals. We first observe that *every singleton set* $\{x\}$ *is a Borel set.* This follows from noting that $\{x\} = \bigcap_{k=1}^{\infty} [x, x + 1/k)$. Then it follows that *every closed interval is a Borel set,* because $[a, b] = [a, b) \cup \{b\}$, with each set being Borel. Similarly, *every open interval is a Borel set,* because $(a, b) = [a, b) - \{a\}$, and differences of Borel sets are Borel.† Since every arbitrary open set of the reals is some countable union of open intervals, it now follows that *every arbitrary open set of the real line is a Borel set.* By complementation we then find that *every arbitrary closed set of the real line is a Borel set.* Furthermore, since $\{x\}$ is a Borel set, it follows that *every countable subset of the real line is a Borel set.* For example, *the set of all rationals is a Borel set.* Then, by complementation, we see that *the set of all irrational numbers is a Borel set.* More generally, *any uncountable set of reals whose complement is countable, is a Borel set.*

In summary, we see that the standard Borel space of the reals is very rich in Borel (i.e. measurable) sets. Actually, the reader may even wonder whether there exist at all any sets that are not Borel? The answer is affirmative. The proof hinges on the fact that the cardinal number of $\mathscr{B}$ can be shown to be $\mathfrak{c}$ (i.e. that of the real line). On the other hand, the cardinal number of the class of all possible subsets of the reals (i.e. the cardinal number of $\mathscr{P}(\mathbf{R})$) is larger than that of the reals (i.e. larger than $\mathfrak{c}$), because of Cantor's theorem (p. 45). Thus, we see that there are "infinitely times more" sets on the real line than there are Borel sets. It is amusing to observe that, nevertheless, it is rather difficult to give an explicit characterization of a set of reals which is not a Borel set.‡

The construction of a Borel structure for the real line $\mathbf{R}$, studied above, can be easily generalized to the n-dimensional§ spaces $\mathbf{R}^n$. Let $x = (\xi_1, \ldots, \xi_n)$ be a point of $\mathbf{R}^n$. We define a "bounded semiclosed rectangle" as the set

$$[\mathbf{a}, \mathbf{b}) = \{x \,|\, \alpha_k \leqslant \xi_k < \beta_k, \quad k = 1, 2, \ldots, n\}, \qquad \text{where} \quad \alpha_k \leqslant \beta_k.$$

Observe that $[\mathbf{a}, \mathbf{b})$ is fully determined by the two points

$$\mathbf{a} = (\alpha_1, \ldots, \alpha_n) \quad \text{and} \quad \mathbf{b} = (\beta_1, \ldots, \beta_n).$$

†In passing we note that every *upper* semiclosed interval is also a Borel set, since $(a, b] = (a, b) \cup \{b\}$.

‡Actually, such a construction can be performed only with a more developed apparatus than what we have so far. We shall come back to this point later.

§The algebraic term "dimension" is used here only for convenience; the linear space structure of $\mathbf{R}^n$ is irrelevant.

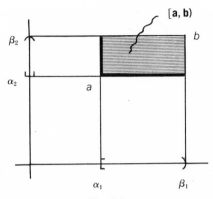

Fig. 7.1.

Figure 7.1 illustrates such a semiclosed rectangle of $\mathbf{R}^2$. Now we define the Borel field on $\mathbf{R}^n$ as the smallest Borel field that contains all semiclosed rectangles.

We observe that singleton sets $\{x\} = \{\xi_1\} \times \{\xi_2\} \times \cdots \times \{\xi_n\}$ are Borel sets. Next, we can consider "hyperplanes" by which we mean "degenerate rectangles." For example,

$$\{\alpha_1\} \times [\alpha_2, \beta_2) \times \cdots \times [\alpha_n, \beta_n)$$

is an "$n-1$-dimensional" hyperplane, and so on. A completely degenerate rectangle is a singleton, i.e. a point. Now, it is easy to see that any hyperplane is a Borel set. This is so because, for example, the $n-1$-dimensional hyperplane described above can be represented as

$$\bigcap_{m=1}^{\infty} [\mathbf{a}, \mathbf{c}_m)$$

with

$$\mathbf{c}_m = \left(\alpha_1 + \frac{1}{m}, \beta_2, \ldots, \beta_n\right).$$

It then follows that closed rectangles $[\mathbf{a}, \mathbf{b}]$ are Borel sets, because they can be represented as the union of the partly closed rectangle $[\mathbf{a}, \mathbf{b})$ and n hyperplanes. Similarly, open rectangles $(\mathbf{a}, \mathbf{b})$ are Borel sets, because any such set is the difference of $[\mathbf{a}, \mathbf{b})$ and the union of n hyperplanes. In consequence of these observations, any open set and (by complementation) any closed set† of $\mathbf{R}^n$ is a Borel set. So are all countable sets and all uncountable sets that have countable complements.

†Here we imagine $\mathbf{R}^n$ equipped with the usual topology so that we really speak of E^n.

After these important examples, we come back to general considerations. We would like to have a systematic procedure which permits us to construct rings, or more important, Borel fields, from an arbitrary given class of "few and simple" sets† of X. This program is facilitated by the following

THEOREM 7.1(3).   *Let $\mathscr{E}$ be an arbitrary given collection of subsets of X. Then there exists a unique ring $\mathscr{R}(\mathscr{E})$ on X such that $\mathscr{R}(\mathscr{E})$ contains all members of $\mathscr{E}$ and such that any other ring on X which contains the collection $\mathscr{E}$, will contain the ring $\mathscr{R}(\mathscr{E})$.*

Proof.   Consider the collection $\{\ldots \mathscr{R}_\alpha \ldots\}$ of all possible rings on X, where each $\mathscr{R}_\alpha$ contains all members of $\mathscr{E}$, i.e. $\mathscr{R}_\alpha \supset \mathscr{E}$. (Such a collection is never empty, because $\mathscr{P}(X)$ is certainly a ring that contains $\mathscr{E}$.) Now construct $\cap_\alpha \mathscr{R}_\alpha$. By Problem 7.1-6, this is also a ring, and it certainly contains $\mathscr{E}$. The ring $\mathscr{R}(\mathscr{E})$ so constructed is the smallest ring containing $\mathscr{E}$, because it is the intersection of all rings that contain $\mathscr{E}$. Thus, it is unique.

The smallest ring $\mathscr{R}(\mathscr{E})$ which contains all members of a given arbitrary collection $\mathscr{E}$ of sets is called the *ring generated by the class $\mathscr{E}$*. The theorem assures us about the existence of such a ring.

In a similar manner, one can construct from any given collection $\mathscr{E}$ the smallest σ-ring or the smallest algebra containing $\mathscr{E}$. To do this, one merely has to take the intersection $\cap_\alpha \mathscr{S}_\alpha$ (or $\cap_\alpha \mathscr{A}_\alpha$) of all σ-rings (or algebras) that contain $\mathscr{E}$. The systems so obtained are called the generated σ-ring $\mathscr{S}(\mathscr{E})$, and the generated algebra $\mathscr{A}(\mathscr{E})$, respectively. Note that if $\mathscr{E}$ contains X itself, then $\mathscr{R}(\mathscr{E})$ contains X, so $\mathscr{R}(\mathscr{E})$ is already an algebra; and $\mathscr{S}(\mathscr{E})$ contains X, so $\mathscr{S}(\mathscr{E})$ is actually a σ-algebra, i.e. a Borel field. However, to obtain from a collection $\mathscr{E}$ a Borel field, it need not be necessary to include X in $\mathscr{E}$ to start with.

Unfortunately, there is no *explicit and simple* prescription‡ to generate a ring from a given $\mathscr{E}$. We give a few examples.

Example ϵ.   Let $X = \{a, b, c\}$. Take $\mathscr{E} = \{\{a, b\}, \{a, c\}\}$. In order to find the generated ring $\mathscr{R}(\mathscr{E})$, we first take the pairwise union

$$\{a, b\} \cup \{a, c\} = \{a, b, c\} = X$$

---

†This procedure is similar to the method of generating a topology for X from an arbitrary given class of sets, via producing first an open subbase, then a base, then a topology.

‡This is rather in contrast to the method of generating a topology τ from a given class of sets.

and the two pairwise differences

$$\{a, b\} - \{a, c\} = \{b\}, \qquad \{a, c\} - \{a, b\} = \{c\}$$

and of course the trivial differences

$$\{a, b\} - \{a, b\} = \{a, c\} - \{a, c\} = \emptyset.$$

In the second step we get as new sets

$$\{b\} \cup \{c\} = \{b, c\}$$

and

$$\{a, b\} - \{b\} = \{a\}.$$

There is no further step necessary; we actually obtained

$$\mathcal{R}(\mathscr{E}) = \{\emptyset, \{a\}, \{b\}, \{c\}, \{a, b\}, \{a, c\}, \{b, c\}, X\}$$

which happens to be the power set $\mathscr{P}(X)$.

Example ζ. Let $X = \mathbf{R}$ and let $\mathscr{E}$ be the collection of all semiclosed intervals $[a, b)$. Observe that the generated σ-ring $\mathscr{S}(\ldots [a, b) \ldots)$ will contain $\mathbf{R}$ itself. Thus, $\mathscr{S}(\ldots [a, b) \ldots)$ is precisely the usual Borel field on the real line.

Example η. We will now show that *the usual Borel field on the reals may be also looked upon as the σ-ring generated by the collection of all* OPEN *intervals* (a, b). To see this, we first recall the nontrivial result that every open interval of the reals is a Borel set. Now, the generated σ-ring $\mathscr{S}(\ldots (a, b) \ldots)$ contains all open intervals (by definition), and it is the *smallest* σ-ring that contains all open intervals. Since the Borel field $\mathscr{B}$ of reals also contains all open intervals, surely

$$\mathscr{B} \supset \mathscr{S}(\ldots (a, b) \ldots). \tag{α}$$

On the other hand, observe that

$$\{a\} = \bigcap_{k=1}^{\infty} \left(a - \frac{1}{k}, a + \frac{1}{k} \right),$$

so that $\{a\} \in \mathscr{S}(\ldots (a, b) \ldots)$. Furthermore, since $[a, b) = (a, b) \cup \{a\}$, we see that $\mathscr{S}(\ldots (a, b) \ldots)$ contains all semiclosed intervals. Since, by the result of Example ζ, the Borel field $\mathscr{B}$ is the *smallest* σ-ring that contains all semiclosed intervals, we certainly have

$$\mathscr{B} \subset \mathscr{S}(\ldots (a, b) \ldots). \tag{β}$$

Combining finally Eqs. (α) and (β), we see that $\mathscr{B} = (\mathscr{S}(\ldots (a, b) \ldots)$, i.e. the usual Borel field of the reals is identical to the σ-ring generated by the collection of all open intervals, as claimed in the first sentence of this example. Incidentally, we now also see that *the smallest Borel field on the reals that contains all open intervals* (a, b) *is precisely the "usual" Borel structure.*

Example ϑ. We leave it to the reader to show that the usual Borel field of the reals may be also looked upon as the σ-ring generated by the collection of all *closed* intervals $[a, b]$ and hence, the smallest Borel field on the reals that contains all closed intervals $[a, b]$ is precisely the "usual" Borel structure.

In view of Examples η and ϑ the reader now may wonder why we used the rather artificial-appearing semiclosed intervals in the *original* definition (p. 299) of the Borel field of the reals, rather than the more natural open intervals. The reason is purely technical: whereas the complement of a *semiclosed* interval is a countable union of *semiclosed* intervals, this is not true for *open* intervals. (Draw a figure to see this.) Of course this does not contradict the fact that complements of open (or of closed) intervals *do* belong to the Borel field.

The considerations of Examples ζ, η, and ϑ generalize easily to $\mathbf{R}^n$. In other words, the usual Borel field $\mathscr{B}$ of $\mathbf{R}^n$ can be generated by either the collection of all semiclosed, or all open, or all closed rectangles. Thus, $\mathscr{B}$ is the smallest Borel field on $\mathbf{R}^n$ which contains all possible types of rectangles. Herein lies the importance of the "usual" Borel fields on $\mathbf{R}$ and $\mathbf{R}^n$.

There exists an important generalization of these results to more general spaces than the discussed Euclidean ones. Without proof we quote the following

THEOREM 7.1(4).   *Let (X, τ) be a locally compact topological space and let $\mathscr{C}$ be the collection of all compact subsets of X. Suppose furthermore that X is a countable union of compact sets. Then the σ-ring $\mathscr{S}(\mathscr{C})$ generated by the class $\mathscr{C}$ is automatically a Borel field on X. It is referred to as the standard (or "usual") Borel structure of (X, τ).*

PROBLEMS

7.1-1. Let $\mathbf{R}$ be the set of reals. Show that the class $\mathscr{R}$ of all subsets of reals of the form

$$\bigcup_{k=1}^{n} [a_k, b_k) \equiv \bigcup_{k=1}^{n} \{x \mid a_k \leqslant x < b_k\}, \qquad a_k, b_k \text{ real}, \quad a_k \leqslant b_k$$

is a ring on $\mathbf{R}$.

7.1-2. The previous problem shows that finite unions of semiopen intervals form a ring. Show that this is not true for finite unions of open (or of closed) intervals.

7.1-3. Let X be an uncountable set. Show that the class $\mathcal{R}$ of all subsets of X which either are countable or have a countable complement, is a ring on X. Is this a σ-ring? Is it an algebra?

7.1-4. Under what circumstances can the class of open sets of a topological space be a ring?

7.1-5. Let $\mathcal{R}$ be a ring on X. Let $\mathcal{A}$ be the class of sets A_α of X such that either $A_\alpha \in \mathcal{R}$ or else $A_\alpha^c \in \mathcal{R}$. Show that $\mathcal{A}$ is an algebra on X.

7.1-6. Let $\mathcal{R}_1$ and $\mathcal{R}_2$ be two rings (σ-rings, algebras) on X. Show that $\mathcal{R}_1 \cap \mathcal{R}_2$ is also a ring (σ-ring, algebra). (*Remark*: These theorems can be extended to arbitrary intersections of families of rings or σ-rings or algebras.)

7.1-7. Let X be an arbitrary set. Let $\mathcal{E}$ be the class of all subsets of X which contain exactly two points. What is the generated ring $\mathcal{R}(\mathcal{E})$? Is this a σ-ring? Is it an algebra?

7.2 MEASURE AND MEASURE SPACES

As already mentioned in the introduction to Chapter 7, measurable spaces (and the special types called Borel spaces) are introduced in order to designate a class of sets for which a "measure" can be defined in a reasonable and useful way.

Quite generally speaking, a measure is nothing but a set function on a ring (i.e. a map defined on the class $\mathcal{R}$ of subsets of some universal set X) with values assumed in the set of the extended reals† $\hat{\mathbf{R}}$ and obeying some basic properties specified below. The largest ring occurring in applications is that of the measurable sets of an arbitrary measurable space $(X, \mathcal{S})$. More often we shall consider a measure defined only on the class of measurable sets (i.e. the Borel sets) of a Borel space $(X, \mathcal{B})$. Sometimes the smaller classes of σ-rings and even arbitrary rings are used as the domain of a measure (which then is often extended to larger classes). We now formulate the following general

DEFINITION 7.2(1). *Let $\mathcal{R}$ be a ring on some set X. Let $\mu : \mathcal{R} \to \hat{\mathbf{R}}$ be a map from the ring to the extended reals which obeys the following axioms:*

  (i) *$\mu(B) \geqslant 0$ for all $B \in \mathcal{R}$.*
  (ii) *$\mu(\emptyset) = 0$.*

---

†Recall that the extended real line $\hat{\mathbf{R}}$ is the union of all real numbers and the two ideal symbols $+\infty$ and $-\infty$, see p. 266.

(iii) *If* $B_k \in \mathcal{R}$ *for* $k = 1, 2, \ldots$ *and* $B_k \cap B_i = \emptyset$ *for* $k \neq i$, *then* $\mu(\bigcup_{k=1}^{\infty} B_k) = \Sigma_{k=1}^{\infty} \mu(B_k)$.

The set function μ with these properties is called a measure on the ring $\mathcal{R}$.

Several remarks are in order. First, since the codomain of μ is $\hat{\mathbf{R}}$, we may have sets whose measure is $+\infty$. Second, for this reason, the infinite series $\Sigma_{k=1}^{\infty} \mu(B_k)$ in axiom (iii) may be divergent,† and then, following the usual convention, we say that "its value is $+\infty$."

The property expressed in axiom (i) simply says that μ is nonnegative. Axiom (ii) tells us that the measure of the empty set is defined to be zero. Axiom (iii) tells us that the measure of a countable union of pairwise disjoint sets is computed additively. (Of course, axiom (iii) together with axiom (ii) assures us that the measure of a finite union $\bigcup_{k=1}^{n} B_k$ is also computed in an additive manner, since we may take all sets B_k with $k > n$ to be the empty set.) The property expressed in axiom (iii) is usually referred to as the *"countably additive"* property of the measure.

In summary, we may say that *a measure μ is an extended real valued, nonnegative, countably additive function defined on a ring‡ and such that* $\mu(\emptyset) = 0$.

Suppose now that we are given a measurable space $(X, \mathcal{S})$ (for the definition, cf. p. 298) and that we define on the σ-ring $\mathcal{S}$ some measure μ. *We then refer to the ordered triple $(X, \mathcal{S}, \mu)$ as a "measure space."* Thus, *a measure space is a measurable space together with a measure.* One often says that "we have a measure on the measure space X," even though this is an abuse of language since the domain of μ is not the set X but rather the class of measurable sets (i.e. the σ-ring $\mathcal{S}$). In most applications we shall be concerned only with the case when the measure space is a Borel space so that we have the system $(X, \mathcal{B}, \mu)$, where the measurable sets are all Borel sets of X. Without causing confusion, we shall refer often to $(X, \mathcal{B}, \mu)$ also by the generic term "measure space." Measure spaces (and in particular measure spaces made out of Borel spaces) are the fundamental structures on which all further developments and applications will be based.

It should be obvious that given a measurable space $(X, \mathcal{S})$ (or Borel space $(X, \mathcal{B})$), we may introduce, in general, several, maybe even infinitely many measures μ, and in each case we obtain a *different* measure space.

†This may happen even if all terms $\mu(B_k)$ are finite.
‡In particular, the ring may be a σ-ring, or a Borel field of a Borel space, or the class of measurable sets of a general measurable space.

Furthermore, if only X is given to start with, many different measurable spaces may be formed from it, and each of which again can be equipped with many different measures. Thus, one always has to clearly specify X, $\mathscr{S}$, (or $\mathscr{B}$) and μ.

In the following we introduce some exotic, but unavoidable *terminology* pertaining to measures and measure spaces.

(a) A set B for which $\mu(B) = 0$, is called a *set of measure zero* or a *null set* (with respect to the given measure μ). Of course, the empty set is always a null set, but in general, there may be many more null sets.

Suppose we have a property of some "object" which holds true for all points $x \in X$ except for points which belong to a set $B \subset X$ of measure zero. We then say that the property in question holds "*almost everywhere*" on X and use the abbreviation "a.e.". For example, a function $f: X \to \mathbf{R}$ is said to be "bounded a.e." iff $|f(x)| \leqslant M$ for all x, except on a subset B of X which has measure zero.

(b) A set B is said to have *finite measure* if $\mu(B) \neq \infty$. A more refined concept is the following. A set B is said to have *σ-finite measure* if there exists a sequence (B_k) of sets in the ring such that $B \subset \bigcup_{k=1}^{\infty} B_k$ and such that $\mu(B_k) < \infty$ for each k. Clearly, a set which has finite measure, has also a σ-finite measure, but not necessarily the other way around.

(c) If the measure of *every* set B of the ring, with the possible exception of $B = X$, is finite (σ-finite), then the measure μ itself is called a *finite* (σ-*finite*) *measure*. If the ring in question is actually the class of measurable sets of a measure space $(X, \mathscr{S}, \mu)$ (specifically of a Borel type measure space $(X, \mathscr{B}, \mu)$), then the space $(X, \mathscr{S}, \mu)$ (or $(X, \mathscr{B}, \mu)$) is said to be a *finite* (σ-*finite*) *measure space* provided μ is a finite (σ-finite) measure.

(d) Suppose X belongs to the ring on which a measure is defined,† and suppose X has a finite (σ-finite) measure. Then the measure μ is called a *totally finite* (*totally σ-finite*) *measure*. If, in particular, the ring is the class of measurable sets of $(X, \mathscr{S}, \mu)$ (specifically of $(X, \mathscr{B}, \mu)$), then the measure space itself is called a *totally finite* (*totally σ-finite*) *measure space*.

At this point we must note that if μ is totally finite (totally σ-finite), then every set of the ring has a finite (σ-finite) measure. (The proof of this statement follows from Theorem 7.2a(3) to be given later on.) Thus, *every totally finite* (*totally σ-finite*) *measure is certainly finite* (σ-*finite*).

---

†That is, the ring must be an algebra. In particular, we are interested in the case when the ring is a Borel field.

It is now high time for a few preliminary examples.

Example α. Let X be the set $\{1, 2, \ldots, k, \ldots\}$ of all positive integers, and let $\mathcal{B}$ be the class of all subsets of X. In view of our previous considerations, this is surely a Borel space. Now define $\mu(B) =$ the number of elements in the set B. It is easily checked that all three axioms of a measure are satisfied. The only set that has measure zero is the empty set. The sets which do not have finite measure are X itself and all infinite subsets. Thus, μ is not a finite measure. But the space is totally σ-finite. To see this, observe that $X = \cup_{k=1}^{\infty}\{k\}$, where $\mu(\{k\}) = 1 < \infty$.

Example β. Let X be an uncountable set and let $\mathcal{B} = \{$all countable sets and their complements$\}$. This is a Borel space (cf. Problem 7.1-3). Let $\mu(B)$ be the number of elements in B. The set function μ so defined is easily seen to be a measure on $(X, \mathcal{B})$. The only null set is $\emptyset$. All finite Borel sets have finite measure, all other Borel sets have infinite measure, in particular, $\mu(X) = \infty$. The measure is not totally σ-finite because X cannot be represented as a countable collection of sets each having only a finite number of points. Moreover, μ is not even σ-finite, because if B is uncountable, it again cannot be contained in a countable union of sets with finitely many points each.

Example γ. Let X be an arbitrary set and let $\mathcal{R} = \{$all finite subsets of $X\}$. (Note that this is, in general, only a ring and not a σ-ring.) Let $f\colon X \to \hat{\mathbf{R}}$ be a nonnegative extended real valued function on X. Define

$$\mu(\{x_1, x_2, \ldots, x_n\}) = \sum_{k=1}^{n} f(x_k)$$

and set also $\mu(\emptyset) = 0$. A simple calculation shows that μ is a measure on $\mathcal{R}$. Clearly, many sets may have measure zero. If f is not bounded, some sets may have infinite measure and μ need not even be σ-finite.

Example δ. An example taken from elementary physics is the following. Let $X = \{x_1, x_2, \ldots, x_n\}$ be a (finite) set of "masspoints" x_k and let the masspoint x_k have the (nonnegative) mass m_k. Declare all subsets of X measurable (i.e. form the Borel field $\mathcal{B} = \mathcal{P}(X)$) and define the mass μ of an arbitrary set $B \subset X$ by $\mu(B) = \Sigma_k m_k$ where the summation ranges over the indices of all points x_k that belong to the set B. The set function μ so defined is obviously a finite measure on the measurable space $(X, \mathcal{B})$. Thus, the concept of "resulting mass of a mass-distribution" is a prototype of measure.

We now proceed to discuss the question of how to *compare measures.*

Let $(X, \mathscr{S})$ be a given measurable space (in particular, a Borel space $(X, \mathscr{B})$) and let μ_1 and μ_2 be two different measures defined on the collection of measurable sets. We say that μ_1 *is inferior to* μ_2 iff for *all* sets B for which $\mu_2(B) = 0$, we have also $\mu_1(B) = 0$. In other words, μ_1 is inferior† to μ_2 if all measurable sets that are null sets for μ_2, are also null sets for μ_1. We then write $\mu_1 \prec \mu_2$. It is rather obvious that this definition gives rise to a partial order relation on the class of all measure spaces (with the same fixed set X and the same collection $\mathscr{S}$ of measurable sets). Note that given a set X, we can have a poset of measurable spaces defined from X, and for each $(X, \mathscr{S})$ we can have a poset of measure spaces $(X, \mathscr{S}, \mu)$.

Two measures μ_1 and μ_2 on $(X, \mathscr{S})$ are called *equivalent measures* iff they have the same null sets. We then write $\mu_1 \sim \mu_2$. Clearly, two measures are equivalent iff $\mu_1 \prec \mu_2$ and $\mu_2 \prec \mu_1$. It is obvious that equivalence of measures gives rise to an equivalence relation on the class of all measure spaces (with the same given X and $\mathscr{S}$).

In the following we wish to discuss how one can construct new measure spaces from given ones.

To start with, let $(X, \mathscr{S}, \mu)$ be a given measure space (in particular, a Borel space with measure) and let A be a measurable subset of X, i.e. $A \in \mathscr{S}$. Let now $\mathscr{S}_A$ be the class of all measurable subsets of A. (That is, $\mathscr{S}_A \subset \mathscr{S}$, consisting of sets $B_A \in \mathscr{S}$ such that $B_A \subset A$). Define $\mu_A(B_A) = \mu(B_A)$ for each $B_A \in \mathscr{S}_A$. In other words, let μ_A be the restriction of μ to the subset $\mathscr{S}_A$ of $\mathscr{S}$. Clearly, $(A, \mathscr{S}_A, \mu_A)$ is a measure space in its own right and will be called a *subspace of the measure space* ‡ $(X, \mathscr{S}, \mu)$.

We now turn in the other direction and discuss briefly how to construct *products of measure spaces*. In order to avoid complications and for the sake of definiteness, we shall restrict ourselves to the case when $(X_1, \mathscr{B}_1, \mu_1)$ and $(X_2, \mathscr{B}_2, \mu_2)$ are given *Borel* spaces equipped with a measure.§ Let $Z = X_1 \times X_2$ be the Cartesian product of the underlying sets. Define $\mathscr{B}$ to be the Borel field on Z which is generated by the collection $\mathscr{E} = \{\dots B_1 \times B_2 \dots\}$, where $B_1 \in \mathscr{B}_1$ and $B_2 \in \mathscr{B}_2$. Then $(Z, \mathscr{B})$

---

†Often the terminology "μ_1 is absolutely continuous with respect to μ_2" is used. The justification of this usage can be given only at a much later stage, and we shall not use this term in the sequel.

‡With a generalization, it is possible to make a measure space of some subsets A of X even if A is not measurable. The corresponding definition is then more similar to the one which applies to the construction of subspaces of topological spaces, cf. Halmos[11], p. 75. We shall not need this generalization in the following.

§The results generalize easily to arbitrary *σ-finite* measure spaces.

is clearly a Borel space. It can be proven that $\mathscr{B}$ (symbolically denoted by $\mathscr{B}_1 \times \mathscr{B}_2$) is actually the collection of sets that are countable unions of all sets of the form $B_1 \times B_2$. (The situation resembles the construction of the product of topological spaces.) Now we proceed to endow $(Z, \mathscr{B})$ with a measure. In order to do this, we must restrict ourselves to the case when both μ_1 and μ_2 are *σ-finite* measures. Then we define μ by setting

$$\mu(B_1 \times B_2) = \mu_1(B_1)\mu_2(B_2) \quad \text{for every} \quad B_1 \in \mathscr{B}_1 \quad \text{and} \quad B_2 \in \mathscr{B}_2.$$

It can be shown that this procedure determines a unique measure† μ for all $B \in \mathscr{B}$. Furthermore, it turns out that the product measure μ (usually denoted by $\mu_1 \times \mu_2$) is itself σ-finite. Thus, from the two given σ-finite measure spaces we constructed the σ-finite product space $(X_1 \times X_2, \mathscr{B}_1 \times \mathscr{B}_2, \mu_1 \times \mu_2)$. The procedure can be generalized to arbitrary (not only finite) collections of factor spaces.

PROBLEMS

7.2-1. Let X be the set of all positive integers and let $\mathscr{R}$ be the class of all finite sets and their complements. Show that this class is a ring but not a σ-ring. Define the set function μ by setting $\mu(A) = 0$ or ∞ according as A is a finite or infinite set of $\mathscr{R}$. Is μ a measure?

7.2-2. Let μ and ν be two measures on a σ-ring. Show that $\mu + \nu$ is also a measure and that $\nu \prec \mu + \nu$.

7.2-3. Let X be some set. Suppose $Y \subset X$ is a (Borel type) measure space $(Y, \mathscr{B}_Y, \mu_Y)$. Show that X can be made into a (Borel type) measure space $(X, \mathscr{B}, \mu)$ by defining $\mathscr{B}$ as the class of all subsets of X such that their intersection with Y is contained in $\mathscr{B}_Y$, and by setting $\mu(B) = \mu_Y(B \cap Y)$ for each $B \in \mathscr{B}$.

7.2a. General Properties of Measures

In this subsection we study basic properties of a measure which follow, mostly, directly from the basic definition of μ and which play a central role in all practical applications. In the following, unless otherwise stated, μ is a measure on some *arbitrary* ring $\mathscr{R}$ whose elements we denote by symbols like $A, B, C, \dots$.

---

†If B is a countable *disjoint* union of sets which have the form $B_1 \times B_2$, then this assertion follows directly from countable additivity and permits the straightforward calculation of $\mu(B)$. For more general sets B, the statement can be proven by using the extension Theorem 7.2a(7) to be given below. But there is no prescription for the actual calculation of $\mu(B)$ for arbitrary $B \in \mathscr{B}$. The proofs of these considerations are far from simple, and, honestly speaking, need much more apparatus than what we so far developed. The interested reader may come back later to these questions and is referred to Halmos[11].

THEOREM 7.2a(1). μ is a monotone function, i.e. if $A \subset B$, then $\mu(A) \leqslant \mu(B)$.

Proof. $B - A \in \mathcal{R}$, so that using additivity, we can write $\mu(B) = \mu(A) + \mu(B - A)$, and both terms on the r.h.s. are of course nonnegative.

THEOREM 7.2a(2). μ is a subtractive function, by which we mean that if $A \subset B$ and $\mu(A) < \infty$, then $\mu(B - A) = \mu(B) - \mu(A)$.

Proof. Looking at the equation in the proof of the preceding theorem, we may subtract $\mu(A) < \infty$ from both sides.

THEOREM 7.2a(3). If $A \in \mathcal{R}$, if (B_k) is a sequence of sets in $\mathcal{R}$, and if $A \subset \cup_k B_k$, then, $\mu(A) \leqslant \Sigma_k \mu(B_k)$.

The proof is not quite trivial (cf. Halmos[11], p. 37) because disjointness of the sets B_k is not stipulated. We note that (as the notation already suggests) the theorem is a fortiori valid for a finite sequence of sets B_k. We also emphasize that, in particular, the theorem holds true if $A = \cup_k B_k$ (improper inclusion), so that we always have

$$\mu\left(\bigcup_k B_k\right) \leqslant \sum_k \mu(B_k)$$

for any countable class B_k ($k = 1, 2, \ldots$) of sets, whether they are pairwise disjoint or not. This is a generalization of countable additivity.

THEOREM 7.2a(4). If $A \in \mathcal{R}$, if (B_k) is a pairwise disjoint sequence of sets in $\mathcal{R}$, and if $\cup_k B_k \subset A$, then $\Sigma_k \mu(B_k) \leqslant \mu(A)$.

Proof. To simplify matters, we make the (unnecessary) assumption that $\mathcal{R}$ is a σ-ring. Then surely $\cup_k B_k \in \mathcal{R}$ and so $\Sigma_k \mu(B_k) = \mu(\cup_k B_k) \leqslant \mu(A)$, where we used monotonity.

THEOREM 7.2a(5). Let (B_n) be an increasing sequence of sets in $\mathcal{R}$ (i.e. let $B_n \subset B_{n+1}$ for $n = 1, 2, \ldots$) and suppose† $\cup_{n=1}^{\infty} B_n \in \mathcal{R}$. Then $\mu(\cup_{n=1}^{\infty} B_n) = \lim_{n \to \infty} \mu(B_n)$.

Proof. Observe that the sets $B_0 \equiv \emptyset$, $B_1 - B_0$, $B_2 - B_1$, $B_3 - B_2$, ... are all disjoint, and that $\mu(\cup_{n=1}^{\infty} B_n) = \mu(\cup_{n=1}^{\infty} (B_n - B_{n-1}))$, so that using countable additivity,

---

†If $\mathcal{R}$ is a σ-ring or Borel field, this part of the assumption is of course redundant.

$$\mu(\bigcup_{n=1}^{\infty} B_n) = \sum_{n=1}^{\infty} \mu(B_n - B_{n-1}) \equiv \lim_{k \to \infty} \sum_{n=1}^{k} \mu(B_n - B_{n-1})$$

$$= \lim_{k \to \infty} \mu(\bigcup_{n=1}^{k} (B_n - B_{n-1})) = \lim_{k \to \infty} \mu(B_k).$$

THEOREM 7.2a(6).   *Let (B_n) be a decreasing sequence of sets in $\mathcal{R}$ (i.e. let $B_n \supset B_{n+1}$ for $n = 1, 2, \ldots$) and suppose$\dagger$ $\bigcap_{n=1}^{\infty} B_n \in \mathcal{R}$. Furthermore, suppose that at least one of the sets B_n has finite measure. Then*

$$\mu(\bigcap_{n=1}^{\infty} B_n) = \lim_{n \to \infty} \mu(B_n).$$

We omit the somewhat technical and uninspiring proof, which can be based on Theorems 7.2a(1), 7.2a(2), and 7.2a(5).

Apart from these simple and very handy theorems, we shall need two, rather sophisticated results concerning the *extension of measures.*

As it will transpire in our later work (especially in applications to functional analysis), it is often unimportant to know "where the measure came from" or what the actual numerical value of $\mu(B)$ is for different, specified measurable sets B. Often all that matters is to know that a measure is defined on our space. Nevertheless, it is worthwhile to know how measures can be constructed in a systematic manner. In many cases it is rather simple to construct explicitly a measure on some simple ring, but it may become necessary to extend this to some Borel field or even to some more general measurable space. This procedure is facilitated by the following

THEOREM 7.2a(7).   *Let μ be a measure on an arbitrary ring $\mathcal{R}$ and let $\mathcal{S}(\mathcal{R})$ be the σ-ring generated by $\mathcal{R}$. Then there exists a measure $\bar{\mu}$ on $\mathcal{S}(\mathcal{R})$ such that $\bar{\mu}(A) = \mu(A)$ for all $A \in \mathcal{R}$ (i.e. $\bar{\mu}$ is an extension of μ). Furthermore, if μ on $\mathcal{R}$ was σ-finite (totally σ-finite), then the same holds for $\bar{\mu}$ on $\mathcal{S}(\mathcal{R})$ and, in addition, the extension $\bar{\mu}$ of μ is unique.*

It is of interest that once we have a measure on a σ-ring (or more specifically, on a Borel field), then, in general, a further extension is possible to an even larger class of sets. This is formalized by the following

THEOREM 7.2a(8).   *Let μ be a measure on a σ-ring $\mathcal{S}$. Consider the collection $\bar{\mathcal{S}}$ of all sets which have the form $B \cup N$, where $B \in \mathcal{S}$ and N is an arbitrary subset of some arbitrary null set of $\mathcal{S}$. Then the collection $\bar{\mathcal{S}}$ is*

$\dagger$Again, for a σ-ring or Borel field, this is automatically true.

a σ-ring. Furthermore, the set function $\bar{\mu}$ on $\bar{\mathscr{S}}$ defined by $\bar{\mu}(B \cup N) = \mu(B)$, is a measure on $\bar{\mathscr{S}}$ and it is clearly an extension of μ.

We omit the very involved proofs of the above two theorems† and restrict ourselves to some comments regarding the last one.

(a) The new σ-ring $\bar{\mathscr{S}}$ contains all sets of $\mathscr{S}$ (take $N = \emptyset$) plus all subsets N of all sets A of $\mathscr{S}$ which had μ-measure zero (take $B = \emptyset$). Of course, some of these additionally constructed sets may have been already present in $\mathscr{S}$.

(b) If $A \in \mathscr{S}$ with $\mu(A) = 0$ and $N \subset A$, then $\bar{\mu}(N) = 0$, because taking $B = \emptyset$ in $\bar{\mu}(B \cup N)$, we get $\bar{\mu}(N) = \mu(\emptyset) = 0$. Now, quite generally, let ν be a measure on an arbitrary ring $\mathscr{R}$, and suppose that $\nu(A) = 0$ and $N \subset A$ always imply that $N \in \mathscr{R}$. Then we say that ν is a *complete measure* on $\mathscr{R}$. It then actually follows that $\nu(N) = 0$, because $\nu(N) \leqslant \nu(A) = 0$ (by Theorem 7.2a(1)) and also, trivially, $\nu(N) \geqslant 0$. We now see that *the measure $\bar{\mu}$ on $\bar{\mathscr{S}}$* (as defined by the Theorem 7.2a(8)) *is a complete measure.* It is also customary to call a measure space which has a complete measure a *complete measure space.* Thus, in essence, Theorem 7.2a(8) tells us that *any measure space can be completed.*

(c) It follows from the construction of $\bar{\mathscr{S}}$ that any given set C of $\bar{\mathscr{S}}$ differs from some set B of $\mathscr{S}$ by at most a set N which has measure zero. That is, given $C \in \bar{\mathscr{S}}$, there exists a set $B \in \mathscr{S}$, such that $\bar{\mu}(C - B) \equiv \bar{\mu}(N) = 0$.

(d) Suppose $\mathscr{S} \equiv \mathscr{B}$ was actually a Borel field, i.e. $(X, \mathscr{B}, \mu)$ was a Borel-type measure space. Then its completion $(X, \bar{\mathscr{B}}, \bar{\mu})$ is clearly also a Borel-type measure space (i.e. $\bar{\mathscr{B}}$ is a Borel field). Nevertheless, it is traditional under these circumstances to reserve the term "Borel set" for members of the original class $\mathscr{B}$, and refer to arbitrary members of $\bar{\mathscr{B}}$ by the generic term "$\bar{\mu}$-measurable sets."

PROBLEMS

7.2a-1. Let μ be a measure on some ring, let $A, B \in \mathscr{R}$. Show that $\mu(A) + \mu(B) = \mu(A \cup B) + \mu(A \cap B)$. (*Hint*: Represent $A \cup B$ as a certain *disjoint* union of sets.) (*Remark*: Rewriting the theorem in the form

$$\mu(A \cup B) = \mu(A) + \mu(B) - \mu(A \cap B),$$

we have displayed one of the most remarkable features of combining measures of sets, which actually explains the nomenclature "measure." Observe that it is

†See, for example, Halmos[11], p. 54.

precisely this formula by which one calculates the area of the union of two arbitrary pointsets in the plane or the joint mass of two sets of masspoints.)

7.2a-2. Let μ be a measure on a σ-ring. Show that the class of all sets of finite measure is a ring. Show that the class of all sets of σ-finite measure is a σ-ring.

7.2a-3. Let $\mathcal{R}$ be a ring and μ a measure on $\mathcal{R}$. For any two sets A and B of $\mathcal{R}$ write $A \sim B$ iff $\mu[(A - B) \cup (B - A)] = 0$. Show that $A \sim B$ is an equivalence relation on $\mathcal{R}$. Let $A \sim B$ and show that $\mu(A) = \mu(B) = \mu(A \cap B)$.

7.2a-4. Let $(X, \mathcal{B}, \mu)$ be a totally finite measure space. Define the function $d: \mathcal{B} \times \mathcal{B} \to \mathbf{R}$ by setting $d(A, B) = \mu(A - B) + \mu(B - A)$ for all $A, B \in \mathcal{B}$. Show that d has all properties of a metric but one. Consider now the equivalence classes of $\mathcal{B}$ defined by the equivalence relation of the preceding problem. Let $\hat{d}([A], [B]) = d(A, B)$, where $[A]([B])$ is an equivalence class with the representative element $A(B)$. Show that the collection of all equivalence classes is a metric space with the metric $\hat{d}$.

7.2a-5. Show that the metric space defined in the preceding problem is complete.

7.2b. Lebesgue Measure

We are now in a position to discuss in some detail the most important (and historically first) measure space and its measure.

Let $X = \mathbf{R}$ be the real line. Let $\mathcal{E}$ be the class of all semiclosed bounded intervals $[a, b)$. This class $\mathcal{E}$ of sets is, of course, not yet a ring, but we can define on $\mathcal{E}$ a set function μ by putting

$$\mu([a, b)) = b - a. \tag{7.1}$$

Note that this map associates to each interval $[a, b)$ its length. Clearly, $\mu \geq 0$. If $b = a$, then $[a, a) = \emptyset$, so that $\mu(\emptyset) = 0$. Now, if $[a, b)$ and $[c, d)$ are disjoint, we *define* $\mu([a, b) \cup [c, d)) = (b - a) + (c - d)$. (This conforms with the idea of length.) Next we observe that the class consisting of all finite disjoint unions of the sets in $\mathcal{E}$ is a ring $\mathcal{R}$. This follows from Problem 7.1-1 and from the easily proven lemma that any finite union of semiclosed intervals can be represented by a disjoint finite union of such intervals. Then, denoting for simplicity the interval $[a_k, b_k)$ by A_k, we consider $A = \bigcup_{k=1}^{n} A_k$ with $A_k \cap A_i = \emptyset$ for $k \neq i$ and set for such an arbitrary $A \in \mathcal{R}$,

$$\mu(A) = \sum_{k=1}^{n} \mu(A_k). \tag{7.2}$$

This definition is perfectly consistent because of the above defined finite additivity of μ. Thus, it is obvious that *we have a measure on the ring $\mathcal{R}$*.

Now we wish to *extend this measure from $\mathcal{R}$ to the class $\mathcal{B}$ of all*

(*usual*) *Borel sets of the real line* (cf. p. 299). Theorem 7.2a(7) assures us
that there *is* such an extension, because $\mathcal{B}$ is the σ-ring generated by the
ring $\mathcal{R}$. (This is so because, as we know from Example ζ on p. 303, already
the class $\mathcal{E}$ generates $\mathcal{B}$.) This extension is unique and σ-finite. (This
follows from the fact that the measure on $\mathcal{R}$ was obviously finite, hence
eo ipso σ-finite.) The entire real line $X = \mathbf{R}$ has infinite measure (because
$\mathbf{R}$ is a countable union of semiclosed disjoint intervals with rational
endpoints, each having a finite length, so the measure on $(\mathbf{R}, \mathcal{B})$ is not
finite). But, as the same argument shows, the measure is actually *totally
σ-finite.*

The totally σ-finite measure on $(\mathbf{R}, \mathcal{B})$ which, by the above argument is
the extension of the previously constructed measure μ on $\mathcal{R}$, is called the
Lebesgue measure for the Borel field of the real line. Since we are no
longer interested in $\mathcal{R}$, we shall denote this extended measure on the
Borel space $(\mathbf{R}, \mathcal{B})$ also by μ. We now study what the value of μ is for
various typical Borel sets of the Borel-type measure space $(\mathbf{R}, \mathcal{B}, \mu)$.

To start with, if x is a point of $\mathbf{R}$, we can write $\{x\} = \bigcap_{n=1}^{\infty} [x, x + 1/n)$,
and the Borel sets $[x, x + 1), [x, x + \frac{1}{2}), \ldots$ form a decreasing sequence.
Therefore, by Theorem 7.2a(6), $\mu(\{x\}) = \lim_{n\to\infty} \mu([x, x + 1/n)) = \lim_{n\to\infty} 1/n = 0$, where we used Eq. (7.1). Thus, every singleton set has
measure zero, and, more generally, using countable additivity, we see that
every countable set of points is a null set. In particular, *the set of all
rationals has measure zero.*

Now, we may write for any closed interval $[a, b] = [a, b) \cup \{b\}$, and
therefore we find that $\mu([a, b]) = b - a$. Similarly, for an open interval,
$(a, b) = [a, b) - \{a\}$ where $[a, b) \supset \{a\}$, so that by the subtractive property
of the measure, $\mu((a, b)) = b - a$. In a similar way we see that
$\mu((a, b]) = b - a$. In summary, *the Lebesgue measure of any type of interval
is precisely the length of the corresponding closed interval.*

We proceed to calculate the measure of more complicated sets. First,
we are interested in *arbitrary open sets.* The measure of such a set may be
calculated with the help of the following lemma: *Any open set of $\mathbf{R}$ is the
union of countably many closed intervals whose interiors are disjoint.*
Now suppose that V is a given open set, which by this lemma can be
represented as

$$V = \bigcup_{k=1}^{\infty} [a_k, b_k] = \bigcup_{k=1}^{\infty} (a_k, b_k) \bigcup_{k=1}^{\infty} \{a_k\} \bigcup_{k=1}^{\infty} \{b_k\}$$

where the three sets on the r.h.s. are disjoint and the last two have
measure zero. Countable additivity then gives

$$\mu(V) = \sum_{k=1}^{\infty} (b_k - a_k). \qquad (7.3)$$

Next we consider *arbitrary closed sets*. Assuming first that C is a bounded closed set, take a bounded open set V such that $C \subset V$. Then $C = V - (V - C)$, so by the subtractive property, $\mu(C) = \mu(V) - \mu(V - C)$. Here both V and $V - C = V \cap C^c$ are open, so that their measures can be calculated by the result of the preceding paragraph. Actually, both terms and thus $\mu(C)$ will be finite. On the other hand, if C is an unbounded closed set, then there exists an unbounded connected open set V such that $C \supset V$, and since then $\mu(V) = \infty$, we have $\mu(C) = \infty$ for an unbounded closed set.

Encouraged by these simple calculations, the reader may wonder whether there exists a prescription for calculating, at least in principle, the Lebesgue measure of an *arbitrary* Borel set of **R**. Surprisingly, the answer is affirmative. Let B be an arbitrary Borel set and consider the class $\{\ldots V_\alpha \ldots\}$ of all open sets V_α that contain the given B. Then, the measure of B can be computed as the infinum of the measures of the sets V_α, i.e.

$$\mu(B) = \inf_{V_\alpha \supset B} \mu(V_\alpha). \qquad (7.4)$$

The *proof* of this statement is based on the observation that (a) the r.h.s. of Eq. (7.4) (determined via Eq. (7.3)) has all the requisite properties of a measure and (b) if B is an open set, Eq. (7.4) reproduces the already known result (7.3); hence, for the class of open intervals (a, b) and so for the class of semiclosed intervals $[a, b)$ it reproduces μ as originally defined. Thus, Eq. (7.4) defines an extension of the measure μ from the ring $\mathcal{R}$ to the Borel field $\mathcal{B}$, and since we know that this extension is unique, $\mu(B)$ as given by Eq. (7.4) is *the* Lebesgue measure on $(\mathbf{R}, \mathcal{B})$, q.e.d. We omit the somewhat lengthy (but not difficult) details of the proof.

In summary, we know all about the Lebesgue measure space $(\mathbf{R}, \mathcal{B}, \mu)$ there is to know. But we may become bolder and try to extend the measure to a larger class of sets on **R**. This is possible, on account of Theorem 7.2a(8). We simply declare all subsets of all Borel null sets to be measurable and hence, to have measure zero (cf. remark (b) following Theorem 7.2a(8)). The measure space† $(\mathbf{R}, \bar{\mathcal{B}}, \bar{\mu}) \equiv (\mathbf{R}, \mathcal{S}, \mu)$ so obtained is slightly bigger than the original space (i.e. $\mathcal{S} \supset \mathcal{B}$), but actually, in view of remark (c), $\mathcal{S}$ is also a Borel field. It is customary to call $\mathcal{S}$ the class of

---

†Without causing confusion, the extension of μ from $(\mathbf{R}, \mathcal{B})$ to $(\mathbf{R}, \mathcal{S})$ is also denoted by μ.

Lebesgue measurable sets of the real line and to refer to μ simply as the *Lebesgue measure on the reals.* By its construction, this measure is the *completion* of the original measure on the "usual" Borel field of the reals,† which, in turn, was the extension of the measure μ defined on the ring $\mathcal{R}$ of finite (disjoint) unions of semiclosed intervals. Thus, *we succeeded in generalizing the simple concept of the "length of an interval" to an enormously large class of sets on the real line.* But the reader should be aware of the fact that not all sets of the real line are Lebesgue measurable: it is possible to construct (with hard labor, though) sets that are not in $\mathcal{S}$, i.e. for which μ is not defined.

In passing we make the comment that, perhaps surprisingly, Lebesgue measurable sets do not differ significantly from open sets. Indeed, it can be shown that *if B is an arbitrary Lebesgue measurable set* (i.e. if $B \in \mathcal{S}$), *then, for any given $\epsilon > 0$, there exists an open set V* (i.e. a set V in the usual topology τ of **R**) *such that $V \supset B$ and $\mu(V) - \mu(B) < \epsilon$. Hence,* (if B is bounded), $\mu(V - B) < \epsilon$. In other words, a Lebesgue measurable set differs from some open set as little as we wish.

We summarize some salient features of μ which will be needed in much of our subsequent work.

The Lebesgue measure μ on the real line is a *complete totally σ-finite measure.* We have $\mu(\mathbf{R}) = \infty$, and also there are many *unbounded*‡ sets which have infinite measure. (For example, every ray $\{x \mid x > a, a$ given real$\}$ has infinite measure.) On the other hand, there do exist unbounded sets which have finite (maybe even zero) measure. (For example, the set of rationals.) Every countable set has zero measure. But the converse does not hold: There *do* exist uncountable sets which are null sets! (For example, let A be the uncountable set of reals between 0 and 1 which, written in the usual decimal form, have no digit equal to 1. This set can be seen to be measurable and to have measure zero.)

We conclude our introduction to the theory of Lebesgue measure on the real line by certain comments which will not play a role in the following but which deal with concepts often met in the literature.

Let μ^* be a set function on the class of *all* subsets of the reals (i.e. on $\mathcal{P}(\mathbf{R})$) defined in the following way: if $\{\ldots V_\alpha \ldots\}$ is the class of all open sets such that $A \subset V_\alpha$ (for all α), then§

---

†Incidentally, it can be shown that this completion is not trivial, i.e. that there exist Lebesgue measurable sets which do not belong to the original Borel field $\mathcal{B}$.

‡From Eqs. (7.4) and (7.3) it should be clear that bounded sets must have finite measure!

§Observe that on the r.h.s. μ denotes the usual Lebesgue measure, so that $\mu(V_\alpha)$ is well defined.

$$\mu^*(A) = \inf_{V_\alpha \supset A} \mu(V_\alpha).$$

The set function μ^* is called the *outer measure* on the reals (more precisely on $\mathscr{P}(\mathbf{R})$) and it has *some*, but not all properties of a measure. Of course, if A happens to be a Lebesgue measurable set, then, by Eq. (7.4), $\mu^*(A)$ gives precisely its Lebesgue measure $\mu(A)$. Next, define the set function μ_* on the class of *all* subsets of the reals by putting, for any $A \in \mathscr{P}(\mathbf{R})$,

$$\mu_*(A) = \sup_{C_\alpha \subset A} \mu(C_\alpha),$$

where $\{\ldots C_\alpha \ldots\}$ is the class of all closed sets of $\mathbf{R}$ which are contained in the given A. This μ_* is called the *inner measure* on the reals (i.e. on $\mathscr{P}(\mathbf{R})$). It can be seen that if A happens to be Lebesgue measurable, then again $\mu(A) = \mu_*(A)$. The statements about μ^* and μ_* have a converse so that the following theorem holds: *A set $S \subset \mathbf{R}$ is Lebesgue measurable iff $\mu^*(S) = \mu_*(S)$, and then $\mu(S) = \mu^*(S) = \mu_*(S)$.* This characterization of the Lebesgue measure plays two roles. First, *it can be used as a criterion for establishing Lebesgue measurability*, and second, *it may serve as an alternative definition of Lebesgue measure*. This new definition of Lebesgue measure (which actually was the *original* definition and the beginning of the whole story of measure theory) has the advantage that it obviates the necessity of first constructing the usual Borel space, building up measure on $\mathscr{B}$, and then finally extending it to the larger class $\mathscr{L}$. All that is needed is a basic definition of measure for all open sets and for all closed sets (which can be given directly from the length of open intervals, cf. Eq. (7.3) and the following paragraph). However, the more modern construction which we followed is less artificial, conforms with the spirit of modern mathematics and appears as a special case of much more general constructions.

We conclude this subsection with the brief discussion of *Lebesgue measure for arbitrary Euclidean spaces* $\mathbf{R}^n$. We start with the collection $\mathscr{E}$ of semiclosed rectangles $[\mathbf{a}, \mathbf{b})$ (see p. 300) and define

$$\mu^n([\mathbf{a}, \mathbf{b})) = \prod_{k=1}^{n} (\beta_k - \alpha_k).$$

Clearly, this quantity is precisely the "volume" of the n-dimensional rectangle. We then may proceed exactly as we did for $\mathbf{R}$. However, the actual labor may be saved by noting that the Borel-type measure space $(\mathbf{R}^n, \mathscr{B}^n, \mu^n)$ so obtained is precisely the product measure space

$$(\mathbf{R} \times \cdots \times \mathbf{R}, \mathscr{B} \times \cdots \times \mathscr{B}, \mu \times \cdots \times \mu),$$

where $\mathscr{B}$ is the Borel field of the reals and μ the corresponding Lebesgue measure. The complete measure space $(\mathbf{R}^n, \mathscr{S}^n, \mu^n)$ is then obtained by the extension Theorem 7.2a(8).

The advantage of looking at $(\mathbf{R}^n, \mathscr{S}^n, \mu^n)$ in this way is that the numerical measure of many sets can be easily computed from the knowledge of Lebesgue measure on $\mathbf{R}$. For example, suppose that $B_1, B_2, \ldots, B_n$ are known to be measurable sets of the real line. Then $B = B_1 \times B_2 \times \cdots \times B_n$ is a measurable set of $\mathbf{R}^n$ and $\mu^n(B) = \mu(B_1)\mu(B_2) \ldots \mu(B_n)$. This consideration shows (among other things) that *the Lebesgue measure of any hyperplane in $\mathbf{R}^n$ is zero*. For example, in $\mathbf{R}^3$, every plane, line,† and point is a null set. The same holds for countable unions of such sets and, of course, for all of their subsets.

Finally, we note the following, not too difficult result. It can be shown that the Lebesgue measure of any arbitrary connected set of $\mathbf{R}^n$ is precisely the usual "volume" of that set, which is defined in elementary geometry as the common limit of the inscribed and circumscribed "boxed volumes" associated with the set. This tells us that μ^n is indeed the powerful generalization of ordinary volume.

PROBLEMS

7.2b-1. Let A be an arbitrary Lebesgue measurable set of the reals. Consider the subset of all irrational numbers contained in A. Show that this set is measurable and calculate its measure. In particular, what is the measure of the set of all irrational numbers?

7.2b-2. Construct a Lebesgue measurable set S in $[0, 1]$ such that for every open set $V \subset [0, 1]$, both $\mu(S \cap V) > 0$ and $\mu(S^c \cap V) > 0$. (*Hint*: Recall that there exists a countable open base. The preceding problem is also helpful.)

7.2b-3. Let S be a Lebesgue measurable set in $[0, 1]$ with the following property: If $x \in S$ and if y is the number obtained from x by changing a *finite* number of digits in the decimal expansion of x, then $y \in S$. Show that $\mu(S)$ is either 0 or 1.

7.2b-4. Show that if A is Lebesgue measurable, then there exist two Borel sets B_1 and B_2 of the real line such that $B_1 \supset A$, $B_2 \subset A$ and $\mu(B_1) = \mu(B_2) = \mu(A)$.

7.2b-5. Let A be a Lebesgue measurable set of the real line. Let B be the set which is obtained from A by adding to each number $x \in A$ the same fixed number c. (That is, $B = \{y \mid y = x + c, x \in A, c \text{ fixed}\}$.) Show that B is measurable and that $\mu(B) = \mu(A)$. (*Remark*: This problem is summarized by saying that Lebesgue measure is *translation invariant*.)

7.2b-6. Give an example of an open, connected, unbounded set of $\mathbf{R}^2$ which has finite and nonzero Lebesgue measure.

---

†Note that in $\mathbf{R}$, the line is *not* a null set, in fact $\mu(\mathbf{R}) = \infty$.

7.2b-7. Let I_1 and I_2 be the unit interval on the X- and Y-axis of $\mathbf{R}^2$, respectively. Let M be a nonmeasurable subset of I_1 (with respect to Lebesgue measure on the reals) and let y be a point of I_2. Is the set $M \times \{y\}$ of $\mathbf{R}^2$ measurable with respect to the product measure? Is it a subset of a null set of the product space? Using the answers to these questions, decide whether the product measure on $\mathbf{R}^2$ is complete? (*Remark*: This problem illustrates that the product of two σ-finite and complete measures need not be complete.)

7.2c. Lebesgue–Stieltjes Measures

In the following we introduce a simple generalization of Lebesgue measure which plays an important role in functional analysis. We shall restrict ourselves to the real line.

Let $\rho : \mathbf{R} \to \mathbf{R}$ be a real valued function defined on the real line which has the following properties:

(a) ρ is *monotone* nondecreasing, i.e. $\rho(x) \geq \rho(y)$ whenever $x \geq y$.

(b) ρ is *right continuous*† in the sense that‡

$$\lim_{\epsilon \to +0} \rho(x + \epsilon) = \rho(x) \quad \text{for all} \quad x.$$

For convenience, we introduce the following standard notation:

$$\rho(x + 0) \equiv \lim_{\epsilon \to +0} \rho(x + \epsilon), \qquad \rho(x - 0) \equiv \lim_{\epsilon \to +0} \rho(x - \epsilon).$$

Let us now consider once again the collection $\mathscr{C}$ of all semiclosed bounded intervals $[a, b)$ of the real line and define on $\mathscr{C}$ the set function μ by putting

$$\mu([a, b)) = \rho(b - 0) - \rho(a - 0). \tag{7.5}$$

Since ρ is monotone, μ is certainly nonnegative. Furthermore, since $[a, a) = \emptyset$, trivially we see that $\mu(\emptyset) = 0$.

Now we proceed exactly as we did when we constructed the Lebesgue measure starting with the definition (7.1). That is, we postulate finite additivity for our new μ, thereby obtaining a measure on the ring $\mathscr{R}$, from which we extend it to the usual Borel field § and then, if desired, a further

---

†Of course, ρ might be continuous in the usual sense (i.e. both right and left continuous), but we do not demand this.

‡We remind the reader that the customary symbolism $\lim_{\epsilon \to +0} f(x + \epsilon) = \alpha$ applied to a function f on the reals means that the sequence of functional values $v_n \equiv f(x + \epsilon_n)$ is convergent for every choice of a decreasing sequence ϵ_n of arbitrary *nonnegative* real numbers which tends to zero ($\epsilon_n \to 0$), and the common limit of all such v_n sequences is the functional value α. Expressions like $\lim_{\epsilon \to -0} f(x + \epsilon)$, etc. are defined in a similar way.

§The extension is unique because, as easily seen, our new measure on $\mathscr{R}$ is σ-finite.

extension can be made to obtain a complete measure space.[†] The measure so defined on the basis of Eq. (7.5) is called the *Lebesgue–Stieltjes measure* on the real line *generated by the* (*right continuous*) *function ρ.* Here we observe that, if in particular, we take $\rho(x) = x$, this (continuous) function will generate precisely the old Lebesgue measure (see Eq. (7.5) and observe that $\rho(x - 0) = \rho(x)$ because of continuity). Thus, the Lebesgue–Stieltjes measures are a *generalization* of the ordinary Lebesgue measure. We now proceed to calculate the Lebesgue–Stieltjes measure for specific sets.

(a) Since

$$\mu(\{x\}) = \mu\left(\bigcap_{n=1}^{\infty} [x, x + \frac{1}{n})\right) = \lim_{n \to \infty} \mu\left([x + \frac{1}{n})\right)$$
$$= \lim_{n \to \infty} \rho\left(x + \frac{1}{n} - 0\right) - \rho(x - 0) = \rho(x + 0 - 0) - \rho(x - 0),$$

we see that for a singleton set

$$\mu(\{x\}) = \rho(x) - \rho(x - 0).$$

Here we observe that (unless ρ happens to be continuous at x in the strict sense), the set $\{x\}$ has a nonzero measure, contrary to the case of ordinary Lebesgue measure. Similarly, countable sets may have nonzero Lebesgue–Stieltjes measure.[‡]

(b) From $\mu([a, b]) = \mu([a, b)) + \mu(\{b\})$
$$= \rho(b - 0) - \rho(a - 0) + \rho(b) - \rho(b - 0) \text{ it follows that}$$
$$\mu([a, b]) = \rho(b) - \rho(a - 0).$$

(c) In a similar manner we find

$$\mu((a, b)) = \rho(b - 0) - \rho(a).$$

(d) Finally,

$$\mu((a, b]) = \rho(b) - \rho(a).$$

We note here that the last three formulae are often written as $\mu([a, b]) = \rho(b + 0) - \rho(a - 0)$, $\mu((a, b)) = \rho(b - 0) - \rho(a + 0)$, and $\mu((a, b]) = \rho(b + 0) - \rho(a + 0)$, respectively, which is permitted because the right continuity of ρ.

[†]This complete measure space will have, in general, a different class of measurable sets than the completed Lebesgue measure had, because the null sets with respect to the new measure will be different from those of the ordinary Lebesgue measure, see below.

[‡]Of course, if ρ is everywhere continuous, then all countable sets are null sets.

We also observe that, in general, the Lebesgue–Stieltjes measure of the four different types of intervals is different, unless ρ happens to be continuous, when actually the measure of all these intervals is simply $\rho(b) - \rho(a)$.

The calculation of the Lebesgue–Stieltjes measure for arbitrary open and closed sets, and then for completely arbitrary measurable sets, is done exactly in the same manner as for the Lebesgue measure.

The Lebesgue–Stieltjes measure is always totally σ-finite. *It may even be totally finite.* For this to hold, it is sufficient that ρ be bounded. Since semicontinuity assures that ρ is certainly bounded at finite points x, the boundedness condition is guaranteed by the simpler requirement that† $\lim_{x \to -\infty} \rho(x)$ and $\lim_{x \to +\infty} \rho(x)$ be finite (i.e. exist as real numbers). It is customary to denote these limits by $\rho(-\infty)$ and $\rho(+\infty)$, respectively, and then we easily see that $\mu(\mathbf{R}) = \rho(+\infty) - \rho(-\infty)$. This shows that $\mathbf{R}$ (hence any measurable subset of $\mathbf{R}$) has a finite measure.

The two main reasons why a Lebesgue–Stieltjes measure may be preferred to ordinary Lebesgue measure are that ($\boldsymbol{\alpha}$) singleton sets may have nonvanishing measure and ($\boldsymbol{\beta}$) the entire real line may have a finite measure. On the other hand, the *Borel sets* for the Lebesgue–Stieltjes measure are the same as for the Lebesgue measure.‡

An often occurring type of a Lebesgue–Stieltjes measure is one which is generated by a *countably piecewise constant* ρ. This means that ρ is constant except at a countable number of points x_k, where it has a jump. Such a "ladder like" ρ is illustrated in Fig. 7.2. We denote the jump at x_k by ρ_k, so that

$$\rho(x_k) = \rho(x_k - 0) + \rho_k$$

(right continuity is assumed). The Lebesgue–Stieltjes measure generated by a countably piecewise constant ρ is called a *discrete measure* μ, and we say that μ is *concentrated* in the points x_k with *weights* ρ_k.

Discrete measures lead to a very big class of null sets. In fact, any measurable set which is contained in one of the open intervals (x_k, x_{k+1}) is a null set. Furthermore, the measure of a singleton is zero unless it is one of the points of concentration, and then $\mu(\{x_k\}) = \rho_k$.

†We take here the opportunity to remind the reader of the meaning of the customary symbolism $\lim_{x \to y} f(x) = \alpha$. Let x_n be an *arbitrary* sequence which converges to y and suppose the sequence of functional values $v_n \equiv f(x_n)$ converges to α (for *all* $x_n \to y$ sequences). Then we simply write $\lim_{x \to y} f(x) = \alpha$. If, on the other hand, $v_n \equiv f(x_n)$ is a positive (negative) unbounded sequence (for arbitrary sequences x_n with $x_n \to y$) then we write $\lim_{x \to y} f(x) = +\infty \, (-\infty)$, respectively.

‡See, however, footnote on p. 321.

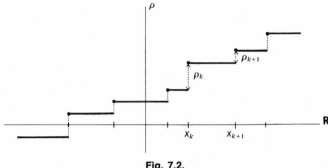

Fig. 7.2.

A particularly simple discrete measure is generated by the *step function* (Fig. 7.3)

$$\rho(x) = \theta(x) = \begin{cases} +1 & \text{if } x \geqslant 0 \\ 0 & \text{if } x < 0. \end{cases}$$

All measurable sets which do not contain the point $x = 0$ have measure zero, $\mu(\{0\}) = 1$, and the same holds for all measurable sets that contain $x = 0$. In particular, $\mu(\mathbf{R}) = 1$.

To conclude this subsection, we should point out that there is nothing sacred about having demanded *right* continuity of a generating function ρ. Left continuity, i.e. the behavior

$$\lim_{\epsilon \to +0} \rho(x - \epsilon) = \rho(x)$$

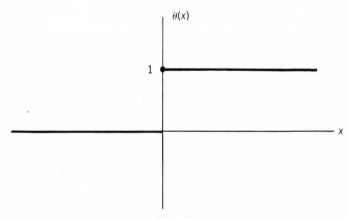

Fig. 7.3.

will also do, provided we formally† change the definition (7.5) to

$$\mu([a, b)) = \rho(b) - \rho(a).$$

We then find similar formal changes for other simple sets, i.e.

$$\mu(\{x\}) = \rho(x + 0) - \rho(x),$$
$$\mu([a, b]) = \rho(b + 0) - \rho(a),$$
$$\mu((a, b)) = \rho(b) - \rho(a + 0),$$
$$\mu((a, b]) = \rho(b + 0) - \rho(a + 0).$$

For discrete measures generated by left continuous functions we define the weights ρ_k by setting‡

$$\rho(x_k) = \rho(x_k + 0) - \rho_k.$$

PROBLEMS

7.2c-1. Prove that two discrete measures μ_1 and μ_2 on the real line are equivalent iff they are concentrated in the same points and their respective weights are not zero. (Why must the second half of the criterion be stated explicitly?)

7.2c-2. Let μ be a discrete Lebesgue–Stieltjes measure on the real line with weights ρ_k at x_k. Show that for any Borel set B we have $\mu(B) = \Sigma_{k_i}\rho_{k_i}$, where the sum ranges over all k_i which correspond to points of concentration that belong to B. (Of course, $\{\ldots x_{k_i}\ldots\} \subset \{\ldots x_k\ldots\}$.)

7.2c-3. Let a Lebesgue–Stieltjes measure on the real line be generated by a (right-continuous) function $\rho(x)$. Show that this function can be uniquely decomposed as $\rho(x) = \rho_c(x) + \rho_d(x)$, where ρ_c is continuous and ρ_d is constant except for a countable set of points of discontinuity. The two parts generate a continuous and a discrete Lebesgue–Stieltjes measure, respectively.

7.2c-4. A measure on the reals with the property that every (usual) Borel set of the line is measurable and the measure of every compact Borel set is finite, is said to be a Borel measure. Demonstrate that every Lebesgue–Stieltjes measure is a Borel measure. Conversely, show that every Borel measure is the extension of a Lebesgue–Stieltjes measure. (*Hint*: Let μ_B be a given Borel measure. Define $\rho(0) = 0$, $\rho(x) = \mu_B([0, x))$ for $x > 0$, $\rho(x) = -\mu_B([x, 0))$ for $x < 0$.) (*Remark*: The notion of Borel measure can be applied, more generally, to $\mathbf{R}^n$, and even to locally compact Hausdorff spaces, where the Borel sets are defined by Theorem 7.1(4).)

---

†Actually, this does not change the numerical value of the measure of $[a, b)$ since now $\rho(x) = \rho(x - 0)$ by left continuity.

‡Otherwise ρ_k would be zero and yet $\mu(\{x_k\}) \neq 0$. Of course, one may define ρ_k in a manner which holds both for the right- and for the left-continuous case, namely by setting $\rho(x_k + 0) = \rho(x_k - 0) + \rho_k$.

7.2d. Signed and Complex Measure

For several applications in integration theory and in functional analysis, a certain generalization of the basic measure concept proves very useful. We shall briefly review this topic.

DEFINITION 7.2d(1).   *Let $\mathfrak{R}$ be a ring on some set X. Let $\mu: \mathfrak{R} \to \hat{\mathbf{R}}$ be an extended real valued function obeying the following axioms:*

(i) *$\mu(B)$ may be either positive or negative or zero, and it may assume* AT MOST *one of the values $+\infty$ or $-\infty$ (but not both).*
(ii) *$\mu(\emptyset) = 0$.*
(iii) *μ is countably additive.*

The set function μ with these properties is called a signed measure on $\mathfrak{R}$.

The novelty here is that μ is not restricted to nonnegative values. The reason why only one of the two possible values $+\infty$ and $-\infty$ is permitted is that the "sum" $(+\infty) + (-\infty)$ is not defined.

In order to avoid confusion, the term "measure" will be reserved for nonnegative measures, and otherwise we always use the explicit qualifier "signed."

For signed measures the concepts finite, σ-finite, totally finite, totally σ-finite are defined in a similar manner as for ordinary measures, except that one considers the values of $|\mu(B)|$ rather than those of $\mu(B)$. (That is, μ is finite if $-\infty < \mu < +\infty$, and so on.)

A physical example of a signed measure is the charge of a (finite) material system.† There are many mathematical reasons why signed measures should be considered. One is the following. If μ_1 and μ_2 are (ordinary) measures on a ring, then $c\mu_1 + d\mu_2$ with $c \geqslant 0$ and $d \geqslant 0$ is also a measure. But this is no longer true if c and d are permitted negative values as well. On the other hand, arbitrary finite real linear combinations $\Sigma_{k=1}^{n} c_k\mu_k$ of finite signed measures on a ring give again a finite signed measure.‡

Naturally, one will be particularly interested in signed measures defined on a measurable space $(X, \mathcal{S})$ (or specifically, on a Borel space $(X, \mathcal{B})$). Suppose we have a signed measure μ on $(X, \mathcal{S})$. It is easy to convince oneself that one can always decompose μ as $\mu^+ - \mu^-$ where μ^+ and μ^-

---

†Make this statement precise by defining the ring and the measure. Why do you have to assume that the system consists of a finite number of charged particles?
‡Thus, we are led to a real linear space whose elements are measures.

are ordinary (nonnegative) measures. In fact, in general, there are many such decompositions. However, there exists a particular decomposition characterized by the property that, whatever property μ had, at least one term in the decomposition will be a finite measure. More explicitly, we have the *Jordan decomposition*:

THEOREM 7.2d(1). *Let μ be an arbitrary signed measure on a measurable space $(X, \mathscr{S})$. Let $B \in \mathscr{S}$ and let $\{\ldots A_\alpha \ldots\}$ be the class of those measurable sets which are contained in B. Then*

$$\mu(B) = \mu^+(B) - \mu^-(B),$$

where

$$\mu^+(B) = \sup_{A_\alpha \subset B} \mu(A_\alpha),$$

$$\mu^-(B) = -\inf_{A_\alpha \subset B} \mu(A_\alpha).$$

At least one of the two measures μ^+ and μ^- is always finite. Furthermore, if μ is (totally) finite or (totally) σ-finite, so are both μ^+ and μ^-.

Remarks: (1) That μ^+ and μ^- are nonnegative measures, is obvious from their definition. That at least one of them is finite, follows easily from the fact that μ cannot have $+\infty$ *and* $-\infty$ in its range.

(2) That part of the theorem which says that if μ is finite then so are both μ^+ and μ^-, is the converse of a special case ($c_1 = -c_2 = 1$) of our preceding result on linear combinations of signed measures.

(3) Apart from handling signed measures, the importance of the theorem lies in the fact that the study of properties of signed measures and the construction of a signed measure on a measurable space can now be reduced to the already known properties of nonnegative measures.

We now proceed to a further generalization of the measure concept. Let $(X, \mathscr{S})$ be a measurable space and let μ_r and μ_i be two signed measures on it. We then call the set function† $\mu : \mathscr{S} \to \mathbf{C}$ defined by

$$\mu = \mu_r + i\mu_i$$

a *complex measure* on $(X, \mathscr{S})$.

A complex measure μ is said to be finite, σ-finite, totally finite, totally σ-finite if these properties hold, respectively, for *both* signed measures μ_r and μ_i. We note in passing that every nonnegative measure may be looked

---

†There is no point in defining complex measure for arbitrary rings, although it could be done.

upon as a special case of a signed measure, so that in the definition of the complex measure, one or both parts may happen to be ordinary measures.

Obviously, any complex measure can be decomposed into ordinary (nonnegative) measures as

$$\mu = \mu_r^+ - \mu_r^- + i\mu_i^+ - i\mu_i^-,$$

where the terms are the corresponding decompositions of μ_r and μ_i. In particular, we may take the unique Jordan decompositions.

One of the attractive properties of complex measures is the following, rather obvious result: Linear combinations $\Sigma_{k=1}^n c_k \mu_k$ of finite complex measures (defined on a given measure space) with arbitrary complex coefficients are again finite complex measures.

There exists the following "converse"

THEOREM 7.2d(2). *Let* $(X, \mathscr{S})$ *be a measurable space and let* $\mu : \mathscr{S} \to \mathbf{C}$ *be a complex valued set function on* $\mathscr{S}$ *with the following properties*:

(i) $\mu(B)$ *is bounded, i.e.* $|\mu(B)| \leq M$ *for all* $B \in \mathscr{S}$.
(ii) $\mu(\emptyset) = 0$.
(iii) μ *is countably additive*.

Then μ *has a unique decomposition* $\mu = \mu_r + i\mu_i$, *where* μ_r *and* μ_i *are both finite signed measures*.

This theorem may be used as a definition of *finite* complex measures.

PROBLEMS

7.2d-1. Let μ be a signed measure on $(X, \mathscr{B})$. Define the function $|\mu| : \mathscr{B} \to \hat{\mathbf{R}}$ by setting

$$|\mu|(B) \equiv \mu^+(B) + \mu^-(B),$$

where μ^+ and μ^- are the terms in the Jordan decomposition of μ. Show that $|\mu|$ is a measure on $(X, \mathscr{B})$. (It is called the "total variation measure.") Show that, in general, $|\mu|(B) \neq |\mu(B)|$. Suppose now that μ is finite. Show that $|\mu|$ is also finite and that, moreover, it is bounded, $|\mu|(B) \leq M$.

7.2d-2. Let μ and ν be two signed measures on $(X, \mathscr{B})$. We say that μ is inferior to ν, $\mu < \nu$, if $\mu(B) = 0$ for every measurable set B for which $|\nu|(B) = 0$. Show that this is a (partial) order relation on the set of equivalence classes of signed measures on $(X, \mathscr{B})$. Observe that this statement would be false if we had defined "$\mu < \nu$ if $\mu(B) = 0$ whenever $\nu(B) = 0$." Using the correct definition of order relation for signed measures, show that μ and $|\mu|$ are always equivalent. Show also that if $\mu < \nu$ in the above defined sense, then $|\mu| < |\nu|$. (Cf. preceding problem.)

8

Theory of Integration

The major application of measure theory (and historically the reason for its development) is its use in the formulation of a very general theory of integration. The Riemann integral, well known from basic calculus, has several shortcomings which severely limit the application of the concept of an integral both in pure mathematics (functional analysis) as well as in many applications to the sciences. The major shortcomings may be summarized as follows:

(a) The whole concept of a Riemann integral applies only to real valued† functions f *which are defined on the real line* **R** (or, more generally on $\mathbf{R}^n$). Actually, the domain of definition of f may be a suitable subset of **R** (or of $\mathbf{R}^n$), but integration is always carried out "over a closed interval of **R**" (or "over a closed connected subset of $\mathbf{R}^n$").

(b) The class of functions for which the Riemann integral may be defined is rather narrow; it consists, essentially, of the class of "piecewise continuous functions."

(c) As a consequence of (b), there are many convergent sequences of Riemann-integrable functions whose (pointwise) limit is not Riemann integrable. But even if the limit is integrable, the process of interchanging integral and limit may lead to a false result. These deficiencies result in serious difficulties not only in functional analysis but even in such simple problems as the termwise integration of infinite series.

---

†The generalization to complex valued functions is trivial: one merely splits $f = f_r + if_i$.

The modern general theory of integration which grew out of the work of Lebesgue, initiated over 70 years ago, overcomes these limitations to a great extent. Remaining first on the real line (or on R^n), the class of integrable functions has been greatly enlarged and the integral can be defined over much more complicated subsets than the closed intervals. The convergence difficulties were considerably reduced. Finally, the very concept of integration has been generalized to suitable classes of real (or complex) valued functions which are defined, not on R (or R^n), but on a completely arbitrary set X, provided the latter is equipped with a measure space structure.†

In this chapter we shall discuss these results, ignoring the historical development and surveying the field in complete generality. Nevertheless, we shall pay special attention to the Lebesgue generalization of the integral on R.

8.1 MEASURABLE FUNCTIONS

Preparatory to the general definition of the integral of a real valued function, we will have to select a broad class of functions for which integration can be usefully defined. We start with the most important (and rather simple) case when the set X on which $f: X \to R$ is defined, has been given a Borel field structure, i.e. when $(X, \mathcal{B})$ is a Borel-type measurable space.

DEFINITION 8.1(1).   *Let $(X, \mathcal{B})$ be a Borel space. Let $f: X \to R$ be a real valued function on X. Then f is said to be a measurable function iff for every (usual) Borel set M of R, the inverse image $f^{-1}(M)$ is a Borel set of $(X, \mathcal{B})$.*

Observe that here and in the following we always assume that the codomain R is equipped with the usual Borel structure, as defined in Section 7.1, p. 299 (see also Examples ζ, η, and ϑ toward the end of Section 7.1). Note also that in Definition 8.1(1) it is only the class of *original* Borel sets of R (and *not* the somewhat larger class of arbitrary Lebesgue measurable sets of R) which plays a role. If we now denote the class of usual Borel sets of R by $\mathcal{B}_R$, Definition 8.1(1) may be summarized

†Under certain circumstances it is even possible to define an integral for functions on some set X whose values lie not in R (or C), but in some other set Y. However, in this case the theory is much less developed, has fewer applications, and will not be considered in the following.

by saying that† $f: (X, \mathscr{B}) \to \mathbf{R}$ is a measurable function iff $M \in \mathscr{B}_\mathbf{R}$ implies $f^{-1}(M) \in \mathscr{B}$.

In some cases of interest the set X on which $f: X \to \mathbf{R}$ is defined, does not have a Borel field structure but we have a general measurable space $(X, \mathscr{S})$ (cf. Definition 7.1(3)). In that case, the definition of a measurable function must be modified.‡ Let $f: (X, \mathscr{S}) \to \mathbf{R}$ be given and define the subset $N(f)$ of X by setting

$$N(f) \equiv \{x \,|\, x \in X \quad \text{and} \quad f(x) \neq 0\}.$$

Then f is said to be a measurable function iff for each $M \in \mathscr{B}_\mathbf{R}$, the set $f^{-1}(M) \cap N(f)$ is a measurable set of $(X, \mathscr{S})$ (i.e. if it belongs to $\mathscr{S}$). It can be seen that if $\mathscr{S}$ happens to be a Borel field, then $N(f)$ is always measurable so that $f^{-1}(M) \cap N(f)$ is measurable provided $f^{-1}(M)$ is; hence the general definition now reduces to the simpler one given above. In the following, we shall not explicitly need this more general definition (because in practically all cases, we are interested in Borel-type measurable spaces only), but we shall indicate in footnotes the corresponding generalizations.

Before proceeding, a number of comments are in order.

(a) The concept of a measurable function is a purely set-theoretic notion (just as was the concept of a measurable set) and does *not* depend on whether any measure μ is defined for the measurable space $(X, \mathscr{S})$ and if so, what the measure μ is. Thus, if $(X, \mathscr{S}, \mu)$ is a measure space and $f: (X, \mathscr{S}, \mu) \to \mathbf{R}$ is a measurable function on X, then for any other measure ν (and corresponding measure space) the function $f: (X, \mathscr{S}, \nu) \to \mathbf{R}$ is also a measurable function on X. (The possibility of confusion arises only because of the notation $f: (X, \mathscr{S}, \mu) \to \mathbf{R}$, instead of which one should write $f: X \to \mathbf{R}$ and specify $\mathscr{S}$.)

(b) On the other hand, the measurability of a given function f depends, naturally, not only on X but also crucially on the prescribed class $\mathscr{S}$ (or $\mathscr{B}$) of measurable sets on X. Thus, using the customary notation, if $f: (X, \mathscr{S}) \to \mathbf{R}$ is measurable and if $\mathscr{S}'$ is another measurable set structure on X, then $f: (X, \mathscr{S}') \to \mathbf{R}$ need not be a measurable function (with respect

---

†The function f is, of course, defined on the underlying set X. However, for the sake of brevity, we shall frequently avail ourselves of the notation $f: (X, \mathscr{B}) \to \mathbf{R}$, which means that $f: X \to \mathbf{R}$ and that on X a Borel field $\mathscr{B}$ is given. A similar notation $f: (X, \mathscr{S}) \to \mathbf{R}$ will be used if X has a more general measurable space structure.

‡The need for modification is connected with the desire to have a reasonable definition of the integral, which is a "generalized sum" and where the value zero plays a somewhat special role.

to this new structure $\mathcal{S}'$). For example, let X be the set of reals and let $\mathcal{B}_R$ be the class of the original Borel sets on the real line. If $f\colon (\mathbf{R}, \mathcal{B}_R) \to \mathbf{R}$ is measurable (with respect to $\mathcal{B}_R$), we shall say, for short that f is a *Borel measurable* function on the reals. If, on the other hand, $\mathcal{B}_L$ is the enlarged (completed) class† of general Lebesgue measurable sets on $\mathbf{R}$, and if $f\colon (\mathbf{R}, \mathcal{B}_L) \to \mathbf{R}$ is measurable (with respect to $\mathcal{B}_L$), then we shall say that f is a *Lebesgue measurable* function on the reals. The two classes of functions do not coincide, the first is somewhat more restrictive, i.e. every Borel measurable function is also Lebesgue measurable, but not conversely.

(c) There is some resemblance between the definition of a measurable function $f\colon (X, \mathcal{B}) \to \mathbf{R}$ and the definition of a continuous function $f\colon (X, \tau) \to \mathbf{R}$. Indeed, making the replacements

$$\text{``Borel set } M \text{ of } \mathbf{R}\text{''} \to \text{``open set } W \text{ of } \mathbf{R}\text{''}$$

and

$$\text{``Borel set } B \text{ of } X\text{''} \to \text{``open set } V \text{ of } X\text{,''}$$

the second definition "goes over" into the first. As a matter of fact, there exists a generalization of the concept of a measurable function. Let $f\colon (X, \mathcal{B}) \to (X', \mathcal{B}')$; we say that f is a *Borel function* iff for every $B' \in \mathcal{B}'$ we have $f^{-1}(B') \in \mathcal{B}$. Starting with this definition, a structure theory of mappings between measurable spaces (including the concept of "isomorphism") can be developed, similarly as was the case for algebraic and topological spaces.‡

Often we have a (universal) set X which has both a prescribed measurable space structure and a topological space structure. We may then compare the class of real valued measurable functions and the class of real valued continuous functions on X. In most interesting cases, *the class of measurable functions is larger than that of the continuous functions.* For example, consider the case when $X = \mathbf{R}$ is the domain and $\mathbf{R}$ is equipped with the usual Borel structure and the usual topology. Now let $f\colon \mathbf{R} \to \mathbf{R}$ be continuous. Then, if $I = (a, b)$ is an arbitrary open interval, $f^{-1}(I)$ is open. We know that every open set is Borel measurable (i.e. belongs to $\mathcal{B}_R$). Furthermore, we recall that the class of Borel sets can be generated by the class of open intervals. From these observations it easily follows that f is Borel measurable. § On the other hand, it is easy to

---

†Recall that $\mathcal{B}_L$ also happens to be a Borel field, but this is irrelevant.

‡Since we are interested only in the application of measure theory to integration, we shall not consider these questions. The interested reader is referred to Halmos [11], Chapter VIII.

§Hence, *eo ipso*, f is Lebesgue measurable.

construct a Borel measurable function which is not continuous. Let $f: \mathbf{R} \to \mathbf{R}$ be given by

$$f(x) = \begin{cases} a & \text{if } x \geq 0 \\ b & \text{if } x < 0, \end{cases}$$

see Fig. 8.1. Clearly, f is not continuous. On the other hand, for *any* set $M \subset \mathbf{R}$ (hence, for any Borel set of $\mathbf{R}$) we obtain the following:

$$\begin{array}{llll}
\text{If} & a \in M, b \notin M, & \text{then} & f^{-1}(M) = \{x \mid x \geq 0\}, \\
\text{If} & a \notin M, b \in M, & \text{then} & f^{-1}(M) = \{x \mid x < 0\}, \\
\text{If} & a \notin M, b \notin M, & \text{then} & f^{-1}(M) = \emptyset, \\
\text{If} & a \in M, b \in M, & \text{then} & f^{-1}(M) = \mathbf{R}.
\end{array}$$

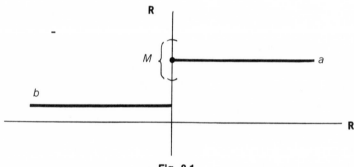

Fig. 8.1.

All sets on the r.h.s. are Borel sets of $\mathbf{R}$, hence f is a Borel measurable function. This shows that the class of Borel (and thus, Lebesgue) measurable functions on the real line is larger than the class of continuous functions, as claimed.

We have one more matter to settle. As everyone knows, in many problems one encounters *extended real valued functions*, i.e. functions whose codomain is not $\mathbf{R}$ but rather $\hat{\mathbf{R}} = \mathbf{R} \cup \{+\infty, -\infty\}$. Let now $f: (X, \mathscr{B}) \to \hat{\mathbf{R}}$. We shall say that f *is measurable iff for every Borel set M of the reals, the set $f^{-1}(M)$ is measurable*† *and if also $f^{-1}(\{+\infty\})$ and $f^{-1}(\{-\infty\})$ are measurable* (*with respect to $\mathscr{B}$, of course*).

We now give a selection of examples for measurable functions.

Example α. Let X be arbitrary and let $\mathscr{B} = \mathscr{P}(X)$. Then, clearly, *every* function $f: (X, \mathscr{B}) \to \hat{\mathbf{R}}$ is measurable.

†If $f: (X, \mathscr{S}) \to \hat{\mathbf{R}}$, then, of course, we demand that $f^{-1}(M) \cap N(f)$ be measurable.

Example β.   Let X be arbitrary and let $\mathscr{B} = \{\emptyset, X\}$. Then only the constant functions are measurable.

Example γ.   Let X be arbitrary and let $A \subset X$. It is customary to define the *characteristic function* χ_A associated with the given set A as the map $\chi_A: X \to \mathbf{R}$ given by

$$\chi_A(x) = \begin{cases} 1 & \text{if } x \in A, \\ 0 & \text{if } x \notin A. \end{cases}$$

Denote, as usual, an arbitrary Borel set of $\mathbf{R}$ by M. Observe that†

If	$1 \in M, 0 \notin M,$	then	$\chi_A^{-1}(M) = A,$
If	$1 \notin M, 0 \in M,$	then	$\chi_A^{-1}(M) = A^c,$
If	$1 \notin M, 0 \notin M,$	then	$\chi_A^{-1}(M) = \emptyset,$
If	$1 \in M, 0 \in M,$	then	$\chi_A^{-1}(M) = X.$

Suppose now that $\mathscr{B}$ is some Borel field on X and that $A \in \mathscr{B}$. Then, together with A, also A^c (and of course, $\emptyset$ and X) are Borel sets, so that, by the above, $\chi_A^{-1}(M)$ is a Borel set. Contrariwise, if A is not in $\mathscr{B}$, the first line in the above table shows that χ_A cannot be measurable. Putting together all these simple facts, we arrive at the following very useful result: *The characteristic function* $\chi_A: (X, \mathscr{B}) \to \mathbf{R}$ *is measurable iff A is measurable*, i.e. *iff* $A \in \mathscr{B}$.

Example δ.   An interesting special case of the preceding example arises in the following way. Let X be the reals, equipped with the usual Borel structure $\mathscr{B}_\mathbf{R}$. Define the function $f: (\mathbf{R}, \mathscr{B}_\mathbf{R}) \to \mathbf{R}$ by setting

$$f(x) = \begin{cases} 1 & \text{if } x \text{ is rational} \\ 0 & \text{if } x \text{ is irrational.} \end{cases}$$

This definition of f amounts to saying that $f = \chi_\mathbf{Q}$, where $\mathbf{Q}$ is the set of all rationals. Since $\mathbf{Q}$ is a Borel set of $\mathbf{R}$, it follows that f is a measurable function. It is worthwhile to note that f is "nowhere continuous."

Example ε.   As we already mentioned previously, every continuous real valued function defined on $\mathbf{R}$ (or on some interval of $\mathbf{R}$) is Lebesgue (and even Borel) measurable. So is any piecewise continuous function, like the one of Fig. 8.1.

In general, the class of measurable functions is so large that it is hard to

---

†Naturally, these results hold not only if $M \in \mathscr{B}_\mathbf{R}$, but for *any* $M \subset \mathbf{R}$. But this is irrelevant for our purposes.

construct an example of a nonmeasurable function. This is so because in most cases the class $\mathscr{S}$ (or $\mathscr{B}$) of $(X, \mathscr{S})$ is very large, so that any "barely reasonable function" will be measurable. Nevertheless, counterexamples can be found, in fact more interesting ones than the case implied by Example β above.

In our future work, it will be important to readily recognize measurable functions. We start with a theorem which serves as a criterion more easily applicable than the original definition:

THEOREM 8.1(1), *If $(X, \mathscr{B})$ is a Borel space and $f: X \to \mathbf{R}$, then f is measurable iff for every real number c, the set*

$$f^{-1}(-\infty, c) \equiv \{x \,|\, x \in X, \quad f(x) < c\}$$

is a measurable set† of $(X, \mathscr{B})$.

Proof. Set, for $x \in X$, $y = f(x)$. Let $M = \{y \,|\, y < c\}$. This open left ray of the reals is a Borel set of $\mathbf{R}$. Now, $f^{-1}(M) = \{x \,|\, f(x) < c\}$, so clearly the criterion of the theorem is a necessary one. To show its sufficiency, suppose that the criterion is satisfied and let $c_1 \leqslant c_2$. Then $\{x \,|\, f(x) < c_2\} - \{x \,|\, f(x) < c_1\} = \{x \,|\, c_1 \leqslant f(x) < c_2\} = f^{-1}([c_1, c_2))$. Thus, for any semiclosed interval $[c_1, c_2)$ one can find two measurable sets such that $f^{-1}(M)$ is their difference. Hence $f^{-1}([c_1, c_2))$ is measurable. Since the σ-ring containing all semiclosed intervals contains all Borel sets of $\mathbf{R}$ (cf. Example ζ of Section 7.1), it immediately follows that $f^{-1}(M)$ is measurable for any Borel set M of $\mathbf{R}$, q.e.d.

It is important to note that if $f^{-1}(-\infty, c) = \{x \,|\, f(x) < c\}$ is measurable for all c, then so are the sets $f^{-1}(-\infty, c] = \{x \,|\, f(x) \leqslant c\}$, $f^{-1}(c, +\infty) = \{x \,|\, f(x) > c\}$, and $f^{-1}[c, +\infty) = \{x \,|\, f(x) \geqslant c\}$, and conversely, the measurability of any of these sets implies the measurability of the other three. Hence, *in the theorem, the set $f^{-1}(-\infty, c)$ may be replaced by any of the other three sets.*

The following few theorems enable us to construct measurable functions from given ones and to recognize measurable functions which are constructed from known ones. In the sequel, (unless explicitly otherwise stated) $f, g, \ldots$ will denote measurable functions on $(X, \mathscr{S})$ to the extended reals $\hat{\mathbf{R}}$. Furthermore, $\alpha, \beta, \ldots$ will stand for any real constant. We start with the rather plausible

---

†If $(X, \mathscr{S})$ is a general measurable space, then the criterion is that $f^{-1}(-\infty, c) \cap N(f)$ be a measurable set of $(X, \mathscr{S})$.

THEOREM 8.1(2).   *The following functions are measurable*:

(a) *The shifted function $f + \alpha$.*
(b) *The pointwise multiple αf.*
(c) *The pointwise sum $f + g$, whenever defined.*†
(d) *The pointwise product fg.*

It follows from this theorem that $f - g$ and more generally any real linear combination $\Sigma_{k=1}^{n} \alpha_k f_k$ of measurable (finite valued) functions is measurable.‡ Also, f^n (n positive integer) is measurable. Let now $\{B_1, B_2, \ldots, B_n\}$ be a finite disjoint class of measurable sets of a measurable space. Let $\alpha_1, \alpha_2, \ldots, \alpha_n$ denote (finite) nonzero real numbers and define the map

$$s : (X, \mathscr{S}) \to \mathbf{R}$$

by setting

$$s(x) = \begin{cases} \alpha_k & \text{if } x \in B_k, \quad k = 1, 2, \ldots, n \\ 0 & \text{if } x \notin \bigcup_{k=1}^{n} B_k. \end{cases}$$

Such a function s will be called a (generalized) *step function*§ on $(X, \mathscr{S})$. Thus, s may be defined as a function which takes on as values a finite number of (nonzero) real numbers, each on some member of a disjoint finite class of measurable sets. If we denote the characteristic function associated with B_k by χ_{B_k}, then, clearly, the step function may be written as

$$s(x) = \sum_{k=1}^{n} \alpha_k \chi_{B_k}(x).$$

Since, by assumption, each B_k is measurable, it follows from Theorem 8.1(2) and from Example γ above that *every step function is measurable.* The simplest step function is, of course, a characteristic function χ_A, provided A is measurable. It is important to note that *real linear combinations $\Sigma_{i=1}^{N} \beta_i s_i$ of step functions and pointwise products* (like $s_1 s_2$) *are also step functions*, hence, by Theorem 8.1(2), they are also measurable. We continue with a useful theorem, but omit the technical proof:

---

†If $f(x) = +\infty$ (or $-\infty$) for some x and $g(x) = -\infty$ (or $+\infty$) for the same x, then $f + g$ cannot be defined.
‡Thus, the class of all measurable finite valued functions on $(X, \mathscr{S})$ is a *linear space*.
§Unfortunately, the terminology is not unique. Many authors call our s a "simple function" and reserve the term "step function" for the case when $\mu(B_k)$ is finite for all B_k in the definition of s. However, with this definition the elementary step function $\theta(x)$ of the physicist and engineer (see Fig. 7.3) could not be called a step function, whereas with our usage it is.

THEOREM 8.1(3). *The function $|f|^\alpha$ is measurable for $\alpha \geq 0$.*

To avoid confusion in the next theorem, we first fix the notation. Let f, g be two real (*finite valued*) functions. Then the function ψ defined for all x by

$$\psi(x) = \max\{f(x), g(x)\},$$

will be denoted by sup (f, g). Similarly, if φ is defined for all x by

$$\varphi(x) = \min\{f(x), g(x)\},$$

we shall denote it by inf (f, g). Now we state

THEOREM 8.1(4). *The functions* sup (f, g) *and* inf (f, g) *are measurable, provided f and g are.*

Proof. Observe that

$$\sup(f, g) = \tfrac{1}{2}(f + g + |f - g|),$$
$$\inf(f, g) = \tfrac{1}{2}(f + g - |f - g|),$$

and apply Theorems 8.1(3) and 8.1(2).

A closely related theorem arises in the following manner. If f is an arbitrary extended real valued function, it is customary to define its *positive part* by

$$f^+ = \tfrac{1}{2}(|f| + f)$$

and its *negative part* by

$$f^- = \tfrac{1}{2}(|f| - f).$$

More explicitly and more precisely†

$$f^+(x) = \begin{cases} f(x) & \text{if } f(x) \geq 0 \\ 0 & \text{if } f(x) < 0 \end{cases}$$

and

$$f^-(x) = \begin{cases} 0 & \text{if } f(x) \geq 0 \\ -f(x) & \text{if } f(x) < 0. \end{cases}$$

Note carefully that both f^+ and f^- are *nonnegative* functions and we have, for any f,

$$f = f^+ - f^-$$

†The detailed form below is indeed "more precise" than the shorter one already given, because it avoids the ambiguity that arises for points x where $f(x)$ *is* $+\infty$ or $-\infty$.

as well as

$$|f| = f^+ + f^-.$$

We now obtain the following

THEOREM 8.1(5). *If f is measurable, then f^+ and f^- are measurable and conversely, the measurability of f^+ and f^- implies that f is measurable.*

Proof. The first part of the theorem follows directly from the definition of f^+ and f^- and from Theorems 8.1(2) and 8.1(3). For the second part, it is sufficient to refer to Theorem 8.1(2).

The next theorem is of a different nature and concerns itself with the limit of a sequence of measurable functions.

THEOREM 8.1(6). *Let (f_n) be a sequence of extended real valued measurable functions and define the function F by $F(x) = \lim_{n \to \infty} f_n(x)$, for all x for which the r.h.s. exists.† Then F is a measurable function on its domain.*

We omit the rather technical proof but point out the most important special case, which is this. *If the sequence (f_n) of measurable functions is pointwise convergent,* i.e. *if $\lim_{n \to \infty} f_n(x) = f(x)$ for all x, then f is measurable.* We conclude with two, rather obvious but quite useful theorems.

THEOREM 8.1(7). *If f is a measurable function on $(X, \mathscr{S})$ and if f_A is the restriction of f to a measurable set A, then f_A is measurable on $(A, \mathscr{S}_A)$.*

THEOREM 8.1(8). *Let f be a measurable function on a complete‡ Borel-type measure space $(X, \mathscr{B}, \mu)$ and let g be equal almost everywhere§ to f. Then g is measurable.*

In passing we note that this is the first occasion when, in connection with a measurable function, we make use of the fact that there is a measure defined for the space whose underlying set X is the domain of definition for f.

In the foregoing, we studied real valued measurable functions. We end this section by pointing out that it is both useful and easy to generalize this

†Observe that the domain of F may be different from the common domain of the functions f_n. Moreover, it is permitted that $\lim_{n \to \infty} f_n(x)$, for some x, is $+\infty$ or $-\infty$, so that, in general, F is also an *extended* real valued function.

‡That is, every subset of a null set is a measurable set, cf. p. 313.

§That is, $g(x) = f(x)$ everywhere except on a set A for which $\mu(A) = 0$.

concept to *complex valued* functions. This is done as follows: *The function* $f: (X, \mathcal{S}) \to \mathbf{C}$ *is said to be measurable iff both* $\mathrm{Re}\, f$ *and* $\mathrm{Im}\, f$ *are measurable.*

PROBLEMS

8.1-1. Let $(X, \mathcal{S})$ be a *general* measurable space. Show that the characteristic function $\chi_A : (X, \mathcal{S}) \to \mathbf{R}$ is measurable iff A is measurable. (*Remark*: In the text we gave the proof for a Borel space only.)

8.1-2. Let $f: (X, \mathcal{S}) \to \mathbf{R}$ be a constant function, $f \not\equiv 0$. Show that f is measurable iff X is measurable. (Why is the condition $f \not\equiv 0$ stated?)

8.1-3. Let $f: \mathbf{R} \to \mathbf{R}$ be a monotone function. Prove that f is Borel measurable.

8.1-4. Show that if f is a measurable function from a Borel space to the reals, then the set $\{x | f(x) = c\}$ is measurable for all c. Is the converse statement true?

8.1-5. Let f be a measurable real valued function defined on a subset $Y \subset X$ of a Borel space $(X, \mathcal{B})$. Using Theorem 8.1(1), show that Y is measurable, i.e. $Y \in \mathcal{B}$.

8.1-6. Let f be a real valued measurable function defined on a Borel space. Show that the graph of f, i.e. the set $\Gamma(f) = \{(x, y) | x \in X, y = f(x)\}$ is a measurable set of the product space $X \times \mathbf{R}$.

8.1-7. Prove the following generalization of Theorem 8.1(8). Let f and g have domains Y_f and Y_g which are subsets of a complete measure space. Suppose Y_f and Y_g differ from one another only by a set of measure zero. Suppose further that $f = g$ almost everywhere on their common domain, i.e. on $Y_f \cap Y_g$. (Functions with these properties are often called equivalent functions.) Show now that if f is measurable, so is g.

8.1-8. Let X be the set of reals, equipped with the Lebesgue measure. Define f by setting

$$f(x) = \begin{cases} 1 & \text{if } x \text{ is irrational} \\ 0 & \text{if } x \text{ is rational.} \end{cases}$$

Use Theorem 8.1(8) to show that f is measurable.

8.1-9. Show that a complex valued function defined on a Borel space is measurable iff for every open set M of the complex plane, the set $f^{-1}(M)$ is measurable.

8.2 DEFINITION OF THE INTEGRAL

We are now finally prepared to introduce the concept of the integral. There are several approaches to this topic, and we shall review two of them.

In both approaches, it will be necessary to restrict ourselves to the most important case when $(X, \mathcal{B}, \mu)$ is a *totally σ-finite* measure space. We shall introduce the integral of certain measurable functions defined only on such spaces. It is possible to define the integral of a certain type of measurable functions on arbitrary measure spaces $(X, \mathcal{S}, \mu)$. However, the procedure is then more involved and for practically all applications our restricted definition suffices.†

The essence of the first approach, to which we now turn, is to consider initially a very small class of functions and then generalize to larger classes. Let s be a step function on‡ $(X, \mathcal{B}, \mu)$, i.e. let

$$s(x) = \sum_{k=1}^{n} \alpha_k \chi_{B_k}(x),$$

where α_k are finite real numbers, $B_k \cap B_i = \emptyset$ $(i \neq k)$ and all B_k are measurable sets. Let us further assume that $\mu(B_k)$ is finite for each $k = 1, 2, \ldots, n$. A step function so restricted will be called a *simple function*.§ Recall that s is measurable.

We now *define the integral of a simple function s over X with respect to the measure μ* by setting

$$\int_X s \, d\mu = \sum_{k=1}^{n} \alpha_k \mu(B_k). \tag{8.1}$$

Observe that the value of this integral is always a *finite* real number. The simplest example is when $s = \chi_A$, with $A \in \mathcal{B}$ and $\mu(A)$ finite. Then

$$\int_X \chi_A \, d\mu = \mu(A).$$

Since the function with domain A and constant value 1 is the restriction of χ_A to A, we may rewrite this result in the suggestive form

$$\int_A d\mu = \mu(A).$$

A specific example in this category is the following. Let $(X, \mathcal{B}, \mu)$ be the real line equipped with the Lebesgue measure and let $s = \alpha$ on some finite interval of length l and $s = 0$ otherwise (cf. Fig. 8.2). Then $\int_{\mathbf{R}} s \, d\mu = \alpha l$. Ob-

---

†The reader who is interested in the most general definition of the integral is referred, for example, to the books by Halmos[11] or Zaanen[43].

‡From now on we shall not repeat that $(X, \mathcal{B}, \mu)$ is totally σ-finite. Actually, we shall often refer to the space simply by X. On the other hand, the reader will notice that several important results are valid under much broader assumptions too.

§See, however, our footnote on nomenclature on p. 335. Many authors call *our* simple function an "integrable simple function." We shall avoid this terminology.

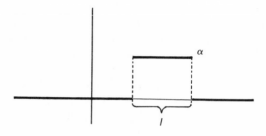

Fig. 8.2.

serve that it does not matter whether the interval is open, closed, or semiclosed.

A more exciting but equally simple example is the following. Let again our space be the measure space of the reals equipped with Lebesgue measure and let

$$s(x) = \begin{cases} 1 & \text{if } x \text{ is rational} \\ 0 & \text{if } x \text{ is irrational}. \end{cases}$$

This is a simple function because $s = \chi_Q$ and $\mu(\mathbf{Q}) = 0$. We then find that

$$\int_\mathbf{R} s \, d\mu = \mu(\mathbf{Q}) = 0.$$

Observe that here we have the first example of a function which is *not* integrable in the classical Riemann sense (because s is nowhere continuous), yet it has a perfectly well-defined integral (with respect to Lebesgue measure) according to our present definition. On the other hand, note also that the function s shown in Fig. 8.2 is also integrable in the Riemann sense† and the value of the new (Lebesgue) integral agrees with that of the classical Riemann integral.

Instead of furnishing further elementary examples, we observe that, if β is a real number and s a simple function, then

$$\int_X \beta s \, d\mu = \beta \int_X s \, d\mu,$$

and if s_1 and s_2 are simple functions, then

$$\int_X (s_1 + s_2) \, d\mu = \int_X s_1 \, d\mu + \int_X s_2 \, d\mu.$$

†Provided the interval is considered a closed one.

We now wish to extend the definition of the integral to a larger class of functions. This is rendered possible by two lemmas which hold for nonnegative measurable functions on totally σ-finite spaces. We have

LEMMA 8.2(1).   *Let f be a nonnegative measurable function on X. Then there exists a sequence* (s_n) *of simple functions on X such that*†

$$s_1(x) \leqslant s_2(x) \leqslant \cdots \leqslant s_n(x) \leqslant \cdots \quad for\ all \quad x \in X$$

and such that‡

$$\lim_{n \to \infty} s_n(x) = f(x) \quad for\ all \quad x \in X.$$

The proof (cf. Goffman and Pedrick[10], p. 121) consists in explicitly producing such a sequence (s_n).

LEMMA 8.2(2).   *If f is a nonnegative measurable function on X and if* (s_n) *and* (t_n) *are two nondecreasing sequences of simple functions both converging pointwise to f, then*

$$\lim_{n \to \infty} \int_X s_n\, d\mu = \lim_{n \to \infty} \int_X t_n\, d\mu.$$

We omit the proof (*ibid.*, p. 122) but emphasize the following. Obviously, if (s_n) is a nondecreasing sequence of simple functions, then the sequence (a_n) defined by

$$a_n \equiv \int_X s_n\, d\mu$$

is a nondecreasing sequence of real numbers. Therefore, (a_n) converges either to a real number or (according to standard terminology), it converges to $+\infty$. The "lim" in the lemma is meant in this sense.

Let now f be a given *nonnegative measurable function* on our $(X, \mathscr{B}, \mu)$. We define its integral by setting

$$\int_X f\, d\mu = \lim_{n \to \infty} \int s_n\, d\mu, \tag{8.2}$$

where (s_n) is some nondecreasing sequence of simple functions converging to f, whose existence is guaranteed by the first lemma. Even though the choice of (s_n) is not necessarily unique, the second lemma assures us that the r.h.s. of Eq. (8.2) is independent of the particular choice, hence Eq. (8.2)

†For simplicity we say that "the sequence (s_n) is nondecreasing."
‡In other words: the sequence (s_n) converges *pointwise* to *f*.

uniquely defines the integral of f. But it must be kept in mind that (in accord with our previous remark) the value of the integral may be $+\infty$. We illuminate these considerations on two examples.

Example α. Let $(X, \mathscr{B}, \mu)$ be totally σ-finite, but *not* totally finite† and let $f(x) = c > 0$ for all $x \in X$. We wish to calculate $\int_X f\, d\mu$. To this end, we chose an increasing sequence (B_n) of measurable sets (i.e. $B_n \subset B_{n+1}$), where each B_n has a finite measure and such that $X = \cup_{n=1}^{\infty} B_n$. (Since the measure space is totally σ-finite, a sequence with these properties can certainly be found.) Now we define

$$s_n(x) = \begin{cases} c & \text{if } x \in B_n, \\ 0 & \text{if } x \notin B_n. \end{cases}$$

Obviously, this is a nondecreasing sequence of simple functions and $\lim_{n \to \infty} s_n = f(x)$. (This follows from noting that $|s_n(x) - f(x)| = 0$ if $x \in \cup_{k=1}^{n} B_k$.) We have

$$\int_X s_n\, d\mu = c\mu(B_n),$$

and thus, from the definition (8.2),

$$\int_X f\, d\mu = c \lim_{n \to \infty} \mu(B_n) = c\mu(X),$$

where in the last step we used Theorem 7.2a(5). Since, by assumption, the measure is not totally finite and $c \neq 0$, we have $\mu(X) = \infty$ and so the value of the integral is actually infinite. A special case of this example arises if we take for X the real line equipped with Lebesgue measure. Then we may choose the sequence $B_n = [-n, n]$ of closed intervals, and set

$$s_n(x) = \begin{cases} c & \text{if } -n \leqslant x \leqslant n \\ 0 & \text{if } |x| > n, \end{cases}$$

so that $\int s_n\, d\mu = c2n$ and hence $\int_{\mathbf{R}} c\, d\mu = \infty$, as expected.

Example β. Let $(X, \mathscr{B}, \mu)$ be the real line with Lebesgue measure and let (see Fig. 8.3)

$$f(x) = \begin{cases} 1 - x & \text{if } 0 < x \leqslant 1 \\ 0 & \text{otherwise.} \end{cases}$$

---

†If the space is totally finite, then f would be a simple function and so we would not have an example of interest.

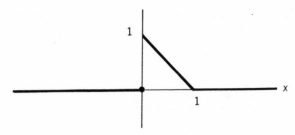

Fig. 8.3.

Take the sequence

$$s_n(x) = \begin{cases} \dfrac{k}{2^n} & \text{if } 1 - \dfrac{k+1}{2^n} < x \le 1 - \dfrac{k}{2^n}, \\ 0 & \text{otherwise.} \end{cases} \qquad k = 0, 1, \ldots, 2^n - 1$$

In detail, (see Fig. 8.4)

$$s_1(x) = \begin{cases} 0 & \text{if } \frac{1}{2} < x \le 1 \\ \frac{1}{2} & \text{if } 0 < x \le \frac{1}{2} \\ 0 & \text{otherwise,} \end{cases}$$

$$s_2(x) = \begin{cases} 0 & \text{if } \frac{3}{4} < x \le 1 \\ \frac{1}{4} & \text{if } \frac{2}{4} < x \le \frac{3}{4} \\ \frac{2}{4} & \text{if } \frac{1}{4} < x \le \frac{2}{4} \\ \frac{3}{4} & \text{if } 0 < x \le \frac{1}{4} \\ 0 & \text{otherwise,} \end{cases}$$

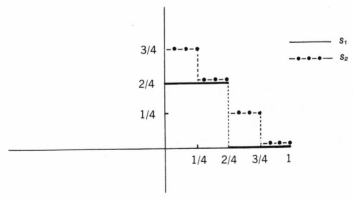

Fig. 8.4.

and so on. This is a nondecreasing sequence of simple functions. Furthermore, for all x where s_n does not vanish, we have

$$\frac{k}{2^n} < 1 - x \leqslant \frac{k+1}{2^n},$$

so that $1 - x = k/2^n + \alpha$, where $\alpha < 1/2^n$. It follows that $|f(x) - s_n(x)| = |1 - x - k/2^n| < 1/2^n$, which means that $s_n(x) \to f(x)$, as desired. Now we calculate

$$\int_{\mathbf{R}} s_n \, d\mu = \sum_{k=0}^{2^n-1} \frac{k}{2^n} \frac{1}{2^n} = \frac{1}{2^{2n}} \sum_{k=0}^{2^n-1} k.$$

This follows from the observation that the value of s_n on the interval $[1 - (k+1)/2^n, \; 1 - (k/2^n)]$ is $k/2^n$ and the measure of this interval (its length) is $1/2^n$. With the elementary formula

$$\sum_{k=0}^{m-1} k = \frac{m}{2}(m-1)$$

we then obtain

$$\int_{\mathbf{R}} s_n \, d\mu = \frac{1}{2} - \frac{1}{2^{n+1}}.$$

Hence, finally, with Eq. (8.2),

$$\int_{\mathbf{R}} f \, d\mu = \lim_{n \to \infty} \left(\frac{1}{2} - \frac{1}{2^{n+1}} \right) = \frac{1}{2}.$$

This agrees with the value of the corresponding Riemann integral of f.

We now proceed and remove the restriction that f be nonnegative. This is easy. Since any measurable extended real valued function can be split into a measurable positive and a negative part (cf. Theorem 8.1(5)), we write $f = f^+ - f^-$ and *set for an arbitrary measurable function f*

$$\int_X f \, d\mu \equiv \int_X f^+ \, d\mu - \int_X f^- \, d\mu. \tag{8.3}$$

Since the integral of f^+ and f^- (both being nonnegative) has been already defined, Eq. (8.3) appears to be an acceptable definition for the general case, which reduces to the already discussed case when $f \geqslant 0$ (because then $f = f^+$ and $f^- = 0$). However, it must be realized that each term in Eq. (8.3) may be $+\infty$. If at least one of the two integrals is finite, no ambiguity arises, and the integral of f may be finite, $+\infty$, or $-\infty$. But if the integrals of both f^+ and f^- happen to be $+\infty$, the r.h.s. is the undefined expression $(+\infty) - (-\infty)$, and strictly speaking, the definition makes no sense. For these reasons, we make the following

DEFINITION 8.2(1).   *A measurable extended real valued function on a totally σ-finite measure space is said to be integrable if both $\int_X f^+ \, d\mu$ and $\int_X f^- \, d\mu$ are finite.*

Several remarks are in order. To start with, the symbol $\int_X f \, d\mu$ is defined for *every* measurable nonnegative (and hence, nonpositive†) function. If, in particular, the value of the integral is finite, then this nonnegative (nonpositive) function is called integrable. If f is measurable but may assume both positive and negative values, then *the integral symbol $\int f \, d\mu$ makes sense only if not both $\int f^+ \, d\mu$ and $\int f^- \, d\mu$ are infinite. Even then, f need not be integrable* because one of the two expressions $\int f^+ \, d\mu$ or $\int f^- \, d\mu$ may be infinite. But if both these integrals are finite, then f is called integrable *and it then follows that the value of its integral is finite.*

In passing we note that a similar situation applies for the elementary Riemann integral. The integral itself is defined for the class of all piecewise continuous functions (then the "Riemann sum" $\Sigma_k f(x_k) \, \Delta x_k$ converges to either a finite number or to $+\infty$, or to $-\infty$), but the term "Riemann integrable function" is used only when the integral has actually a finite value.

Returning to our general theory of integration, we realize that the concept of integral cannot be even thought of in connection with non-measurable functions. The class of measurable functions, and even the class of those for which Eq. (8.3) is unambiguously defined, is very large. However, the class of *integrable* functions is considerably smaller. For example, as we already saw, the "very well-behaved" constant function $f(x) = c$ is not integrable on X if the space is not totally finite (because then $\int c \, d\mu = \infty$). In general, integrability with respect to a given measure is quite a serious restriction on the function.‡

Finally, we must comment on some unfortunate nonuniformity in terminology. Many, more recent, books call the functions which we decided to call integrable, "*summable functions*," and use the term "integrable" for the more general case when at least one of the integrals $\int f^+ \, d\mu$ or

---

†If f is nonpositive, i.e. $f(x) \leq 0$, then, clearly, $\int f \, d\mu = \int (-1)(-f) \, d\mu = -\int (-f) \, d\mu$, where $-f \geq 0$.

‡The reader is reminded that we are considering integration on a *totally σ-finite* (Borel-type) measure space. For arbitrary measure spaces $(X, \mathcal{S}, \mu)$ the class of integrable functions is again a subclass of the measurable functions and is selected by somewhat different criteria. But, even in this case, the integral of an integrable function is always finite, by way of the pertinent definition.

$\int f^- \, d\mu$ is finite (i.e. for the case when $\int f \, d\mu$ is unambiguously defined by Eq. (8.3) but may be $+\infty$ or $-\infty$). We shall *not* use this terminology,† and reserve the term integrable to be used only in the sense of our Definition 8.2(1). Nevertheless, we shall freely avail ourselves of writing the symbol $\int f \, d\mu$ (whenever unambiguously defined) even though we face the possibility that its value may be not necessarily finite but could be $+\infty$ or $-\infty$.

We give a few trivial illustrations. In each case we take for our measure space the real line equipped with Lebesgue measure.

Example γ. Let

$$f(x) = \begin{cases} -1 & \text{if } -1 \le x < 0 \\ +1 & \text{if } \ \ 0 \le x \le +1 \\ 0 & \text{if } |x| > 1. \end{cases}$$

Here

$$f^+(x) = \begin{cases} +1 & \text{if } 0 \le x \le +1 \\ 0 & \text{otherwise} \end{cases}, \qquad f^-(x) = \begin{cases} +1 & \text{if } -1 \le x < 0 \\ 0 & \text{otherwise} \end{cases}$$

hence

$$\int_R f^+ \, d\mu = 1, \qquad \int_R f^- \, d\mu = 1$$

so that f is integrable ‡ and $\int_R f \, d\mu = 0$.

Example δ. Let

$$f(x) = \begin{cases} -1 & \text{if } -1 \le x < 0 \\ +1 & \text{if } \ \ x \ge 0 \\ 0 & \text{if } x < -1. \end{cases}$$

Then one easily sees that

$$\int_R f^+ \, d\mu = +\infty, \qquad \int_R f^- \, d\mu = 1$$

so that f is not integrable, but the symbol $\int_R f \, d\mu$ still makes sense and its value is $+\infty$.

Example ε. Let

$$f(x) = \begin{cases} -1 & \text{if } x < 0 \\ +1 & \text{if } x \ge 0. \end{cases}$$

---

†Most theorems of fundamental importance apply anyway to only the class of "summable" functions.

‡Naturally, our f is actually a simple function, so that the decomposition into f^+ and f^- is really unnecessary.

Then both

$$\int_{\mathbf{R}} f^+ \, d\mu = +\infty \quad \text{and} \quad \int_{\mathbf{R}} f^- \, d\mu = +\infty,$$

so f is not integrable and $\int_{\mathbf{R}} f \, d\mu$ makes no sense.

We now turn to a completely different, *alternative definition of the integral* (on totally σ-finite measure spaces) which ties up better with elementary knowledge and which also makes the evaluation of the integral somewhat easier.

Let f be a measurable *nonnegative* real function† on $(X, \mathscr{B}, \mu)$. Define the *ordinate set* F of f to be that subset of $X \times \mathbf{R}$ which consists of all points (x, y) such that for each $x \in X$ we have $0 \leqslant y \leqslant f(x)$. Thus,

$$F = \{(x, y) | x \in X, \quad 0 \leqslant y \leqslant f(x)\}. \tag{8.4}$$

The set F is a measurable subset of $X \times \mathbf{R}$, with respect to the product Borel field $\mathscr{B} \times \mathscr{B}_{\mathbf{R}}$. We shall verify this statement only under the additional assumption that f is actually a simple function of the form $\sum_{k=1}^{n} \alpha_k \chi_{B_k}$. In that case,

$$F = \bigcup_{k=1}^{n} (B_k \times [0, \alpha_k]),$$

where B_k is a measurable set of $(X, \mathscr{B}, \mu)$ and the interval $[0, \alpha_k]$ is a Lebesgue measurable (Borel) set of the reals. Since, by the definition of the product of measurable spaces (Section 7.2, p. 309), the Borel field $\mathscr{B} \times \mathscr{B}_{\mathbf{R}}$ on $X \times \mathbf{R}$ is generated by sets of the form $B \times M$, where $B \in \mathscr{B}$ and $M \in \mathscr{B}_{\mathbf{R}}$, it is clear that F, being a finite union of such sets, belongs to the class $\mathscr{B} \times \mathscr{B}_{\mathbf{R}}$, so that F is a measurable set of the measurable space $(X \times \mathbf{R}, \mathscr{B} \times \mathscr{B}_{\mathbf{R}})$, q.e.d.‡

We next recall (p. 310) that the product measurable space $(X \times \mathbf{R}, \mathscr{B} \times \mathscr{B}_{\mathbf{R}})$ can be turned into a measure space by endowing it with the product measure.§ If we denote this measure on the product space $X \times \mathbf{R}$ by ν, then we can write $\nu = \mu \times m$, where m is the Lebesgue measure on the real line.

We now define the integral of f on X by setting

$$\int_X f \, d\mu \equiv \nu(F). \tag{8.5a}$$

---

†For simplicity we assume that f is finite everywhere.

‡For the generalization to the case when f is not simple, we must again represent f as the limit of a nondecreasing sequence of simple functions. For details, see Halmos[11], p. 143, paragraphs (5b), (5c), and (5d).

§Observe that both $(X, \mathscr{B})$ and $(\mathbf{R}, \mathscr{B}_{\mathbf{R}})$ are σ-finite.

Since we know that the ordinate set F of f is a measurable set of $(X \times \mathbf{R}, \mathscr{B} \times \mathscr{B}_{\mathbf{R}}, \mu \times m)$, the r.h.s. is uniquely determined.
Finally, if f is not nonnegative, we again put

$$\int_X f \, d\mu = \int_X f^+ \, d\mu - \int_X f^- \, d\mu = \nu(F^+) - \nu(F^-), \qquad (8.5b)$$

where $F^+(F^-)$ is the ordinate set of $f^+(f^-)$. Naturally, $\int_X f \, d\mu$ is well defined only if at least one of the two functions f^+ and f^- has a finite integral, i.e. if $\nu(F^+)$ and $\nu(F^-)$ are not both infinite.

We still have to show that this new definition of the integral is equivalent to the previously given one. As a start, we take a simple function $s = \sum_{k=1}^n \alpha_k \chi_{B_k}$, so that

$$f = \bigcup_{k=1}^n (B_k \times [0, \alpha_k]),$$

where the terms in the union are nonintersecting measurable sets of $X \times \mathbf{R}$. Hence, using the definition of the product measure and the fact that $m([0, \alpha_k]) = \alpha_k$, we get, with additivity,

$$\nu(F) = \sum_{k=1}^n \nu(B_k \times [0, \alpha_k]) = \sum_{k=1}^n \mu(B_k) m([0, \alpha_k]) = \sum_{k=1}^n \alpha_k \mu(B_k).$$

Thus, $\int_X s \, d\mu = \sum_{k=1}^n \alpha_k \mu(B_k)$, exactly as we found with the original definition of the integral. From here it is then not difficult to proceed to arbitrary (nonnegative) measurable functions and show the equivalence of the two definitions.

The pictorial content of the new definition of the integral becomes evident if one takes the special (but most important) case where X is the real line with Lebesgue measure, i.e. if one considers the integral of real functions on the real line with respect to Lebesgue measure. It then turns out that $\nu(F)$ is a simple generalization of what, in elementary calculus, is referred to as the "*area under the curve*." Indeed, let f be a continuous function on a bounded closed interval $I = [a, b]$ and zero outside I. Then, as Fig. 8.5 shows, F is an (uncountable) union of vertical lines which together make up the area bounded by the graph of f, the interval I, and two vertical lines at a and b. The product measure ν is precisely the product of Lebesgue measure with itself and therefore $\nu(F)$ is nothing but the area of the closed connected bounded set F.

A trivial example of applying this method is given in Problem 8.2-3. An equally elementary application is to recalculate, with this method, the

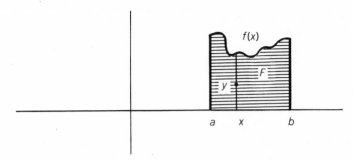

Fig. 8.5.

Lebesgue integral of

$$f(x) = \begin{cases} 1 & \text{if } x \text{ is rational,} \\ 0 & \text{if } x \text{ is irrational.} \end{cases}$$

Here the ordinate set is given by

$$F = \{(x, y) | x = \text{rational}, \quad 0 \le y \le 1 \quad \text{and} \quad x = \text{irrational}, \quad y = 0\}$$

$$= (\mathbf{Q} \times [0, 1]) \cup (\mathbf{Q}^c \times \{0\}).$$

Thus

$$\int_{\mathbf{R}} f(x) \, d\mu = \nu(F) = m(\mathbf{Q}) \cdot 1 + m(\mathbf{Q}^c) \cdot 0 = 0 \cdot 1 + \infty \cdot 0 = 0,$$

as found before.

To complete our understanding of the basic definition of the integral, we add now a simple but very useful remark. Suppose f is integrable on $(X, \mathcal{B}, \mu)$ and let E be a *measurable* subset of X. Then, as we know already, the restriction of f to E, or equivalently, the function $f\chi_E$, is also measurable on E. In fact, it is integrable on X. To see this, we note that the ordinate sets of its positive and negative parts are

$$F(f^{\pm}\chi_E) = \{(x, y) | x \in E, \quad 0 \le y \le f^{\pm}(x) \quad \text{and} \quad x \notin E, \quad y = 0\}.$$

These measurable sets are obviously subsets of the corresponding ordinate sets $F(f^{\pm})$. Since, by assumption, these have both finite ν-measure, the subsets $F(f^{\pm}\chi_E)$ have *eo ipso* finite ν-measure. Hence, if we define

$$\int_E f \, d\mu = \int_X f\chi_E \, d\mu, \tag{8.6}$$

then we are assured that this integral of the restriction of f to E is finite, which means that *f is integrable on the set E.*

So far we have considered the integral of an extended real valued function. We conclude this section with some generalizations of the integral concept.

The *generalization of integrals for complex valued functions* is rather trivial. Let

$$f: (X, \mathcal{B}, \mu) \to \mathbf{C}$$

be a complex valued function on the measure space X. We split f into its real and imaginary part, $f = u + iv$, and recall that f is said to be measurable if both u and v are measurable. Similarly, we say that f is integrable if both u and v are integrable; and if this is the case, we can define, without running into an ambiguity,

$$\int_X f d\mu = \int_X u \, d\mu + i \int_X v \, d\mu. \tag{8.7}$$

Clearly, the integral is a (finite) complex number.

We now turn our interest to a less trivial generalization which plays an important role in some branches of functional analysis. Suppose we have a *signed measure* μ on a measurable space $(X, \mathcal{B})$ which has the Jordan decomposition $\mu = \mu^+ - \mu^-$. Let f be an (extended) real valued function on X which is measurable (with respect to $\mathcal{B}$). Set

$$\int_X f d\mu = \int_X f d\mu^+ - \int_X f d\mu^-. \tag{8.8}$$

The question arises whether the r.h.s. is well defined. This of course is contingent upon the integrability (in the usual sense) of f with respect to the (ordinary) measures μ^+ and μ^-. It can be shown that f is integrable with respect to both μ^+ and μ^- provided it is integrable with respect to the so-called "total variation measure" defined by $|\mu| = \mu^+ + \mu^-$. (For details concerning the measure $|\mu|$, cf. Problem 7.2d-1.) Thus, if f is integrable with respect to $|\mu|$, the r.h.s. of Eq. (8.8) makes sense and so it defines *the integral of f with respect to the signed measure μ*. This integral is, of course, finite.

There is no problem for integrating a complex valued measurable function with respect to a signed measure. Clearly, we then simply have to split f on the r.h.s. of Eq. (8.8) into its real and imaginary part and make sure that both u and v are integrable with respect to $|\mu|$.

Finally, we can go one step further and define the integral with respect to a *complex measure*. Let $\mu = \mu_r + i\mu_i$ be a complex measure on $(X, \mathcal{B})$ (where μ_r and μ_i are, of course, signed measures) and let f be an (extended) real valued function on X, measurable with respect to $\mathcal{B}$. We then set

$$\int_X f \, d\mu = \int_X f \, d\mu_r + i \int_X f \, d\mu_i = \int_X f \, d\mu_r^+ - \int_X f \, d\mu_r^-$$
$$+ i \int_X f \, d\mu_i^+ - i \int_X f \, d\mu_i^- . \tag{8.9}$$

According to our preceding analysis, if f is integrable with respect to both $|\mu_r|$ and $|\mu_i|$, then the r.h.s. is unambiguous and so Eq. (8.9) defines *the integral of f with respect to a complex measure μ.* It is, naturally, finite.

Once again, we may generalize for the case when f is a complex valued measurable function on $(X, \mathcal{B})$. If f is complex and μ is complex, we have

$$\int f \, d\mu = \int u \, d\mu_r + i \int v \, d\mu_r + i \int u \, d\mu_i - \int v \, d\mu_i$$
$$= \int u \, d\mu_r^+ - \int u \, d\mu_r^- + i \int v \, d\mu_r^+ - i \int v \, d\mu_r^-$$
$$+ i \int u \, d\mu_i^+ - i \int u \, d\mu_i^- - \int v \, d\mu_i^+ + \int v \, d\mu_i^- . \tag{8.10}$$

This is well defined and finite if u and v are measurable with respect to $|\mu_r|$ and $|\mu_i|$.

PROBLEMS

8.2-1. Let $(X, \mathcal{B}, \mu)$ be the positive real numbers equipped with the Lebesgue measure. Show that the functions

$$s_n(x) = \begin{cases} \dfrac{k}{2^n} & \text{if } \dfrac{2^n}{k+1} < x \leqslant \dfrac{2^n}{k}, \qquad k = 0, 1, \ldots, n2^n - 1 \\[2mm] n & \text{if } x \leqslant \dfrac{1}{n} \end{cases}$$

form a monotone sequence of simple functions which converges everywhere to $f(x) = 1/x$. Compute now the Lebesgue integral of $1/x$ over the set of the positive reals. Is $1/x$ Lebesgue integrable?

8.2-2. Let $(X, \mathcal{B}, \mu)$ be the subset of the reals belonging to the interval $[0, 1]$, equipped with the Lebesgue measure. Show that the functions

$$\begin{cases} n - \dfrac{1}{n} & \text{if } x \leqslant \dfrac{1}{n^2} \\[2mm] \dfrac{n^2 - k - 2}{n} & \text{if } \dfrac{n^2}{(n^2 - k)^2} < x \leqslant \dfrac{n^2}{(n^2 - k - 1)^2}, \qquad k = 0, 1, \ldots, n^2 - n - 1, \end{cases}$$

form a monotone sequence of simple functions which converges everywhere to $f(x) = 1/x^{1/2}$. Compute now the Lebesgue integral of $1/x^{1/2}$ over $[0, 1]$. Is $1/x^{1/2}$ Lebesgue integrable? Is it Riemann integrable? (*Remark*: It will help to know that

$$\sum_{k=1}^{m} \frac{1}{(n^2 - k - 1)(n^2 - k)} = \frac{m}{(n^2 - 1)(n^2 - 1 - m)}.$$

Other sums will also occur, but you will be able to evaluate their limits using estimates and the above formula. Of course, you may conduct your calculations in any other way and use tables of sums for series.)

8.2-3. Let $(X, \mathcal{B}, \mu)$ be the measure space of the reals equipped with the Lebesgue measure. Let $f: \mathbf{R} \to \mathbf{R}$ be the function given by $f(x) = x$. Decompose f into its positive and negative part (draw figures) and "compute" $\int_{\mathbf{R}} x\, d\mu$, using the second definition of the integral. Is $f(x)$ integrable? Does $\int_{\mathbf{R}} x\, d\mu$ make sense at all? Now calculate the same integral over the closed bounded interval $[-a, a]$.

8.2-4. Calculate the Lebesgue integral of $f(x) = 1/x^{1/2}$ over $[0, 1]$ (as you did in Problem 8.2-2), but use now the definition of the integral based on the measure of the ordinate set. (*Hint*: Observe that one can write $F = \bigcup_{k=1}^{\infty} A_k$, where A_k is the portion of the plane which is bounded by the segment $[0, 1]$ of the X-axis, the vertical of length one erected at $x = 1$, the graph of $f(x)$ for $1/k^2 \le x \le 1$, the horizontal straight line at $y = k$ for $0 \le x < 1/k^2$, and the segment $[0, k]$ of the Y-axis. Realize also that, for continuous functions on an interval, the area under the graph can be computed by the familiar Riemann integral.)

8.2-5. A Lebesgue integral over a set E where, at an "endpoint" $f(x)$ becomes unbounded, is often called an improper Lebesgue integral. By generalizing the insight you gained in the previous problem, show that such an integral can be calculated by the following prescription. Let $f \ge 0$ and set

$$f_k(x) = f(x) \quad \text{if} \quad x \in E \quad \text{and} \quad f(x) \le k,$$
$$f_k(x) = k \qquad \text{if} \quad x \in E \quad \text{and} \quad f(x) > k.$$

Then

$$\int_E f\, d\mu = \lim_{k \to \infty} \int_E f_k\, d\mu.$$

If f is not nonnegative, the positive and negative parts must be dealt with separately, of course. (*Remark*: Even if f_k is integrable for all k, obviously f need not be integrable.)

8.2-6. Use the method of the previous problem to calculate the Lebesgue integral of $f(x) = x$ over the positive real line.

8.2-7. Let $(X, \mathcal{B}, \mu)$ be the measure space where $X = $ set of all positive integers, $\mathcal{B} = $ class of all subsets, $\mu(B) = $ the number of points in B. Let $f: (X, \mathcal{B}, \mu) \to \mathbf{R}$ and write, for simplicity, f_k for the value of f at the integer k. Use both definitions of the integral to show that f is integrable with respect to μ iff $\Sigma_{k=1}^{n} |f_k|$ is finite. Calculate the integral $\int_X f\, d\mu$. (*Hint*: In both approaches you will have to split f in a positive and negative part. In the first approach, you will have to find first a monotone sequence of simple functions which converges to $f^{\pm}$. This is quite easy.) (*Remark*: The result of this problem shows that, as expected, the sum of function values can be considered as the simplest example of an integral, i.e. it is the "discrete" case of an integral.)

8.2-8. Let $\mathscr{I}$ be the set of all irrationals in $[0, 1]$. Let $f: \mathbf{R} \to \mathbf{R}$ be given by $f(x) = x^2$. Calculate the Lebesgue integral $\int_{\mathscr{I}} x^2 \, d\mu$.

8.3 GENERAL PROPERTIES OF THE INTEGRAL

In this section we are going to establish the fundamental properties of the integral. The theorems will be formulated for extended real valued functions but, with obvious modifications whenever necessary, analogous results hold for the integrals of complex valued functions. In general, we shall not name (or indicate on the integration symbol) the space X over which we integrate because this is usually obvious. If the integration is over a subset E of X, it will be understood that E is measurable (cf. the discussion of Eq. (8.6) in the preceding section). We shall frequently use the notation

$$f(x) \overset{\circ}{=} g(x) \quad \text{or} \quad f(x) = g(x) \quad \text{a.e.}$$

which means that f is almost everywhere equal to g on their common domain, i.e. equal everywhere except on a set A which has measure zero.

THEOREM 8.3(1).   *If* $f \overset{\circ}{=} g$, *then integrability of one of the two functions implies the integrability of the other and actually*

$$\int f \, d\mu = \int g \, d\mu.$$

Proof.   The theorem is almost obvious if one thinks of the definition of the integral in terms of the ordinate set.

An important implication of this simple theorem is that *integrability of a function and the value of the integral are not affected if one alters the function in any arbitrary way on a set of measure zero. Furthermore, when formulating or deriving theorems on integration, we may disregard any null set.*† Finally, our theorem implies that *we may speak of the integral* $\int_X f \, d\mu$ *even if the domain of* f *is not* X *but only the set* $X - N$, *where* N *is*

---

†This harmless remark will be very important in several subsequent theorems.

an arbitrary null set. This is so because on N we may define f in any way we wish.†

COROLLARY TO THEOREM 8.3(1).   *A function f equals the null function almost everywhere iff $\int |f|\, d\mu = 0$.*

Proof.   (a) Assume that $f \stackrel{\circ}{=} 0$, i.e. that $f(x) \neq 0$ only on some null set. By our preceding result, this set can be disregarded, hence $\int |f|\, d\mu = 0$. (b) If $\int |f|\, d\mu = 0$, then consideration of the ordinate set reveals that $f \stackrel{\circ}{=} 0$.

THEOREM 8.3(2).   *If f is integrable on X and N is a set of measure zero, then $\int_N f\, d\mu = 0$.*

The *proof* follows trivially from observing that, as we pointed out above, we may alter f on the whole set N to be identically zero, without changing the value of the integral. It is amusing to observe that, even if $f = +\infty$ *everywhere on N*, its integral on N is still zero!

THEOREM 8.3(3).   *Let c be a constant and f, g be integrable functions. Then*

(a) cf *is integrable and* $\int cf\, d\mu = c \int f\, d\mu$,
(b) $f + g$ *is integrable and* $\int (f + g)\, d\mu = \int f\, d\mu + \int g\, d\mu$.

The *proof* of (a) is trivial. Concerning (b), we omit the details of this plausible statement and observe only that the desired result may be obtained by considering the positive and negative parts separately, representing them in terms of simple functions, and using the validity of the theorem for such functions which has been already pointed out on p. 340.
    An important but obvious consequence of Theorem 8.3(3) can be formally stated by

COROLLARY TO THEOREM 8.3(3).   *All linear combinations of integrable functions on a given measure space are integrable and therefore, defining sum and scalar multiple of functions in the usual pointwise manner, we can say that the set of integrable functions forms a linear space.*

The following two theorems compare integrals.

---

†The reader will appreciate the depth of this simple statement if he recalls the situation concerning Riemann integrals on the real line, where no such gimmick makes sense!

THEOREM 8.3(4).   *If f and g are integrable and* $f(x) \geq g(x)$ a.e., *then* $\int f \, d\mu \geq \int g \, d\mu$.

Proof. Recall our comment on disregarding null sets. Then, assuming first that f and g are nonnegative, the theorem follows easily from considering the corresponding monotone sequences of simple functions. If f and g have arbitrary signs, observe that $f^+ + g^- \geq f^- + g^+$.

THEOREM 8.3(5).   *If f is integrable, then*

$$\left| \int f \, d\mu \right| \leq \int |f| \, d\mu.$$

Proof. Since $\int f^{\pm} \, d\mu \geq 0$, we have

$$-\int f^+ \, d\mu - \int f^- \, d\mu \leq \int f^+ \, d\mu - \int f^- \, d\mu \leq \int f^+ \, d\mu + \int f^- \, d\mu.$$

This can be written as

$$-\int |f| \, d\mu \leq \int f \, d\mu \leq \int |f| \, d\mu,$$

which is equivalent to the statement of the theorem.

The next theorem relates integrability and absolute integrability, assuring us that these properties are equivalent. More precisely, we have

THEOREM 8.3(6).   *A measurable function f is integrable iff* $|f|$ *is integrable.*

Proof. Recalling that $|f| = f^+ + f^-$ and that integrability of f implies that of f^+ and f^-, Theorem 8.3(3) immediately leads to the first half of the statement. The converse follows similarly, by writing $f = f^+ - f^-$.

The importance of this theorem is appreciated if one considers the particular case of integrating a real function on the real line. If one used the classical Riemann integral, the theorem would not be valid for certain "improper" Riemann integrals. For example, the (improper) Riemann integral of $f(x) = \sin x / x$ is

$$\int_0^{+\infty} \frac{\sin x}{x} \, dx \equiv \lim_{a \to \infty} \int_0^{+a} \frac{\sin x}{x} \, dx = \lim_{a \to \infty} \text{Si } a = \frac{\pi}{2},$$

so that, in this customary sense, $\sin x / x$ is Riemann integrable. On the other hand,

$$\int_0^{n\pi} \left| \frac{\sin x}{x} \right| dx = \int_0^\pi \frac{\sin x}{x} dx - \int_\pi^{2\pi} \frac{\sin x}{x} dx + \int_{2\pi}^{3\pi} \frac{\sin x}{x} dx - \cdots$$

$$\pm \int_{(n-1)\pi}^{n\pi} \frac{\sin x}{x} dx = 2(\mathrm{Si}\,\pi + \mathrm{Si}\,3\pi + \cdots + \mathrm{Si}\,n\pi),$$

which goes to $+\infty$ when $n \to \infty$, so that

$$\int_0^\infty \left| \frac{\sin x}{x} \right| dx = \lim_{a \to \infty} \int_0^a \left| \frac{\sin x}{x} \right| dx = +\infty,$$

hence $|\sin x / x|$ is not Riemann integrable. This unpleasant phenomenon cannot occur with the more general concept of integration; for example, if Lebesgue integration is used, then neither $|\sin x / x|$ *nor* $\sin x / x$ is Lebesgue integrable. (This is so because the ordinate set of both the positive and negative parts of $\sin x / x$ is infinite.)

The next theorem makes an important statement about the integrability of products:

THEOREM 8.3(7).   *Let f be integrable, let g be measurable and bounded a.e. Then fg is integrable.*

Proof.   By assumption, $|g| \le K$ (we may ignore the null set where this does not hold), so that $|fg| \le K|f|$. By virtue of Theorem 8.3(6), $|f|$ is integrable, hence surely measurable. Now recall that the product $|fg| = |f||g|$ must be measurable. Obviously, the ordinate set of $|fg|$ will have a (product) measure not greater than that of the ordinate set of $K|f|$. Since $K|f|$ is integrable, the measure of its ordinate set is finite, hence *eo ipso* the same holds for that of $|fg|$, i.e. $|fg|$ is integrable. Using again Theorem 8.3(6), it follows that fg is integrable, q.e.d.

Observe that the boundedness of g is essential: in general, fg is not integrable even if *both f* and *g* are integrable. This contrasts the fact that products of measurable functions are always measurable, and it illustrates our previous remark that the class of integrable functions is considerably smaller than that of the measurable functions.

The following theorem relates integrals defined over different sets:

THEOREM 8.3(8).   *Let A_1 and A_2 be measurable, disjoint sets. Let $A = A_1 \cup A_2$. Suppose f is integrable on A. Then f is also integrable over* *both A_1 and A_2 and, in fact,*

$$\int_A f\,d\mu = \int_{A_1} f\,d\mu + \int_{A_2} f\,d\mu.$$

Proof. Observe that χ_{A_1} and χ_{A_2} are measurable and bounded on A. Since f is integrable on A, using Theorem 8.3(7) we see that $\chi_{A_1}f$ and $\chi_{A_2}f$ are integrable on A. But

$$\int_A \chi_{A_k}f\,d\mu \equiv \int_{A_k} f\,d\mu \qquad \text{for } k = 1, 2.$$

Hence, f is integrable over both A_1 and A_2. Furthermore, observe that under the given conditions,

$$\chi_A = \chi_{A_1 \cup A_2} = \chi_{A_1} + \chi_{A_2}.$$

Therefore,

$$\int_A f\,d\mu \equiv \int_A f\chi_A\,d\mu = \int_A f\chi_{A_1}\,d\mu + \int_A f\chi_{A_2}\,d\mu = \int_{A_1} f\,d\mu + \int_{A_2} f\,d\mu,$$

q.e.d.

Clearly, *the additivity of the integral over nonintersecting sets* as expressed by the theorem, can be generalized, by induction, to an arbitrary finite number of sets. Actually, the theorem holds even for the countable case: *If $A = \bigcup_{k=1}^{\infty} A_k$, with each A_k measurable and $A_i \cap A_j = \emptyset$ for $i \neq j$, and if f is integrable over A, then it is integrable over each set A_k and*

$$\int_A f\,d\mu = \sum_{k=1}^{\infty} \int_{A_k} f\,d\mu.$$

However, to prove this assertion, the not yet stated Theorem 8.3(11) is needed (see Problem 8.3-14).

To proceed, we wish to relate integrability and boundedness. First we observe that if f assumes infinite values $\pm\infty$ on a subset E of X which is not a null set, then f cannot be integrable on X, since the ordinate set of f^+ and/or f^- will have an infinite measure. Therefore, *if f is integrable over a set, it must be finite almost everywhere on the set.* This is a necessary criterion of integrability. The question of a sufficient criterion of a *similar* nature is settled, in a sense, by the following

THEOREM 8.3(9).   *If E is a measurable set which has finite measure and if f is an* a.e. *bounded measurable function on E, then f is integrable over E.*

Proof.   If E has finite measure, then χ_E is integrable, so that $f \equiv f\chi_E$ is the product of an integrable function and of an a.e. bounded measurable function; hence, by Theorem 8.3(7) it is integrable.

This sufficiency criterion is rather reassuring because the class of measurable functions is very large and the theorem *does* permit f to become unbounded "quite badly." For example, if E is the subset [0, 1] of the reals equipped with Lebesgue measure, and if we define

$$f(x) = \begin{cases} +\infty & \text{if } x \text{ is rational} \\ c & \text{if } x \text{ is irrational,} \end{cases}$$

then f is integrable, and in fact $\int_{[0,1]} f(x)\, d\mu = c$. On the other hand, the theorem *fails* to hold if E has infinite measure. In particular, we cannot apply the criterion if the integral is over all of X, where X is only a (totally) σ-finite (but not totally finite) measure space. This occurs often, like for the case of Lebesgue integration on the real line.

Furthermore, the reader should be aware of not confusing unboundedness and "becoming infinite." In fact, Theorem 8.3(9) may fail if f is not bounded a.e. *even if* it becomes infinite only on a null set or not at all.† For example, the function $f(x) = 1/x$ is not Lebesgue integrable over the interval [0, 1] (cf. Problem 8.2-1; the formal value of the integral is $+\infty$), because, even though $f = \infty$ only at the single point $x = 0$, given any number $N \geq 0$, we have $f(x) > N$ for the whole interval [0, 1/N]. Thus, f is unbounded on a set of nonzero measure, and the theorem does not apply.

On the other hand, Theorem 8.3(9) is only a sufficient, but not necessary criterion. For example, $f(x) = 1/x^{1/2}$ is Lebesgue integrable on [0, 1], (see Problem 8.2-4) even though f is again unbounded on a set of finite measure.

The next three theorems make statements on sequences of functions in relation to integrability. We precede these theorems with the negative remark that, unfortunately, pointwise convergence of a sequence of integrable functions to an integrable function does *not*, in general, imply the convergence of the corresponding sequence of integrals. That is, if $f_n(x) \to f(x)$ with all f_n and f being integrable, it does *not* follow that $\lim_{n \to \infty} \int f_n\, d\mu = \int f\, d\mu$. For example, let E be the open interval $(0, 1)$,

---

†Thus, our theorem is *not* a converse of the preceding observation.

equipped with Lebesgue measure. Let

$$f_n(x) = \begin{cases} n(n+1) & \text{if } \dfrac{1}{n+1} < x < \dfrac{1}{n}, \\ 0 & \text{otherwise.} \end{cases}$$

Clearly (draw a figure), $\int_{(0,1)} f_n \, d\mu = 1$ for each n, and $f_n(x) \to f(x) \equiv 0$, so that $\int_{(0,1)} f \, d\mu = 0$. But we see that $\lim_{n\to\infty} \int f_n \, d\mu \neq \int f \, d\mu$.

The first result, giving us a useful criterion in this area, is called the *monotone convergence theorem*:

THEOREM 8.3(10).   *If* (f_n) *is a monotone sequence of (nonnegative) integrable functions (in the sense that* $f_1(x) \leqslant f_2(x) \leqslant \cdots$*) and if* f *is the pointwise limit of* f_n *almost everywhere (i.e. if* $f(x) = \lim_{n\to\infty} f_n(x)$*a.e.), then*

$$\int f \, d\mu = \lim_{n\to\infty} \int f_n \, d\mu.$$

We omit the somewhat technical proof, but make three comments.

(a) The premise of the theorem certainly ensures that $f(x)$ is measurable (cf. Theorem 8.1(6)), but it does *not* imply that $f(x)$ is integrable. Therefore, it may happen that $\int f \, d\mu = +\infty$. Then, of course, $\lim_{n\to\infty} \int f_n \, d\mu$ is also $+\infty$.

(b) It is obvious that if (f_n) is *nonpositive* and *nonincreasing* (i.e. $f_1(x) \geqslant f_2(x) \geqslant \cdots$), the theorem also holds. Now $\int f \, d\mu$ may happen to be $-\infty$.

(c) In the formulation of the theorem, the term "nonnegative" may be actually omitted, which can be seen if one considers the sequence $(f_n - f_1)$ and the limit function $f - f_1$. Similarly, for nonincreasing sequences, the condition of being nonpositive may be omitted.

The essence of the above theorem is that, under the stated conditions, the integration and the limit may be interchanged. But the theorem has only a limited usefulness because, first of all, f need not be integrable and second, monotonity of (f_n) and integrability of each f_n are rather restrictive conditions. Therefore, the following theorem, often called the *Lebesgue dominated convergence theorem*, is much more powerful:

THEOREM 8.3(11).   *Let* (f_n) *be an arbitrary sequence of measurable functions. Let, for every* n*,* $|f_n(x)| \leqslant g(x)$ *a.e., where* g *is some integrable function. Let* $f(x) = \lim_{n\to\infty} f_n(x)$ *a.e. Then* f *is integrable and*

$$\int f \, d\mu = \lim_{n \to \infty} \int f_n \, d\mu.$$

Thus, when the given conditions hold, we can interchange the limit and the integral, and, moreover, the last sentence of the theorem assures us that the integral will be finite. The condition on (f_n) is rather mild and easy to check,† hence the usefulness of the theorem. To avoid confusion, the reader should observe that the premise of the theorem tacitly *assumes* that the sequence (f_n) is (pointwise) convergent, and the statement holds for this limit function.

In most applications, only the following, restricted form of the Lebesgue theorem is needed:

THEOREM 8.3(11a).   *Let (f_n) be an arbitrary sequence of measurable functions and let E be a set which has finite measure. Let the sequence (f_n) be bounded* a.e. *in the sense that* $|f_n(x)| \le K$ *a.e. Let* $f(x) = \lim_{n \to \infty} f_n(x)$ *a.e. Then f is integrable and*

$$\int_E f \, d\mu = \lim_{n \to \infty} \int_E f_n \, d\mu.$$

The validity of this result follows from the general case, because if $\mu(E) < \infty$, then the constant function $g(x) = K$ is integrable.

We illustrate the power of the theorem on the following example. Let E be the interval $[0, 1]$ of the reals, equipped with Lebesgue measure. Imagine the rationals in E be enumerated in some way, labeling them as $r_1, r_2, \ldots, r_n, \ldots$. Let

$$f_n(x) = \begin{cases} 1 & \text{if } x = r_1, r_2, \ldots, r_n \\ 0 & \text{otherwise.} \end{cases}$$

Clearly, this is a sequence of measurable functions. Furthermore, $|f_n(x)| \le 1$. The sequence is pointwise convergent; indeed,

$$\lim_{n \to \infty} f_n(x) = f(x) = \begin{cases} 1 & \text{if } x \text{ is rational} \\ 0 & \text{if } x \text{ is irrational.} \end{cases}$$

Our theorem assures us that f is integrable. Furthermore, for each n, $\int_{[0,1]} f_n(x) \, d\mu = 0$, because $f_n = 0$ with the exception of finite many points.

---

†It is interesting to note here that the criterion $|f_n(x)| \le g(x)$ actually implies that each f_n is integrable, cf. Problem 8.3-2.

Hence, the theorem also tells us that $\int_{[0,1]} f \, d\mu = 0$. Of course, this last result is not new to us and we obtained it previously in an elementary manner. However, let us compare the whole situation with the case when not Lebesgue but rather Riemann integration is used. Then each f_n is Riemann integrable, because it is a piecewise continuous function and $\int_0^1 f_n \, dx = 0$. But $\lim_{n \to \infty} f_n = f$ is *not* Riemann integrable, in fact, the integral of f cannot be defined. The question of interchanging limit and integration does not even arise. One merit of the Lebesgue theorem is that, under mild assumptions, it assures that the pointwise limit of sequences is an integrable function.

We conclude the study of the properties of the integral by briefly quoting the results concerning the relation of integrals on product spaces to those on the factor spaces.

Suppose $(X, \mathcal{A}, \mu)$ and $(Y, \mathcal{B}, \nu)$ are (σ-finite) measure spaces. Let $(Z, \mathcal{C}, \lambda)$ be the product measure space, i.e.

$$Z = X \times Y, \qquad \mathcal{C} = \mathcal{A} \times \mathcal{B}, \qquad \lambda = \mu \times \nu.$$

Let f be an integrable function on $(Z, \mathcal{C}, \lambda)$, and, as usual, denote its value for the point $z = (x, y) \in Z$ by the symbol $f(x, y)$. If we fix y and let x go through all over X, we obtain a function f_y on $(X, \mathcal{A}, \mu)$. It can be shown that, for almost all $y \in Y$, the functions f_y are integrable on $(X, \mathcal{A}, \mu)$, i.e.

$$-\infty < \int_X f(x, y) \, d\mu < +\infty \quad \text{for almost all fixed } y.$$

Similarly, if we fix x and let y vary freely over Y, the functions f_x on $(Y, \mathcal{B}, \nu)$ are integrable for almost all choices of $x \in X$, i.e.

$$-\infty < \int_Y f(x, y) \, d\nu < +\infty \quad \text{for almost all fixed } x.$$

More important than these results is, however, *Fubini's theorem*:†

THEOREM 8.3(12). *Let f be integrable on $(Z, \mathcal{C}, \lambda)$. Then the following hold*:

(a) *The function h on $(X, \mathcal{A}, \mu)$, defined by*

$$h(x) = \int_Y f(x, y) \, d\nu$$

†Often the already quoted results are also referred to as a part of Fubini's theorem.

is integrable, i.e. for the "iterated integral"† we have

$$-\infty < \int_X \left(\int_Y f(x, y) \, dv \right) d\mu < +\infty.$$

(b) *The function g on $(Y, \mathcal{B}, \mu)$, defined by*

$$g(x) = \int_X f(x, y) \, d\mu$$

is integrable, i.e. for the "iterated integral" we have

$$-\infty < \int_Y \left(\int_X f(x, y) \, d\mu \right) dv < +\infty.$$

(c) *The values of the two iterated integrals are equal and in fact coincide with the value of the integral of f on $(Z, \mathcal{C}, \lambda)$, i.e.*

$$\int_Z f(x, y) \, d\lambda = \int_Y \left(\int_X f(x, y) \, d\mu \right) dv = \int_X \left(\int_Y f(x, y) \, dv \right) d\mu.$$

In passing we note that the integral of f on $(Z, \mathcal{C}, \lambda)$ is often called the "double integral" of $f(x, y)$ and it is customary to use the notation

$$\int_Z f \, d\lambda \equiv \iint_Z f(x, y) \, d\mu \, dv.$$

Fubini's theorem is very powerful because under the single assumption of f being integrable on $(Z, \mathcal{C}, \lambda)$, it permits the calculation of this integral by the method of iteration, performing the two successive integrations in either order. As a bonus, Fubini's theorem also provides a method to construct from the given f, several integrable functions on X and Y.

On the other hand, care must be exercised, because *there is no converse theorem*: the existence and equality of the two iterated integrals is a necessary but *not sufficient* criterion for the integrability of f on Z. Thus, calculating the iterated integrals and finding their values finite and equal to each other, does not permit the conclusion that $\int_Z f \, d\lambda$ exists. In other words, if f is not integrable on Z, it still may happen that one or even both of the iterated integrals exist. In the latter case they may be either equal or not. Of course, if we find that only one of the two integrals is finite or if they do not agree, we can surely conclude that f is *not* integrable.

---

†There are several subtleties here. First, note that because of our preparatory remarks, $h(x)$ is defined a.e. on $(X, \mathcal{A}, \mu)$, hence we *may* consider its integral. Second, it is tacitly asserted that $h(x)$ is measurable and that $\int_X h(x) \, d\mu$ is well defined (not $+\infty - (+\infty)$); and third, the explicit statement tells us that this integral is finite.

PROBLEMS

8.3-1. Prove the following: If f is integrable and positive a.e. on a measurable set E, and if $\int_E f \, d\mu = 0$, then $\mu(E) = 0$.

8.3-2. Prove the following theorem: If f is measurable and g is integrable and if $|f| \le |g|$ a.e., then f is integrable. (*Hint*: Refer to the ordinate sets.)

8.3-3. Let f be a measurable complex valued function. Show that f is integrable iff $|f|$ is integrable.

8.3-4. Let f be a real integrable function and show that $|\int f \, d\mu| = \int |f| \, d\mu$ iff either $f \ge 0$ a.e. or $f \le 0$ a.e.

8.3-5. Let f be a complex valued integrable function and show that $|\int f \, d\mu| = \int |f| \, d\mu$ iff $f(x)$ has the form $f(x) \overset{\circ}{=} c|f(x)|$, where c is a *constant* with $|c| = 1$.

8.3-6. Prove the following theorem: If f is integrable and $\int_F f \, d\mu = 0$ for every measurable set F, then $f = 0$ a.e. (*Hint*: Consider the sets $E = \{x | f(x) > 0\}$ and $G = \{x | f(x) < 0\}$ and use the Corollary to Theorem 8.3(1).)

8.3-7. Let (f_n) be a sequence of nonnegative integrable functions. Suppose that

$$\sum_{n=1}^{\infty} \int f_n \, d\mu < \infty. \qquad (\alpha)$$

Using Theorem 8.3(9) show that $\sum_{n=1}^{\infty} f_n(x)$ converges a.e. to an integrable function $f(x)$ and

$$\int f \, d\mu \equiv \int \sum_{n=1}^{\infty} f_n \, d\mu = \sum_{n=1}^{\infty} \int f_n \, d\mu.$$

(*Remark*: This theorem shows that, under the stated circumstances integration can be done termwise. As a matter of fact, the restriction concerning the non-negativeness of f_n can be removed provided the criterion (α) is replaced by

$$\sum_{n=1}^{\infty} \int |f_n| \, d\mu < \infty. \qquad (\alpha')$$

This is so because absolute convergence implies convergence.)

8.3-8. Let f_n be the "n-tooth saw-function" on the real interval $[0, 1)$, defined by

$$f_n(x) = \begin{cases} 0 & \text{if } \dfrac{k}{2n} \le x < \dfrac{k+1}{2n}, & k = 0, 2, 4, \dots, 2n-2 \\ 1 & \text{if } \dfrac{k}{2n} \le x < \dfrac{k+1}{2n}, & k = 1, 3, 5, \dots, 2n-1. \end{cases}$$

Show that the "infinitely fine saw-function" given by $f(x) = \lim_{n \to \infty} f_n(x)$ is Lebesgue integrable and calculate its integral. Is f Riemann integrable?

8.3-9. Let (f_n) be a sequence of integrable functions on a measure space and suppose there exists some integrable function f such that $\lim_{n \to \infty} \int |f_n - f| \, d\mu = 0$. We then say that (f_n) *converges in the mean* to f. Show that a sequence (f_n) satisfying the criteria of the Lebesgue dominated convergence theorem converges in the mean. (In other words, prove that if (f_n) satisfies the Lebesgue criteria, then it converges not only pointwise (as assumed) but also in the mean, and the pointwise limit is equal to the limit in the mean.)

8.3-10. The notion of "convergence in the mean," as introduced in Problem 8.3-9, is rather subtle and must not be confused with pointwise convergence a.e. Neither implies the other. This and the next problem illustrate these points.

Let $(X, \mathcal{B}, \mu)$ be the measure space of positive integers (cf. Problem 8.2-7). Let

$$f_n(k) = \begin{cases} \dfrac{1}{k} & \text{if } 1 \leq k \leq n, \\ 0 & \text{if } k > n. \end{cases}$$

Show that although this sequence converges pointwise everywhere to the constant function $f(k) = 1/k$, it does not converge to f in the mean. What went wrong? (That is, which part of the Lebesgue criteria is not satisfied?)

8.3-11. Let $(X, \mathcal{B}, \mu)$ be the measure space of the reals with the Lebesgue measure. Let

$$f_1 = \chi_{[0,1]}, f_2 = \chi_{[0,1/2]}, f_3 = \chi_{[1/2,1]}, f_4 = \chi_{[0,1/3]}, f_5 = \chi_{[1/3,2/3]}, f_6 = \chi_{[2/3,1]}, \ldots,$$

where $\chi_{[a,b]}$ is the characteristic function of the interval $[a, b]$. Show that although this sequence does not converge pointwise a.e. to any function, yet it converges in the mean to the zero function on X.

8.3-12. Let f be integrable. Show that there exists a sequence (u_n) of simple functions which converges in the mean to f, i.e. there exists a simple sequence (u_n) such that

$$\lim_{n \to \infty} \int |u_n - f| \, d\mu = 0.$$

(*Hint*: Use the Lebesgue dominated convergence theorem, split $f = f^+ - f^-$, and recall the original definition of the integral of a nonnegative function. Be very careful in justifying all details of the calculation.)

8.3-13. Show that the theorem which you derived in the previous problem holds true even if f is a *complex* valued integrable function. (Then, of course, u_n is a complex simple function, i.e. one whose real and imaginary parts are simple functions.)

8.3-14. Prove the following generalization of Theorem 8.3(8). If f is integrable over A and if $A = \bigcup_{k=1}^{\infty} A_k$ with $A_k \cap A_j = \emptyset$, A_k measurable, then f is integrable over each A_k and

$$\int_A f \, d\mu = \sum_{k=1}^{\infty} \int_{A_k} f \, d\mu.$$

(*Hint*: Write $B_n = \bigcup_{k=1}^{n} A_k$, define $f_n = f\chi_{B_n}$, and apply the Lebesgue Theorem 8.3(11) as well as Theorem 8.3(8).)

8.3-15. Let f and g be integrable functions on X and Y, respectively. Define $\psi(x, y) = f(x)g(y)$. Show that ψ is integrable on $X \times Y$ and that

$$\int_Z \psi \, d\lambda = \int_X f \, d\mu \cdot \int_Y g \, d\nu.$$

(*Hint*: The first part of the statement is the more difficult to prove. It is probably easier to use the definition of the integral in terms of the measure of the ordinate set, rather than the one based on simple functions. The second part of the statement follows easily from Fubini's theorem.)

8.4 COMMENTS ON LEBESGUE AND LEBESGUE–STIELTJES INTEGRALS

The most frequently occurring generalized integral is, of course, the Lebesque integral of functions on the reals, defined over subsets of the reals equipped with Lebesque measure. Most of our previous examples were taken from problems of this kind. In this section we wish to record some points of special interest concerning Lebesgue integration.

First we have a remark on notation. It is customary to denote the Lebesgue integral of f over **R** or over a subset E of **R** by the symbol

$$\int_E f(x) \, d\mu(x) \quad \text{or simply} \quad \int_E f(x) \, dx.$$

In particular, if E is an interval, often the same notation is used as if we had a Riemann integral, i.e. we write

$$\int_a^b f(x) \, dx \quad \text{and} \quad \int_{-\infty}^{+\infty} f(x) \, d\mu(x) \quad \text{or even} \quad \int_{-\infty}^{+\infty} f(x) \, dx.$$

At this point we recall that (whenever it exists at all) the Riemann integral of a function over a (bounded) interval is always meant to refer to a *closed* interval $[a, b]$. For Lebesgue integration, no such restriction of the concept exists: at the endpoints we may alter f in any way we wish (the set $\{a, b\}$ has measure zero), so that Lebesgue integration makes sense for closed, open, or semiclosed intervals and the value of the integral is in each case the same.

Next, we wish to compare Riemann and Lebesgue integrability. *If f is Riemann integrable† on a closed and bounded interval $[a, b]$, then f is also Lebesgue integrable on $[a, b]$ and the two integrals have the same*

---

†Incidentally, f is Riemann integrable on $[a, b]$ iff it is continuous almost everywhere (relative to Lebesgue measure).

value. This can be seen rather easily by considering the Lebesgue integral as the measure of the ordinate set and comparing with the standard definition of the Riemann integral which, in the given case, reduces to the "area under the curve."

Of course, the converse does not hold: we already met examples of functions which were Lebesgue integrable on some interval $[a, b]$, but for which the Riemann integral could not be defined. Lebesgue, but not Riemann, integrable functions on intervals may arise either in trivial cases (e.g., if f has countably many points where it is infinite), or when one considers a wildly discontinuous function, or, more significantly, when limits of sequences of Riemann integrable functions are concerned. (See, for example, the illustration that followed Theorem 8.3(11a), or Problem 8.3-8.)

The above-mentioned relation between Riemann and Lebesgue integrable functions for bounded intervals fails to hold if one considers unbounded intervals. The Riemann integral of a function on an unbounded interval (if it exists) is the so-called "improper" Riemann integral, defined by

$$\int_a^\infty f\,dx = \lim_{b\to\infty} \int_a^b f\,dx.$$

It is well known that even if f has an improper Riemann integral, $|f|$ need not have one (see the example on p. 356). On the other hand, as we recall, Lebesgue integrability of f always implies that of $|f|$. Therefore, on unbounded intervals there can exist functions which have an (improper) Riemann integral but which are not Lebesgue integrable.†

Naturally, in contrast to Riemann integrals, a Lebesgue integral is meaningful even if taken over sets which are not intervals (or unions of intervals). One rarely meets this situation, but we give an example. Let E be the interval $[0, 1]$ and let A be that subset of E which consists of all numbers $0 < x < 1$ except the sequence $0.01, 0.01001, 0.010010001, \ldots$. Let $f: A \to \mathbf{R}$ be given by $f(x) = x$. To calculate $\int_A f(x)\,d\mu(x)$, we only note that we can replace f by the function $g : [0, 1] \to \mathbf{R}$ given by $g(x) = x$, since the domain of f and g differ only by a set of measure zero and, on their

†In passing we note here that there is no need to talk about "improper Lebesgue integrals." For *all* measurable sets of the reals, the Lebesgue integral is defined by the same procedure. Nevertheless, the term "improper Lebesgue integral" is sometimes used for the case when f becomes unbounded at an endpoint (which may be either $x = \pm\infty$ or even a finite point). The evaluation of such Lebesgue integrals is very simple, cf. Problem 8.2-5.

common domain, they agree everywhere. Hence,

$$\int_A f \, d\mu(x) = \int_0^1 g(x) \, dx = \frac{1}{2}.$$

While we are at the comparison of Lebesgue and Riemann integrals, it may be interesting to consider the following argument. The familiar way to "compute" the Riemann integral over a closed and bounded interval $[a, b]$ is to take a partition

$$a = x_1 < x_2 < x_3 \cdots < x_n = b,$$

set the length of $(x_\nu, x_{\nu+1})$ equal to $\Delta_\nu x$, choose a point ξ_ν such that $x_\nu < \xi_\nu < x_{\nu+1}$, compute the Riemann sum

$$\sum_{\nu=0}^n f(\xi_\nu)\Delta_\nu x,$$

and then take the limit for an "infinite refinement" of the partition of $[a, b]$. Now, it can be shown that the Lebesgue integral may be "computed" in a rather similar fashion. One starts with the *codomain* of f, which is an interval† $[\alpha, \beta]$ of the "Y-axis." Then one takes a partition

$$\alpha = f(x_1) < f(x_2) < f(x_3) < \cdots < f(x_n) = \beta,$$

and sets $(f(x_\nu), f(x_{\nu+1})) = \Delta_\nu y$. Next, one finds the inverse image of the interval $\Delta_\nu y$ on the real axis. Denoting this subset of **R** by A_ν and choosing a point $f(\eta_\nu)$ in the codomain such that $f(x_\nu) < f(\eta_\nu) < f(x_{\nu+1})$, one then computes the "Lebesgue sum"

$$\sum_{\nu=1}^n f(\eta_\nu)\mu(A_\nu).$$

The limit of this sum for an infinite refinement of the partition of $[\alpha, \beta]$ can be shown to give the Lebesgue integral. The process is visualized in Fig. 8.6. Comparing with the Riemann integral, we see that *we now approach the problem from the opposite end*. The subdivisions are made in the codomain, not in the domain. Therefore, what becomes significant on the real axis, is not the length of an interval but rather the measure of some set. The figure also reveals that, for continuous functions the two constructions are identical.‡

†We assume that f is bounded.

‡We remark here that the sketched method may be generalized for the definition of the integral of functions on arbitrary measure spaces, but this approach does not have much merit.

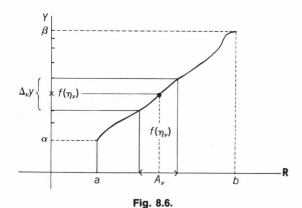

Fig. 8.6.

As a final remark on Lebesgue integrals we note the following. Let f be Lebesgue integrable over a bounded closed interval $[a, b]$. Let us define the "*primitive function of f*" as the map $\varphi : [a, b] \to \mathbf{R}$ given by

$$\varphi(x) = \varphi(a) + \int_a^x f \, d\mu(x), \qquad (8.11)$$

where $\varphi(a)$ is an arbitrary real number. With the Radon–Nikodym theorem (to be discussed in the next section) it can be shown that φ *is continuous and differentiable* a.e. in $[a, b]$, and that actually

$$\frac{d\varphi}{dx} = f \quad \text{a.e.} \qquad (8.11a)$$

In the present context, we may use these results to conclude that the usual method of *partial integration* can be used for Lebesgue integrals. That is, if f and g are Lebesgue integrable on $[a, b]$ and φ and ψ denote their primitive functions as defined by Eq. (8.11), then we have the relation

$$\int_a^b \varphi(x)g(x) \, d\mu(x) = [\varphi(x)\psi(x)]_a^b - \int_a^b f(x)\psi(x) \, d\mu(x). \qquad (8.12)$$

In Section 8.5 we shall show that the familiar *substitution law* may be also used for the evaluation of Lebesgue integrals.

Regarding Lebesgue integration on $\mathbf{R}^n$, there is nothing basically new to add. Of course, Lebesgue integrals over hypersurfaces will be always zero, because they have zero Lebesgue measure. One important comment is that repeated use of Fubini's theorem greatly simplifies the evaluation of Lebesgue integrals in $\mathbf{R}^n$.

We now turn our interest to *Lebesgue–Stieltjes integrals* over Lebesgue–Stieltjes measurable sets of **R**. If the Lebesgue–Stieltjes measure is generated by a function $\rho(x)$, it is customary to use the symbols

$$\int_E f\, d\mu_\rho \quad \text{or} \quad \int_E f(x)\, d\mu_\rho(x) \quad \text{or even†} \quad \int_E f(x)\, d\rho(x)$$

to denote the integral of f over the Lebesgue–Stieltjes measurable set E with respect to the Lebesgue–Stieltjes measure generated by ρ. In particular, if E is an interval, the symbols

$$\int_a^b f\, d\mu_\rho, \qquad \int_a^b f(x)\, d\mu_\rho(x), \qquad \int_a^b f(x)\, d\rho(x), \qquad \text{etc.}$$

are often used. It can be shown that if $[a, b]$ is a closed and bounded interval over which f is Riemann–Stieltjes integrable with respect to a function ρ of bounded variation,‡ then f is also Lebesgue–Stieltjes integrable with respect to the Lebesgue–Stieltjes measure μ_ρ generated by ρ and the two integrals have the same value.

It will be useful to remember how the concept of "primitive function" generalizes for the case of Lebesgue–Stieltjes integration. If f is Lebesgue–Stieltjes integrable, then the function

$$\varphi(x) = \varphi(a) + \int_a^x f\, d\rho(x) \tag{8.13a}$$

is called the primitive of f and the Radon–Nikodyn theorem (cf. next section) tells us that φ is continuous and differentiable a.e. Furthermore we have

$$\frac{d\varphi}{d\rho} = f \quad \text{a.e.} \tag{8.13b}$$

Particularly interesting are Lebesgue–Stieltjes integrals corresponding to a discrete measure. Since such measures lead to many null sets, it will be possible to integrate "very badly behaved" functions. Apart from discrete measures, Lebesgue–Stieltjes integrals may have another attractive feature, namely when $\mu_\rho(\mathbf{R})$ is finite. Then certain functions (such as a constant) which are not Lebesgue integrable over **R**, become Lebesgue–Stieltjes integrable. On the other hand, since for a

---

†This last notation is somewhat misleading, and is borrowed from the standard practice for Riemann–Stieltjes integrals.

‡The reader unfamiliar with the classical theory of Riemann–Stieltjes integration may consult any standard text on classical analysis.

Lebesgue–Stieltjes measure, singleton sets may have nonvanishing measure, unboundedness of a function in only finite many points may cause f to become not integrable.

An interesting example of a Lebesgue–Stieltjes integral is given by the discrete measure generated by the step function

$$\rho(x) = \theta(x) = \begin{cases} +1 & \text{if } x \geq 0 \\ 0 & \text{if } x < 0. \end{cases}$$

Let f be an extended real valued function, measurable with respect to the complete measure space corresponding to μ_θ and subject to the only restriction that it be finite at $x = 0$. Then it is integrable and actually†

$$\int_{-\infty}^{+\infty} f \, d\theta(x) = f(0).$$

This follows immediately from the fact that we may disregard all subsets of **R** except the singleton $\{0\}$, which has measure one. Integrals of this type are very familiar in physics and are called "δ-function integrals." They are written in the form

$$\int_{-\infty}^{+\infty} f(x) \delta(x) \, dx = f(0),$$

where the "Dirac delta function" is stipulated to be zero everywhere except at $x = 0$ where it assumes the value ∞ in such a way that $\int_{-\infty}^{+\infty} \delta(x) \, dx = 1$. Of course, such a "function" does not exist, since (interpreting the integral as a Riemann or even ordinary Lebesgue integral, as is understood) its value ought to be zero. We now see that there is no real need for such an imaginary object. Formally, we may write $d\theta(x) = \delta(x) \, dx$.

More generally, if ρ generates a discrete measure concentrated in the points x_k with weights ρ_k, and if f is measurable with respect to‡ μ_ρ and finite at all x_k, we have

$$\int_{-\infty}^{+\infty} f \, d\rho(x) = \sum_k f(x_k) \rho_k. \tag{8.14}$$

Thus, *weighted sums* at countably many points of even very badly behaved functions may be represented by a suitable Lebesgue–Stieltjes integral.

†Following standard practice, we use the notation $d\theta(x)$ instead of the more reasonable $d\mu_\theta$ symbol.

‡That is, with respect to the complete measure space corresponding to μ_ρ.

PROBLEMS

8.4-1. Show that if $f: \mathbf{R} \to \hat{\mathbf{R}}$ is continuous on $\mathbf{R}$ and if $N(f) = \{x \mid f(x) \neq 0\}$ is a bounded set of $\mathbf{R}$, then f is Lebesgue integrable on $\mathbf{R}$.

8.4-2. Let $f: [0, 1] \to \mathbf{R}$ be given by

$$f(x) = \begin{cases} 5 & \text{if } x \text{ is irrational} \\ 2 & \text{if } x \text{ is rational.} \end{cases}$$

Calculate the Lebesgue integral $\int_0^1 f(x)\, dx$.

8.4-3. Let f and g be Lebesgue integrable over $\mathbf{R}$. Define the *convolution* of f and g by setting

$$(f * g)(x) = \int_{-\infty}^{+\infty} f(x - y)g(y)\, dy.$$

Show that $f * g$ is Lebesgue integrable over $\mathbf{R}$. (*Hint*: Recall Theorem 8.3(6), estimate $\int_{-\infty}^{+\infty} |f * g|\, dx$ by Theorem 8.3(5), use the substitution law of integration, apply the result of Problem 8.3-15 for the special case when $X = Y = \mathbf{R}$, and finally appeal again to Theorem 8.3(6).)

8.4-4. Let X be the positive real axis, and let μ_ρ be the measure generated by the function $\rho(x) = 1 - e^{-x}$. Let

$$f(x) = \begin{cases} +\infty & \text{if } x \text{ is integer} \\ c > 0 & \text{if } x \text{ is not integer.} \end{cases}$$

Calculate $\int_0^\infty f(x)\, d\mu_\rho(x)$. Now consider the function $f(x) = x$. Is this integrable over X? (*Hint*: Represent $f(x)$ as the limit of the sequence

$$f_n(x) = \begin{cases} x & \text{if } 0 \leqslant x \leqslant n \\ n & \text{if } x > n, \end{cases}$$

and consider the ordinate sets. When doing this for the interval $0 \leqslant x \leqslant n$, it may help to observe that

$$\sum_{l=0}^{k} e^{-l\alpha}(l + 1) = \sum_{l=0}^{k} e^{-l\alpha} - \frac{d}{d\alpha} \sum_{l=0}^{k} e^{-\alpha},$$

where $\sum_{l=0}^{k} e^{-l\alpha}$ is a finite geometrical series.)

8.5 THE RADON–NIKODYM THEOREM

In our previous work, we studied the generalization of the "definite Riemann integral." We now turn our attention to the generalization of the "indefinite integral" concept.

Let us start by surveying familiar ground. Suppose f is continuous on a bounded interval $[a, b]$ and hence, f is Riemann (and, of course, Lebesgue) integrable. It is well known that if we define the "primitive function" or the "indefinite integral" of f by setting

$$\rho(x) = \rho(a) + \int_a^x f(x)\, dx, \qquad a \le x \le b, \tag{8.15}$$

where $\rho(a)$ is an arbitrary constant, then $\rho(x)$ is continuous and differentiable on $[a, b]$, and in fact $d\rho/dx = f$. We can then express the (definite) integral of f in terms of its primitive:

$$\int_a^b f\, dx = \rho(b) - \rho(a). \tag{8.15a}$$

Suppose that, in addition to continuity, f is also nonnegative. Then we see from Eq. (8.15) that $\rho(x) \ge \rho(y)$ if $x \ge y$, (because the definite integral in Eq. (8.15) is positive and a monotone function of the length of the interval). Thus, ρ is a nondecreasing continuous function and as such, it will generate a Lebesgue–Stieltjes measure ν on $[a, b]$. We then have $\nu([a, b]) = \rho(b) - \rho(a)$, so that in view of Eq. (8.15a) we have

$$\nu([a, b]) = \int_a^b f\, dx.$$

Furthermore, since without restricting generality, we may take $\rho(a) = 0$, we may write

$$\nu([a, x]) = \rho(x) \quad \text{for all} \quad a \le x \le b.$$

We now inquire about the converse of these results. Suppose there is *given* a continuous Lebesgue–Stieltjes measure ν on the class of all intervals of the form $[a, x]$ with a fixed, $a \le x \le b$, and having the values $\nu([a, x]) = \rho(x)$. Is there a function f on $[a, b]$ such

$$\int_a^x f(x)\, dx = \nu([a, x]) \quad \text{for all} \quad a \le x \le b?$$

The answer is affirmative,† since indeed the function f defined a.e. by

$$f(x) = \frac{d\rho}{dx} \equiv \frac{d\nu([a, x])}{dx}$$

satisfies the demand.

In the following we shall study how these simple facts generalize.

Let $(X, \mathcal{B}, \mu)$ be a totally σ-finite measure space and f be integrable on X. Let us define the set function

$$\nu \colon \mathcal{B} \to \hat{\mathbf{R}}$$

†Provided the measure ν satisfies a mild technical criterion, cf. Theorem 8.5(2) below.

by setting

$$\nu(B) = \int_B f \, d\mu \qquad (8.16)$$

for every measurable set B. It is by obvious analogy that we may call ν the "*indefinite integral of f.*"

Suppose that, for the time being, f is a.e. nonnegative. Then, clearly, $\nu(B) \geqslant 0$ for all $B \in \mathscr{B}$ and we also see that $\nu(\emptyset) = 0$. Finally, if (B_k) is a countable sequence of disjoint measurable sets, then

$$\nu\left(\bigcup_{k=1}^{\infty} B_k\right) = \sum_{k=1}^{\infty} \nu(B_k).$$

This follows from noting that

$$\nu\left(\bigcup_{k=1}^{\infty} B_k\right) = \int_{\bigcup_k B_k} f \, d\mu = \sum_{k=1}^{\infty} \int_{B_k} f \, d\mu = \sum_{k=1}^{\infty} \nu(B_k),$$

where we used the most general form of Theorem 8.3(8). Combining these simple results, we see that ν *is a measure on* $\mathscr{B}$. Actually, it is a *totally finite* measure, because $\nu(X) = \int_X f \, d\mu < \infty$ by assumption. Our next observation is that ν is inferior to the given measure μ, i.e. $\nu < \mu$. This follows from noting that if $\mu(B) = 0$, then $\nu(B) = \int_B f \, d\mu = 0$. Furthermore, if f is not only nonnegative, but actually positive definite a.e. on X, then we also have $\mu < \nu$. This follows from noting that if $\nu(B) = \int_B f \, d\mu = 0$, then, f being positive, $\mu(B)$ must be zero (cf. Problem 8.3-1). Thus, *for a.e. positive f, the two measures ν and μ are equivalent.*

Now we remove the restriction of nonnegativity of f, i.e. we only assume that f is integrable. Countable additivity and $\nu(\emptyset) = 0$ still follows from the definition (8.16), but we can no longer conclude that $\nu(B) \geqslant 0$. But we also see that $-\infty < \nu(B) < +\infty$, even for $B = X$, so that putting together all this information, we conclude that ν *is a totally finite signed measure.* If we set $f = f^+ - f^-$ and define $\nu^+(B) = \int_B f^+ \, d\mu$ and $\nu^-(B) = \int_B f^- \, d\mu$, then each of these is a (nonnegative) measure and $\nu(B) = \nu^+(B) - \nu^-(B)$. As a matter of fact, this is the Jordan decomposition of ν, which follows from the fact that $\sup_{B \supset A} \nu(A) = \nu^+(B)$ and a similar result for $\nu^-(B)$. Finally, as it was for $f \geqslant 0$, we again have $\nu < \mu$, in the sense as is used for signed measures, cf. Problem 7.2d-2. Indeed, by definition, $|\nu|(B) = \nu^+(B) + \nu^-(B)$ and so $|\nu|(B) = \int_B |f| \, d\mu$. Therefore, if $\mu(B) = 0$, we have $|\nu|(B) = 0$, which means $\nu^+(B) = \nu^-(B) = 0$ so that $\nu(B) = 0$. But, of course, we now surely cannot have μ equivalent to ν.

We summarize these results in

THEOREM 8.5(1).   *Let f be integrable on the totally σ-finite measure space $(X, \mathcal{B}, \mu)$. Then $\nu(B) = \int_B f \, d\mu$ defines a new, signed measure on $(X, \mathcal{B})$. This ν is totally finite and $\nu \prec \mu$. Its Jordan decomposition is given by $\nu^+(B) = \int_B f^+ \, d\mu$, $\nu^-(B) = \int_B f^- \, d\mu$.*

One of the customary applications of this theorem is its use for generating new and very good measures ν^+, ν^-, ν from a given measure μ, by choosing suitable integrable functions f.

Again, we may ask about a converse result. Given a measure μ on $(X, \mathcal{B})$, is it possible to generate every inferior totally finite signed measure ν by integrating some function on X with respect to μ? This question is answered by the celebrated Radon–Nikodym theorem:

THEOREM 8.5(2).   *Let $(X, \mathcal{B}, \mu)$ be a totally σ-finite measure space. Let ν be a totally finite signed measure, inferior to μ, i.e. $\nu \prec \mu$. Then there exists a function f which is integrable on X and is such that $\nu(B) = \int_B f \, d\mu$ for every $B \in \mathcal{B}$. This f is almost unique in the sense that, if another function g also yields $\nu(B) = \int_B g \, d\mu$, then $g \overset{\circ}{=} f$ with respect to μ.*

For the rather technical proof the reader is referred, for example, to Halmos [11], p. 129. We only have two comments.

(a) If ν is actually a nonnegative measure, then the associated function f is also nonnegative.

(b) The theorem has a generalization for the case when ν is not totally finite but only totally σ-finite. Then (if the other conditions hold) it is still possible to find a measurable and almost everywhere finite f such that $\nu(B) = \int_B f \, d\mu$, but f will not be integrable. Total finiteness of ν is a sufficient and necessary criterion for f to be integrable.

We illustrate the Radon–Nikodym theorem on a rather simple example. Let X be the set of all positive integers, $X = \{1, 2, \ldots, n, \ldots\}$, let $\mathcal{B}$ be the class of all subsets of X. Let (α_k) be a sequence of positive definite numbers such that $\Sigma_{k=1}^{\infty} \alpha_k < \infty$. Define μ on $\mathcal{B}$ by setting $\mu(B) = \Sigma_{k \in B} \alpha_k$. It is easy to check that this is a measure on $(X, \mathcal{B})$ (actually, it is a slight generalization of the measure discussed in Problem 8.2-7). In the present case, the measure of the singleton set $\{n\}$ is $\mu(\{n\}) = \alpha_n$. Our measure happens to be totally finite, since $\mu(X) = \Sigma_{k=1}^{\infty} \alpha_k < \infty$ by assumption. Any function $\psi : X \to \hat{\mathbf{R}}$ is measurable, and, at least formally, we have

$$\int_B \psi \, d\mu = \sum_{k \in B} \psi(k) \alpha_k,$$

in particular,

$$\int_X \psi \, d\mu = \sum_{k=1}^{\infty} \psi(k)\alpha_k.$$

(This statement can be verified, for example, by generalizing the method of Problem 8.2-7.) Furthermore, if ψ is such that $\Sigma_{k \in B} |\psi(k)|\alpha_k < \infty$, then ψ is integrable over B.

Now define on $(X, \mathcal{B})$ another, similar measure ν by selecting another positive definite sequence (β_k) which satisfies $\Sigma_{k=1}^{\infty} \beta_k < \infty$ and setting $\nu(B) = \Sigma_{k \in B} \beta_k$.

The two measures μ and ν happen to be equivalent, because in each case the only null set is the empty set. Thus, $\nu < \mu$ (and also $\mu < \nu$, but for the present, this is irrelevant). Therefore, by the Radon–Nikodym theorem, there should exist an integrable function f on $(X, \mathcal{B}, \mu)$ such that $\nu(B) = \int_B f \, d\mu \equiv \Sigma_{k \in B} f(k)\alpha_k$ for any $B \in \mathcal{B}$. Indeed, consider the function f defined on $(X, \mathcal{B}, \mu)$ by

$$f(k) = \frac{\beta_k}{\alpha_k}.$$

This f is integrable because

$$\sum_{k=1}^{\infty} |f(k)|\alpha_k = \sum_{k=1}^{\infty} \beta_k < \infty;$$

and we find

$$\int_B f \, d\mu = \sum_{k \in B} \frac{\beta_k}{\alpha_k} \alpha_k = \sum_{k \in B} \beta_k = \nu(B),$$

as we wanted. In the present case, f is uniquely determined, because the only null set of $(X, \mathcal{B}, \mu)$ is the empty set.

Since $\mu \approx \nu$, in our case we also have $\mu < \nu$. Therefore, there must exist an integrable function g on $(X, \mathcal{B}, \nu)$ such that $\mu(B) = \int_B g \, d\nu$. Naturally, $g = \alpha_k/\beta_k$ serves the purpose.†

To illustrate comment (b) made after the Radon–Nikodym theorem, let us modify our example by choosing the positive numbers β_k so that $\Sigma_{k=1}^{\infty} \beta_k = \infty$. Then the measure ν is only totally σ-finite: there exist some infinite sets such that $\mu(B) = \Sigma_{k \in B} \beta_k = \infty$. Nevertheless, the function $f(k) = \beta_k/\alpha_k$ still is a solution for $\nu(B) = \int_B f \, d\mu \equiv \Sigma_{k \in B} f(k)\beta_k$, but obviously, f is not integrable over sets for which $\nu(B) = \infty$. In particular, $\int_X f \, d\mu = +\infty$. On the other hand, $f(k)$ is everywhere finite, because $\alpha_k \neq 0$, $\beta_k \neq \infty$.

---

†Observe that $g = 1/f$. This illustrates the general theorem expressed in the last part of Problem 8.5-3.

Returning to general considerations, we point out that there exists a very suggestive and useful notation and terminology for the function f that occurs in the integrand of the Radon–Nikodym theorem. The function f *is called the Radon–Nikodym derivative of ν with respect to μ* and we write, symbolically,

$$f = \frac{d\nu}{d\mu} \quad \text{or} \quad d\nu = f\,d\mu. \tag{8.17}$$

With this notation, we have

$$\nu(B) = \int_B \frac{d\nu}{d\mu}\,d\mu.$$

Of course, in general, this is nothing more than a formal, symbolic notation. However, if, in particular, $(X, \mathscr{B})$ is the real line with its usual Borel structure and μ_σ and ν_ρ are Lebesgue–Stieltjes measures generated by σ and ρ respectively, then it turns out that

$$\frac{d\nu_\rho}{d\mu_\sigma} = \frac{d\rho}{d\sigma}. \tag{8.18a}$$

More specifically, if $\sigma(x) = x$, i.e. if $\mu_\sigma = \mu$ is the standard Lebesgue measure, then

$$\frac{d\nu_\rho}{d\mu} = \frac{d\rho}{dx}, \tag{8.18b}$$

i.e. *the Radon–Nikodym derivative of the Lebesgue–Stieltjes measure ν_ρ coincides* a.e.† *with the ordinary derivative of $\rho = \rho(x)$ with respect to x.* These statements follow rather easily from our earlier observation that, for a.e. continuous functions on bounded intervals, the Lebesgue–Stieltjes integral and the Riemann–Stieltjes integral coincide. Equation (8.18b) is well illustrated in the simple case we considered at the beginning of this section.

It is very useful to know that, under rather broad conditions, the familiar formal manipulations with differentials hold true also for Radon–Nikodym derivatives. For example, $d(\nu_1 + \nu_2)/d\mu = d\nu_1/d\mu + d\nu_2/d\mu$. More important is the fact that the *chain rule* holds:

THEOREM 8.5(3).   *Let $(X, \mathscr{B})$ be a Borel space. If ν is a totally σ-finite (signed) measure, and μ and λ are totally σ-finite measures, and if*

---

†The "a.e." clause applies because f, i.e. the Radon–Nikodym derivative, is uniquely determined only a.e.

$\nu < \mu < \lambda$. then

$$\frac{d\nu}{d\lambda} = \frac{d\nu}{d\mu} \frac{d\mu}{d\lambda} \quad \text{a.e. with respect to } \lambda.$$

In detail, this simply means that if $\nu(B) = \int_B f \, d\mu$, and if $\mu(B) = \int_B g \, d\lambda$ with $g \geq 0$, then fg is measurable with respect to λ and $\nu(B) = \int_B fg \, d\lambda$. An immediate consequence of this is

THEOREM 8.5(4).   Let μ and λ be totally σ-finite measures and suppose $\mu < \lambda$. Let f be finite valued a.e. Then

$$\int f \, d\mu = \int f \frac{d\mu}{d\lambda} \, d\lambda.$$

This relation is the generalization of the well-known "substitution law" of Riemann integration and may be extremely useful in the calculation of generalized integrals.

PROBLEMS

8.5-1. Let μ and ν be two totally finite measures on a measurable space and suppose $\nu < \mu$. Define $\bar{\mu} = \mu + \nu$ and show that there exists a function f such that $\nu(B) = \int_B f \, d\bar{\mu}$ for every measurable set B and such that $0 \leq f(x) < 1$ a.e. with respect to μ.

8.5-2. Let μ and ν be two totally finite measures on a measurable space and suppose $\mu \approx \nu$. Show that

$$\frac{d\mu}{d\nu} = 1 \bigg/ \frac{d\nu}{d\mu} \quad \text{a.e.}$$

(*Hint*: Use the theorem formulated in Problem 8.3-15 and note that, quite generally, if $\int_B \psi \, d\lambda = \lambda(B)$ for all measurable B, then $\psi = 1$ a.e. (Why?) Make it clear where in the proof the condition $\mu \approx \nu$ is used.) (*Remark*: The above theorem is true even if μ and ν are just σ-finite, or even σ-finite signed measures, but the proof is then quite different.)

8.5-3. Putting together the two theorems on indefinite integrals as given in the text and using also the result of Problem 8.5-2 (in its more general form), justify the following statements:

Let μ and ν be two totally σ-finite measures on a measure space $(X, \mathcal{B})$. The necessary and sufficient condition for $\nu < \mu$ is that there exist an almost everywhere uniquely determined finite valued nonnegative, measurable function f with domain X such that

$$\nu(B) = \int_B f \, d\mu$$

for every measurable set B. If also $\mu < \nu$ (i.e. if actually $\mu \approx \nu$), then the function f is positive a.e., and

$$\mu(B) = \int_B \frac{1}{f}\, d\nu.$$

If both μ and ν are totally finite, then both f and $1/f$ are integrable on X. (*Remark*: This theorem is the probably most often used form of the Radon–Nikodym theorem.)

8.5-4. Show that the Lebesgue–Stieltjes measure generated by $\rho(x) = 1 - e^{-x}$ is equivalent to the ordinary Lebesgue measure. (*Hint*: In view of the previous problem, you must find an a.e. finite and positive measurable function whose Lebesgue integral reproduces the Lebesgue–Stieltjes measure generated by ρ for all finite intervals on the real line. Explain why this is the correct procedure.)

8.5-5. Recalculate Problem 8.4-4 with the use of the "substitution law" for integrals. (Do not forget to justify that the substitution law applies in this case.) You will be certainly impressed how useful the substitution law can be when evaluating Lebesgue–Stieltjes integrals. Incidentally, can this method be followed if the Lebesgue–Stieltjes measure is discrete?

Appendices

Appendix I: Some Inequalities

In connection with metric spaces and some other related topics, frequent use is made of certain simple algebraic inequalities. For easy reference, we list these relations in this appendix.

In the following, a_k, b_k denote arbitrary complex numbers. Sums Σ_k are either finite sums $(k = 1, 2, \ldots, n)$ or infinite sums, provided obvious convergence properties hold.

(a) *Minkowski inequality.* Let p be a real number and $1 \leqslant p < \infty$. Then

$$\left[\sum_k |a_k + b_k|^p \right]^{1/p} \leqslant \left[\sum_k |a_k|^p \right]^{1/p} + \left[\sum_k |b_k|^p \right]^{1/p}.$$

(b) *Hölder inequality.* Let p be a real number and $1 < p < \infty$. Let $q = p/(p - 1)$, i.e.

$$\frac{1}{p} + \frac{1}{q} = 1.$$

(Note that, by this definition, $1 < q < \infty$.) We then have

$$\sum_k |a_k b_k| \leqslant \left[\sum_k |a_k|^p \right]^{1/p} \left[\sum_k |b_k|^q \right]^{1/q}.$$

The special case of the Hölder inequality for $p = q = 2$ is usually called the *Cauchy inequality* (or Cauchy–Buniakowsky–Schwarz inequality),

$$\left(\sum_k |a_k b_k| \right)^2 \leqslant \left[\sum_k |a_k|^2 \right] \left[\sum_k |b_k|^2 \right].$$

Appendix III: Annotated Reading List

In the following we give a somewhat capriciously chosen, to a large extent arbitrarily selected list of text- and reference books which the reader may profitably use to fill many gaps in our exposition and to broaden his knowledge. Obviously, the inclusion or omission of a book from this list does not constitute any authoritative judgment relative to its merits. At any rate, these were the books from which the present author, during years past, attempted to learn the subject material covered in this volume and from which, he feels he profited immensely.

The listing is in alphabetical order, and a brief discussion of the topics covered by the book, as well as some hints concerning the level of sophistication and the style of exposition, is given in each case. These comments are highly personal and must not be construed as an evaluation or a critique of the quoted work. Following each description, we give, in parentheses, the chapter numbers of the present book for the study of which the quoted reference appears to be most relevant.

Here is the list.†

[1] N. AKHIEZER AND I. M. GLAZMAN: *Theory of Linear Operators in Hilbert Space.* (Ungar, 1961.) An extremely careful, highly pedagogical presentation of Hilbert space theory. A special feature of the book is that it gives much attention to the often neglected topics of unbounded operators and to questions of domains. (10, 12, 13)
[2] D. M. BURTON: *Introduction to Modern Abstract Algebra.* (Addison-Wesley, 1967.) A clear, rather simple, modern introductory exposition of the fundamental algebraic

†For books that have several editions, we give only the bibliographical data of the latest edition known to us. For books that were written in a foreign language, we give bibliographic reference only to the translation in the English language.

structures and their major properties. It features a good selection of illuminating examples and problems. (3, 4)

[3] C. CHEVALLEY: *Fundamental Concepts of Algebra.* (Academic Press, 1956.) This is a very modern, advanced level introductory text to the fundamental algebraic structures, written in the now classic "abstract" style of the leading French mathematical school. (3, 4)

[4] P. M. COHN: *Lie Groups.* (Cambridge University Press, 1968.) This small volume is a concise introduction, requiring few prerequisites from algebra or topology. Even though Lie groups were not considered in our treatise, the book of Cohn is recommended as a starter for the reader's future studies in this direction. (4, 5, 6)

[5] J. DIEUDONNÉ: *Foundations of Modern Analysis.* (Academic Press, 1966.) This serious, but not very difficult text provides a solid background for classical analysis, treated in a modern way, emphasizing the *conceptual* aspects, and approaching the subjects in the now well-accepted "axiomatic" manner. It also includes a very modern approach to differential calculus and to analytic functions. (1, 2, 5, 6, 9, 10, 13)

[6] N. DUNFORD AND J. T. SCHWARTZ: *Linear Operators*, Vols. I, II, and III. (Interscience, 1963–1971.) A monumental, authoritative and encyclopedic classic on many branches of functional analysis, with a strong emphasis on linear operators and their spectral theory. While leading into frontier areas, these books manage to present even the simplest basic concepts (topological spaces, measure theory, etc.) in a compact yet clear form. Considerable effort is spent on the application of theoretical structures to applied problems of mathematical analysis and even to physics. (9, 10, 11, 12, 13)

[7] R. E. EDWARDS: *Functional Analysis.* (Holt, Reinhart & Winston, 1965.) This is a detailed, scholarly, sophisticated, carefully presented treatise on a large variety of topics, such as topological vector spaces, Banach and Hilbert spaces, dual space theory, distributions, linear maps of topological vector spaces, theory of compact operators, and so on. There is a strong emphasis on methods of computation and even on direct applications to nontrivial problems of practical analysis. (9, 10, 11, 12, App. II)

[8] B. EPSTEIN: *Linear Functional Analysis.* (Saunders, 1970.) A well-written, simple, first introduction to the topology of metric-, normed-, and inner product spaces, to Lebesgue integration on the real line, and to the elements of the theory of linear operators. Applications to classical analysis (for example, to integral equations) are often given. (5, 7, 8, 9, 10, 11, 12)

[9] G. FALK: *Algebra.* (Article in Vol. II of the "Encyclopedia of Physics," S. Flügge (Ed.), Springer, 1955.) This is a simple but quite detailed introduction to classical algebraic systems, concentrating on linear associative algebras. It appeals strongly to the fundamental needs of the physicist and it also gives numerous applications to the use of algebras in, essentially classical, structural problems of physics. (4)

[10] C. GOFFMAN AND G. PEDRICK: *First Course in Functional Analysis.* (Prentice-Hall, 1965.) An often terse, rigorous, not too difficult but sometimes a little "dry" textbook on the graduate level. It covers a large variety of topics of topological linear spaces, Banach and Hilbert spaces, integration theory, Banach algebras, etc. There are many applications to "hard analysis," and a number of rarely discussed topics is also included. (4, 7, 8, 9, 10, 11, 12)

[11] P. R. HALMOS: *Measure Theory.* (Van-Nostrand, 1950.) An easily readable, detailed, up-to-date and comprehensive monograph on most aspects of measure theory and integration, including many applications. (7, 8)

[12] P. R. HALMOS: *Naïve Set Theory.* (Van-Nostrand, 1960.) The title of this delightful text

is a modest understatement: in reality it is a "naïve" introduction to axiomatic set theory. Special emphasis is given to the theory of ordinal and cardinal numbers. (1, 2)

[13] M. HAMERMESH: *Group Theory and its Application to Physical Problems.* (Addison-Wesley, 1962.) A pedagogically outstanding coverage of basic group theory and of the simpler aspects of representation theory. It gives a fairly detailed survey of not only the important finite groups, but also of the classical Lie groups. Many straightforward applications to physics are discussed in detail. (4)

[14] S. HELGASON: *Differential Geometry and Symmetric Spaces.* (Academic Press, 1962.) In this rather advanced monograph the reader will find important material on the theory of Lie algebras and topological groups, as well as an introduction to modern differential geometry—a topic which we did not touch upon in the present book but which can be easily mastered with the tools now available to the reader. (4, 6, 10)

[15] G. HELLWIG: *Differential Operators of Mathematical Physics.* (Addison-Wesley, 1964.) Following a general introduction to Hilbert spaces, the detailed theory of differential operators on such spaces is developed, in an easy-to-follow yet comprehensive way. Many applications to both classical and modern problems are treated. (10, 12, 13)

[16] G. HELMBERG: *Introduction to Spectral Theory in Hilbert Space.* (North-Holland—Interscience, 1969.) A very detailed, carefully written and well-readable (even though occasionally pedantic) textbook on Hilbert spaces and operators. It complements well Ref. [1]. (10, 12, 13)

[17] N. JACOBSON: *Lectures in Abstract Algebra*, Vols. 1, 2, 3. (Van-Nostrand, 1953.) This textbook is a very detailed, systematic, highly rated work, covering all fundamental concepts, in a sufficiently modern language. (3, 4)†

[18] N. JACOBSON: *Lie algebras.* (Interscience, 1962.) A scholarly and well-readable, but not always easy account of all basic aspects and of many advanced topics in the theory of Lie algebras. Lie groups are not covered. (4)

[19] J. M. JAUCH: *Foundations of Quantum Mechanics.* (Addison-Wesley, 1968.) This excellent book starts with a brief but modern introduction to the basic ideas of integration theory, Hilbert space, linear functionals, operators, and spectral representations. In the main body of the book, where the formalism of modern quantum mechanics is systematically developed, the reader will find the most beautiful applications of the simpler parts of functional analysis and group theory. (10, 12, 13)

[20] T. F. JORDAN: *Linear Operators for Quantum Mechanics.* (Wiley, 1969.) The first half of this book gives a review of operators on Hilbert spaces and their spectral theory. The second half applies the framework to clarify, in an almost elementary yet very instructive way, the structure of the mathematical apparatus of quantum mechanics. (12, 13)

[21] J. L. KELLEY: *General Topology.* (Van-Nostrand, 1955.) This is a standard, detailed, yet very readable text and reference on all aspects of topological concepts. Highly recommended. (5, 6)

[22] K. KURATOWSKI: *Introduction to Set Theory and Topology.* (Addison-Wesley, 1962.) A masterly introduction to the central ideas of set theory and general topology, presented

Parts of Vol. 2 will also be useful to supplement material of our Chapters 10 and 12. Volume 3 is devoted to topics that are of lesser interest to the physicist.

with an eye to applications in modern analysis. Many well-chosen and simple examples enliven the reading. (1, 2, 5, 6)

[23] A. G. KUROSH: *The Theory of Groups.* (Chelsea, 1956.) A clear, very elegant exposition, covering both basic and quite advanced material on groups. (4)

[24] L. LIUSTERNIK AND V. SOBOLEV: *Elements of Functional Analysis.* (Ungar, 1961.) Topological spaces, linear functionals, spectral theory of bounded selfadjoint operators, and some topics in nonlinear functional analysis are covered, with a healthy emphasis on possible applications to "hard analysis." (6, 9, 10, 11, 12, 13)

[25] E. R. LORCH: *Spectral Theory.* (Oxford University Press, 1962.) This delightful and most informative text has a slightly misleading title: it is actually a self-contained introduction to Banach spaces and Banach algebras. While it is not an elementary survey, it presents the material so clearly and with such pedagogical skill that it can be highly recommended to all readers. (9, 10, 11, 12, 13)

[26] G. W. MACKEY: *Induced Representations of Groups and Quantum Mechanics.* (Benjamin, 1968.) Even though in our work we did not cover the theory of group representations, we highly recommend the serious study of Mackey's book, because it will not only introduce the reader to a modern and important topic, but also because it will illustrate the confluence of many ideas we dealt with. (4, 7, 12)

[27] S. MAC LANE AND B. BIRKHOFF: *Algebra.* (Macmillan, 1970.) A very modern and up-to-date classic on all fundamental and structural aspects of algebraic systems. Some of it is not easy reading, but worthwhile the effort. (1, 2, 3, 4)

[28] G. McCARTY: *Topology.* (McGraw-Hill, 1967.) A delightful, modern introduction to general topology and to topological groups. No prerequisites are needed to enjoy this work. Strongly recommended. (1, 2, 4, 5, 6)

[29] TH. O. MOORE: *Elementary General Topology.* (Prentice-Hall, 1964.) A very easy-to-read introduction to topological concepts, with a large selection of do-it-yourself exercises and problems. (5, 6)

[30] M. A. NAIMARK: *Normed Rings.* (Nordhoff, 1959.) An authoritative, monumental classic, covering in an encyclopedic manner vast areas of the theory of topological linear spaces, in particular Banach and Hilbert spaces. It presents a deep discussion of the theory of linear operators, from the viewpoint of Banach algebras. (8, 9, 10, 11, 12, 13)

[31] J. VON NEUMANN: *Mathematical Foundations of Quantum Mechanics.* (Princeton University Press, 1955.) This work is a renowned classic which every student of theoretical physics should read carefully. Apart from material directly relevant to the foundations of quantum mechanics, this book contains the first, and crystal clear, presentation of abstract Hilbert space theory. (9, 10, 12, 13)

[32] L. PONTRJAGIN: *Topological Groups.* (Gordon and Breach, 1966.) This is a famous treatise on the classical theory of topological groups, in particular Lie groups. It does not cover much of representation theory. It is not easy reading, but it does not demand prerequisites either. (4, 5, 6)

[33] E. PRUGOVEČKI: *Quantum Mechanics in Hilbert Space.* (Academic Press, 1971.) The purpose of this volume is to develop the tools from functional analysis that are needed for a rigorous formulation of nonrelativistic quantum theory and to illustrate these methods on relatively simple problems. In its sophistication and in its depth, this book is halfway between the similar works Refs. [19] and [20]. Hilbert space, measure and integration, spectral theory are discussed in considerable detail. (7, 8, 9, 10, 12, 13)

[34] F. RIESZ AND B. SZ.-NAGY: *Functional Analysis.* (Ungar, 1955.) This is a monumental classic on a vast field of topics in functional analysis. It covers, with great mathematical rigor and yet in an enjoyable style, Lebesgue integration, L^p spaces, general Banach- and Hilbert spaces, operator theory, spectral theory, transformation groups, and many other topics. (8, 9, 10, 11, 12, 13)

[35] I. E. SEGAL AND R. A. KUNZE: *Integrals and Operators.* (McGraw-Hill, 1968.) A very modern, elegant, yet not too difficult advanced textbook on integration theory with many applications, discussing Banach and Hilbert spaces and spectral analysis in depth. (7, 8, 9, 10, 13)

[36] G. F. SIMMONS: *Introduction to Topology and Modern Analysis.* (McGraw-Hill, 1963.) A truly outstanding, well-written, serious, even exciting introduction to general topology, topological linear spaces, (bounded) operators, and operator algebras. It does not require any previous acquaintance with the subject matter. It provides the student with motivation and with a clear, intuitively developed line of reasoning. (1, 2, 5, 6, 9, 10, 11, 12)

[37] V. I. SMIRNOV: *A Course of Higher Mathematics,* Vol. V. (Pergamon Press—Addison-Wesley, 1964.) This book covers the theory of Lebesgue integration, normed linear spaces, Hilbert spaces, and the theory of (bounded) operators. It is a very thorough and detailed text, but sometimes the continuity of development is obscured, some unusual terminology is employed (translator's fault) and a rather cumbersome notation is used. (5, 6, 8, 9, 10, 12, 13)

[38] M. H. STONE: *Linear Transformations in Hilbert Space.* (Amer. Math. Soc. Coll. Publ., 1958.) A detailed, authoritative, scholarly classic on the theory of operators in Hilbert spaces. The second half of this highly recommended book is devoted to examples and to applications on the fundamental level. (10, 11, 12, 13)

[39] A. E. TAYLOR: *Introduction to Functional Analysis.* (Wiley, 1964.) This is a well-known, rather advanced, yet easily readable broad-scoped introduction to the most important branches of the theory of topological linear spaces, operators, and their spectral analysis. It also contains an introduction to general topology and to the theory of integration. (5, 6, 8, 9, 10, 11, 12, 13)

[40] B. Z. VULIKH: *Introduction to Functional Analysis for Scientists and Technologists.* (Pergamon Press, 1963.) An elementary introduction to a surprisingly broad field of topics, such as metric- and normed spaces, Hilbert space, (bounded) linear operators and functionals, approximation methods. Applications to simple problems in analysis are strongly emphasized. (5, 6, 9, 10, 12)

[41] B. L. VAN DER WAERDEN: *Modern Algebra,* Vols. 1, 2. (Ungar, 1953.) Even though this is quite an old work (first German edition: 1931), it is strongly recommended. It presents a scholarly treatment of fundamental algebraic systems: finite groups, rings, modules, linear spaces, algebras, the theory of polynomials, ideals, and many direct applications. (3, 4)

[42] E. P. WIGNER: *Group Theory and its Applications to Quantum Mechanics of Atomic Spectra.* (Academic Press, 1959.) A well-known, excellently written classic. Provides a masterly introduction to linear spaces, operators, several classical groups, the basic theory of their representations, and illuminates, in a now historical perspective, the role which group theory played in the early applications of quantum theory to simple systems. (4, 12)

[43] A. C. ZAANEN: *Linear Analysis.* (Interscience–North-Holland, 1953.) This is a

comprehensive handbook of "soft analysis." It contains a discussion of measure and integration, of bounded operators on Banach and Hilbert spaces, and a very valuable study of the theory of linear integral equations.   (7, 8, 9, 10, 11, 12)

[44] M. ZAMANSKY: *Linear Algebra and Analysis.* (Van-Nostrand, 1969.) A rigorous first introduction to sets, functions, algebraic systems, topology, topological linear spaces, and the theory of integration.   (1, 2, 3, 4, 5, 6, 8)

[45]† G. FANO: *Mathematical Methods of Quantum Mechanics.* (McGraw-Hill, 1971.) Similar to Ref. [20], but perhaps giving more detailed, direct applications to quantum mechanics.   (10, 12, 13)

[46]† M. REED AND B. SIMON: *Methods of Modern Mathematical Physics.* Vol. 1. (Academic Press, 1972.) A rather rigorous and in-depth discussion of many topics in functional analysis.   (7, 8, 9, 10, 12, 13)

[47]† R. B. ASH: *Measure, Integration, and Functional Analysis.* A modern, rigorous, well readable introduction to integration theory and some parts of functional analysis, paying special attention to the interplay between measure theory and topology.   (5, 6, 7, 8, 9, 10, 12)

†This book has been added at proofreading stage and therefore it does not appear in this list at the appropriate alphabetical position.

Appendix IV: Frequently Used Symbols

Below we collect those symbols and notations which appear to have a fixed meaning throughout this book. They are listed essentially in the order as they appear in the book. No attempt at completeness was made.†

(a) LETTERS WITH FIXED MEANING

N: set of natural numbers, $\{0, 1, 2, \ldots\}$.

N⁺: set of positive integers, $\{1, 2, \ldots\}$.

Z: set of all integers, $\{\ldots, -2, -1, 0, +1, +2, \ldots\}$.

Q: set of all rationals.

R: set of all real numbers; also the real line.

R⁺: set of positive real numbers.

C: set of all complex numbers.

E: set of even integers, $\{\ldots, -4, -2, 0, +2, +4, \ldots\}$.

O: set of odd integers, $\{\ldots, -3, -1, +1, +3, \ldots\}$.

Rⁿ: set of all real n-tuples; also the n-dimensional real vector space over the reals.

Cⁿ: set of all complex n-tuples; also the n-dimensional complex vector space over the field of complex numbers.

∅: the empty set.

---

†This will be a good place to advise the reader that the word "iff" is not a misprint but an abbreviation for "if and only if."

(b) SETS AND MAPS

$\{a, b, \ldots, k\}$, or $\{x \,|\, P(x)\}$, or $\{\ldots A_\alpha \ldots\}$:   sets.†
$a \in A$:   element a belongs to set A.
$A \subset X$: A is a subset of X (proper or improper).
$A \cup B$, or $\cup_\alpha A_\alpha$, etc.:   union of sets.
$A \cap B$, or $\cap_\alpha A_\alpha$, etc.:   intersection of sets.
A^c: complement of the set A.
$A - B$:   difference of two sets.
$A \times B$:   Cartesian product of sets.
$\mathscr{P}(X)$:   power set of X.
(a, b):   ordered pair.
$(a_1, a_2, \ldots, a_n)$:   ordered n-tuple.
$(a_1, a_2, \ldots, a_k, \ldots)$ or (a_k):   sequence.
(x_α):   family (indexed by elements α of some A)
$x \, R \, y$: x stands in relation R to y.
$x \sim y$: x is equivalent to y.
π: a partition.
$[x]$:   an equivalence class with representative element x.
X/E: quotient set of X by (or modulo) the equivalence relation E.
$x \leqslant y$ or $x < y$:   an order relation.
$\inf(A)$ or $\inf(\{a, b \ldots\})$:   infimum of a set.
$\sup(A)$ or $\sup(\{a, b \ldots\})$:   supremum of a set.
$\aleph_0$:   aleph null, cardinal number of countably infinite sets.
$\mathfrak{c}$:   cardinal number of the continuum (real line).
$f: X \to Y$: f is a map from X into Y.
$x \mapsto y$:   element x is mapped onto element y.
$f \circ g$:   composite of map g with map f.
$X \simeq Y$:   set isomorphism (set equivalence).
$\ker(f)$:   equivalence kernel of a map f.
$X \cong Y$:   order isomorphism (similarity) of sets.
$\inf_{x \in X} f$ (sometimes $\inf f(x)$):   infimum of a function.
$\sup_{x \in X} f$ (sometimes $\sup f(x)$):   supremum of a function.

(c) ALGEBRA

$\square, *, \circ, \triangle$:   composition law (like $a \,\square\, b = c$).
$(X, *)$:   algebraic system with underlying set X and composition law $*$.

---

†Greek labels, like in A_α, refer to arbitrary indexing sets. Latin indices, like in A_k, refer to a countable indexing set.

$(A, \square) \approx (B, \triangle)$ or simply $A \approx B$:   algebraic isomorphism.

Im (f):   image of morphism.

Ker (f):   kernel of morphism.

G: group.

Z_n: cyclic group of order n.

$G \times G', R \times R'$, etc.:   direct product of algebraic systems.

Aut (G):   group of automorphisms.

$\Gamma g, g\Gamma$:   left (right) cosets of g by Γ.

$\Gamma + g$:   additive coset.

G/Γ: coset space, in particular quotient group.

e: unit element of group.

o: zero element of additive group (or of ring, of linear space =null vector, of algebra).

ϵ: unit (identity) element of ring (or algebra).

$\mathscr{I}$:   ideal.

$\mathscr{I} + x$:   ring coset (or algebra coset) of x by $\mathscr{I}$.

$\mathscr{R}/\mathscr{I}$:   quotient ring.

$\mathscr{M} \oplus \mathscr{N}$:   direct sum of linear spaces (or manifolds).

$\{\ldots e_\alpha \ldots\}$:   basis of linear space.

$\{\ldots e_k \ldots\}$:   countable basis.

$\{e_1, \ldots, e_n\}$:   finite basis.

$\mathscr{L}/\mathscr{M}$:   quotient space of linear space $\mathscr{L}$ by manifold $\mathscr{M}$.

$\mathscr{M} + x$:   coset of x by manifold $\mathscr{M}$.

$\mathscr{C}(\mathscr{L}_1, \mathscr{L}_2)$:   linear space of all linear transformations from $\mathscr{L}_1$ to $\mathscr{L}_2$.

$\mathscr{C}(\mathscr{L}, \mathscr{L}) = \mathscr{C}(\mathscr{L})$:   algebra of all linear operators on $\mathscr{L}$.

$\mathscr{C}(\mathscr{L}, K) = \mathscr{L}^*$:   (algebraic) dual space of linear space $\mathscr{L}$ over the field K.

$\mathscr{A} \oplus \mathscr{B}$:   direct sum of algebras.

$\mathscr{A}/\mathscr{I}$:   quotient algebra.

$[x, y]$:   Lie product.

ad_x: adjoint map on Lie algebra.

Ad L: adjoint Lie algebra.

(d) TOPOLOGY

τ: a topology.

(X, τ):   topological space.

$d(x, y)$:   a metric (distance).

(X, d):   metric space.

E^n: Euclidean real n-space (in particular, E is the real line with the usual topology).

C^n: unitary (complex) n-space.

E_p^n, C_p^n: generalization of E^n and C^n.

E^∞, C^∞:   infinite dimensional Euclidean (unitary) space of square summable sequences.

E_p^∞, C_p^∞:   generalization of above.

$C[a, b]$:   metric space of continuous (complex) functions on interval $a \leqslant x \leqslant b$.

$S_r(x_0)$:   open ball with radius r about x_0.

$S_r[x_0]$:   closed ball with radius r about x_0.

$(a, b), [a, b), (a, b], [a, b]$:   intervals on the real line.

(A, τ_A):   subspace of a topological space with the relative topology τ_A.

$(X \times Y, \tau_X \times \tau_Y)$:   topological product of two topological spaces.

$\mathcal{B}$:   a base for a topology.

$N(x)$:   a neighborhood of a point x.

$\mathring{A}$ or Int A: interior of A.

$\bar{A}$:   closure of A.

bA: boundary of A.

$\lim_{n \to \infty} x_n$:   limit of sequence (x_n).

$x_n \to x$:   sequence (x_n) converges to limit x.

$(X, \tau) \approx (X', \tau')$ or simply $X \underset{h}{\approx} X'$:   homeomorphism.

X/f: topological quotient space associated with set $X/\ker(f)$.

$\varphi_{a \to b}$:   image of a path from a to b.

$\pi(X)$:   homotopy (fundamental) group.

$\hat{\mathbf{R}} = \mathbf{R} \cup \{+\infty, -\infty\}$:   extended real line.

R^*:   projective real line.

(X^*, τ^*):   one-point compactification of (X, τ).

(e) MEASURE THEORY AND INTEGRATION

$\mathcal{R}$:   a ring of sets.

$\mathcal{S}$:   a σ-ring of sets.

$\mathcal{A}$:   an algebra of sets.

$\mathcal{B}$:   a Borel field (σ-algebra).

$\mathcal{R}(\mathcal{E})$:   ring generated by set $\mathcal{E}$.

$(X, \mathcal{B})$:   Borel space.

$(X, \mathcal{S})$:   a general measurable space.

μ: measure on an arbitrary ring.

$(X, \mathcal{S}, \mu)$:   a general measure space.

$(X, \mathcal{B}, \mu)$:   a measure space made from a Borel space.

a.e.:   almost everywhere.

$f \overset{\circ}{=} g$ or $f = g$ a.e.: f equals g almost everywhere.

χ_A: characteristic function of set A.

$\mu_1 < \mu_2$: μ_1 is inferior to μ_2.

$\mu_1 \sim \mu_2$: μ_1 is equivalent to μ_2.

$(A, \mathscr{S}_A, \mu_A)$: subspace of a measure space.

$(X \times Y, \mathscr{B}_X \times \mathscr{B}_Y, \mu_X \times \mu_Y)$: product of two measure spaces.

$\bar{\mu}$: an extension of μ.

ρ: a right- (or left-) continuous function generating a Lebesgue–Stieltjes measure.

μ^+, μ^-: positive (negative) part of a signed measure.

μ_r, μ_i: real (imaginary) part of a complex measure.

$\int_X f \, d\mu$: integral of f on X with respect to measure μ.

f^+, f^-: positive (negative) part of a function.

$\dfrac{d\mu}{d\nu}$: Radon–Nikodym derivative.

(f) FUNCTIONAL ANALYSIS

$\mathscr{L}$: a topological linear space.

$\mathscr{L}_1 \underset{t}{\approx} \mathscr{L}_2$: topological isomorphism.

$\|x\|$: norm of a vector x.

l_n^p: n-dimensional (complex) normed space (Banach p-space) of n-tuples (special case: l_n^2, n-dimensional Hilbert space).

l^p: normed (complex) linear space (Banach p-space) of p-summable sequences (special case: l^2, sequential Hilbert space).

$L^p(X, \mu)$: normed (complex) linear space (Banach L^p-space) of pth power absolute integrable functions on $(X, \mathscr{B}, \mu)$. (Special case: $L^2(X, \mu)$, Hilbert space of square integrable functions.)

$L^p(-\infty, +\infty)$ or $L^p(a, b)$: same as above, for the real line (or an interval of reals) with Lebesgue measure.

p, q: conjugate numbers, $1/p + 1/q = 1$, $1 < p < \infty$, $1 < q < \infty$.

$B(X)$: Banach space of bounded (complex) functions on X. (Special case: $B[a, b]$.)

$C(X)$: Banach space of (complex) continuous and bounded functions on X. (Special case: $C[a, b]$.)

$\mathscr{B}$: generic symbol for Banach space.

$\mathscr{H}$: generic symbol for Hilbert space.

$\langle x, y \rangle$: inner product.

$x_n \rightarrow y$: weak convergence of vector sequence (x_n) to vector y.

$\mathscr{M}^\perp$: orthogonal complement of $\mathscr{M}$.

$\bigoplus_{k=1}^{\infty} \mathscr{H}_k$:   direct sum of Hilbert spaces.

$\int_{\oplus} \mathscr{H}_\lambda \, d\rho(\lambda)$:   direct integral of Hilbert spaces.

$\|F\|$:   norm of a linear transformation F.

$\langle x|y \rangle$:   ket-bra notation for continuous linear functionals.

$\mathscr{A}(\mathscr{B})$:   Banach algebra of all bounded linear operators on the Banach space $\mathscr{B}$.

$A_n \Rightarrow A, A_n \to A, A_n \rightharpoonup A$:   uniform, strong, and weak convergence of operator sequences (A_n).

$\bar{T}$:   closure of an operator.

$T^\dagger$:   adjoint of an operator on $\mathscr{H}$.

$A \oplus B$ or $\bigoplus_k A_k$, etc.:   direct sum of operators.

$\rho(A)$:   resolvent set of A.

$\sigma(A)$:   spectral set of A.

$P\sigma(A), C\sigma(A), R\sigma(A)$:   pointspectrum, continuous spectrum, and residual spectrum of A.

$\mathscr{R}(\lambda)$:   resolvent operator at λ.

$\mathscr{D}^*, \mathscr{S}^*$:   space of distributions (tempered distributions).

Index to Volume 1

Numbers in **boldface type** refer to pages on which a longer discussion of the subject begins.